Ac Electric Machines and Their Control

David A. Torrey, Ph.D., P.E.
Advanced Energy Conversion, LLC, Schenectady, NY
Union Graduate College, Schenectady, NY

E-Man Press LLC

Copyright © by E-Man Press LLC, 2010

All rights reserved. No part of this publication may be reproduced, stored in a retrieval system, or transmitted, in any form by any means, electronic, mechanical, photocopying, recording, or otherwise, without the prior written permission of the publisher.

The author and publisher have taken care in the preparation of this book, but make no expressed or implied warranty of any kind and assume no responsibility for errors or omissions. No liability is assumed for incidental or consequential damages in connection with or arising out of the use of the information or programs contained herein.

Library of Congress Preassigned Control Number Data

Torrey, David Allan
    Ac Electric Machines and Their Control / David A. Torrey
        p. cm.
    Library of Congress Control Number: 2010924755
    Includes bibliographic references and index.
    ISBN 978-0-9826926-0-8

Printed in the United States of America.

To S, J, and J

Contents

Preface		xi
1	**Preliminaries**	**1**
1.1	Introduction	1
1.2	Maxwell's Equations and the Quasistatic Limit	2
1.3	Magnetic Circuits	8
1.4	Materials Used in Electric Machines	13
	1.4.1 Saturation and Hysteresis	14
	1.4.2 Permanent Magnets	16
	1.4.3 Equivalence Between Permanent Magnets and Windings	19
	1.4.4 Losses	22
	1.4.5 Eddy Currents and Laminations	23
1.5	Mechanical Elements	24
1.6	Summary	25
2	**Electromechanical Energy Conversion**	**29**
2.1	Introduction	29
2.2	Electromechanical Coupling	30
2.3	Electromagnetic Forces	37
2.4	Energy, Force, and Torque	43
2.5	Reciprocity	50
2.6	Energy Conversion Cycles	51
2.7	Conditions for Average Power Production	55

	2.8 Summary	61

3 Windings   63

    3.1 Introduction 63
    3.2 Phases, Poles, Slots, and Pitches 64
    3.3 Types of Windings 66
    3.4 Integral Slot Double-layer Windings 70
    3.5 Fractional Slot Double-layer Windings 79
    3.6 Slot and Skew Winding Factors 84
    3.7 Winding Connections 89
    3.8 Summary 92

4 Inductances, Back Emfs, and Space Harmonics   93

    4.1 Introduction 93
    4.2 Air Gap Modeling 94
    4.3 Winding Inductances 97
        4.3.1 Air Gap Inductances 98
        4.3.2 Slot Inductances 104
        4.3.3 Other Inductances 107
    4.4 Back Emfs 108
    4.5 Space Harmonics 115
    4.6 Summary 118

5 Elements of Electric Machine Design   119

    5.1 Introduction 119
    5.2 Electromagnetic Design 121
        5.2.1 Electric Loading 122
        5.2.2 Magnetic Loading 125
        5.2.3 The Output Equation 126
        5.2.4 Detailed Magnetic Design 127
    5.3 Mechanical Design 128

| | 5.4 | Thermal Design . . . . 132 |
| | 5.5 | Summary . . . . 138 |

6 Transformations 139

	6.1	Introduction 139
	6.2	Three Phases to Two Phases ($abc \leftrightarrow \alpha\beta 0$) 140
	6.3	Rotation ($\alpha\beta 0 \leftrightarrow dq0$) 142
	6.4	Useful Combinations 144
	6.5	Summary 145

7 Inverters 147

	7.1	Introduction 147
	7.2	An Inverter Phase Leg 148
		7.2.1 Basic Principles 149
		7.2.2 Snubber Circuits and Clamps 150
		7.2.3 Interfacing to Controllable Switches 152
	7.3	Single-Phase Inverters 154
	7.4	Three-Phase Inverters 157
	7.5	Multilevel Inverters 157
	7.6	Voltage Waveform Synthesis Techniques 160
		7.6.1 Harmonic Elimination 160
		7.6.2 Harmonic Cancellation 161
		7.6.3 Pulse-Width Modulation 161
	7.7	Current Waveform Synthesis Techniques 168
		7.7.1 Hysteresis and Sliding-Mode Control 170
		7.7.2 Predictive Current Regulation 171
	7.8	Summary 171

8 Current Regulators 173

| | 8.1 | Introduction . . . . 173 |
| | 8.2 | The Structure of Current Regulators . . . . 173 |

8.3	The Current Regulator Dynamics	176
8.4	The Modulator	182
8.5	Overmodulation	186
8.6	An Example Current Regulator	195
8.7	A Space-Vector Predictive Current Regulator	202
8.8	Current Regulation in the dq Reference Frame	204
8.9	Summary	205

9 Induction Machine Models — 207

9.1	Introduction	207
9.2	A Physical Model	208
9.3	The Phase Equivalent Model	212
	9.3.1 Model Development	212
	9.3.2 Model Analysis	217
9.4	Two Phase Equivalent Models	220
	9.4.1 The $\alpha\beta$ Model	220
	9.4.2 The dq Model	225
9.5	Using the dq Induction Machine Model	229
9.6	Summary	230

10 Control of Induction Machines — 239

10.1	Introduction	239
10.2	Scalar Control	240
10.3	Field Oriented Control	242
	10.3.1 Torque Expressions	243
	10.3.2 Field Orientation	245
	10.3.3 Implementation of Field Oriented Control	247
	10.3.4 Stator Field Orientation	252
	10.3.5 Air Gap Field Orientation	255
10.4	Direct Torque Control	259
10.5	Summary	262

11 Permanent Magnet Machine Models 265

- 11.1 Introduction 265
- 11.2 A Physical Model 266
  - 11.2.1 Sinusoidal Windings 270
  - 11.2.2 Trapezoidal Windings 271
- 11.3 Two Phase Equivalent Models 274
  - 11.3.1 The $\alpha\beta$ Model 274
  - 11.3.2 The dq Model 277
- 11.4 Cogging Torque 283
  - 11.4.1 Overview 283
  - 11.4.2 Flux Tube Preliminaries 285
  - 11.4.3 The Stator Slot Shape Function 287
  - 11.4.4 The Air Gap Shape Function 289
  - 11.4.5 The Magnet Shape Function 289
  - 11.4.6 The Magnet Magnetization Function 290
  - 11.4.7 Energy and Cogging Torque 291
  - 11.4.8 Skew 293
  - 11.4.9 Model Results 294
- 11.5 Summary 297

12 Control of Permanent Magnet Machines 303

- 12.1 Introduction 303
- 12.2 Sinusoidal Permanent Magnet Machines 304
  - 12.2.1 Maximizing Torque Per Ampere 304
  - 12.2.2 Direct Torque Control 306
  - 12.2.3 Field Weakening 307
- 12.3 Trapezoidal Permanent Magnet Machines 310
- 12.4 Summary 318

13 Variable-reluctance Machines 323

- 13.1 Introduction 323

	13.2	Modeling VRM Systems		324
		13.2.1	Energy Conversion	327
	13.3	Control of VR Motors		329
		13.3.1	The Algorithm	331
	13.4	Control of VR Generators		338
		13.4.1	Excitation and Generation	339
		13.4.2	Control Implications	345
		13.4.3	VRG Control	346
		13.4.4	Control Implementation	352
	13.5	Summary		354

A Symbols Used — 357

B Signal Analysis — 363

B.1 Introduction . . . 363
B.2 Fourier Series . . . 363
    B.2.1 The Basic Series . . . 364
    B.2.2 Symmetry Conditions . . . 366
B.3 The Decomposition of Waveforms . . . 369
B.4 Measures of Distortion . . . 371
B.5 Summary . . . 371

C Matlab Simulation Code — 373

C.1 Current Regulator . . . 373
C.2 Induction Machine . . . 387
C.3 Simulation of a Brushless Dc Machine System . . . 392

Index — 407

Preface

This book addresses the electromechanics and control of ac electric machines. It is the result of an opportunity to develop a sequence of three graduate courses for Union Graduate College, Schenectady, New York, that deal with electrical energy conversion. After some iteration, the sequence now consists of the following courses:

1. *Electromechanics and Ac Machines*: consideration of electromechanical energy conversion principles and their application to ac electric machines.

2. *Electronic Power Conversion*: consideration of the form and function of electronic power converters, with some consideration of their dynamic models.

3. *Modeling and Control of Power Converters and Electric Machines*: development of dynamic models for power converters and electric machines, and how those models can be used in their control.

The material presented here has evolved over three years and several course offerings. Undergraduate students, graduate students, and working professionals have found the material accessible.

Given the large number of machines books available, one might legitimately ask why another book, such as this one, is necessary. In fairness, the material contained in this book is covered elsewhere. However, it is not packaged as it is here. Many electric machinery books tend to focus on the steady state behavior of electric machines. Treatment of induction machines, for example, focuses on the single-phase equivalent circuit model. Treatment of synchronous machines focuses on phasor diagrams as applied to power systems with (nominally) constant frequency. While this material is useful, I find it limiting in two ways. First, it tends to deemphasize the fields within the machine in favor of the electrical terminals outside of the machine. Second, a significant number of applications of electric machines exist where operation at nominally constant speed is the exception, not the rule.

There are principally two parts to the book. Chapters 1 through 5 deal with the internal issues of electric machines. This treatment starts with magnetic circuits and materials, and builds through force and torque development. Once the fundamentals of electromechanical energy conversion are covered, we delve into windings, inductances, back emfs, and the interaction of space harmonics with current harmonics in producing ripple torque. This section of the book concludes with a discussion of how one might go about designing an electric machine. This discussion is not intended to be a comprehensive design guide, but a productive way of summarizing the material covered to that point, by placing it in the context of the machine design problem. The material in Chapters 1 through 5 has supported the design of a number of successful permanent magnet and variable-reluctance machines.

Chapters 9 through 13 treat the modeling and control of induction, permanent magnet, and variable-reluctance machines. Dynamic models are developed for each of these machines to prepare for a discussion of their control. For induction and permanent magnet machines, control includes field orientation and direct torque control, and is equally applicable to motor and generator applications. For variable-reluctance machines, separate consideration is given to both motoring and generating because each needs to be handled in a slightly different manner.

Chapters 6 through 8 cover material that is needed to properly introduce the dynamic control of electric machines. Chapter 6 discusses the transformations that allow one to view the dynamics of ac electric machines from a more productive perspective, namely that of an observer traveling through the air gap at the same speed as the excitation. Chapter 7 introduces the structure and operation of the inverters used to create an ac source with adjustable amplitude and frequency for dynamic control of ac machines. Chapter 8 considers current regulation, developing the current regulator dynamics and the principles of space vector modulation.

Given the packaging of material in the book, it might be useful to consider how it can be used. In a ten-week term at Union Graduate College, in a first course that carries no prerequisites other than graduate standing, I cover the following material:

- All of Chapter 1.

- Chapter 2 with the exception of Section 2.3.

- All of Chapters 3 and 4.

- Chapter 6, to set up the transformations needed to build the single-phase equivalent circuit model of the induction machine.

- Sections 9.2 and 9.3.

- If time allows, Sections 11.2.2 and 12.3 on permanent magnet machines with trapezoidal windings. The inverter and current regulation are handled at a high level.

In a second ten-week course that assumes students have had prior exposure to power electronics and electric machines, I cover the following material:

- Modeling and control of power converters.

- All of Chapters 6 (again) and 8.

- Chapter 9, with the exception of Section 9.3.

- All of Chapters 10 through 12.

The material on modeling and control of power converters starts with Part II of *Principles of Power Electronics* by J. G. Kassakian, M. F. Schlecht, and G. C. Verghese. This is the same text I use for teaching power electronics, the second course in the three-course sequence. The emphasis is on the development of averaged, state-space, and sampled-data models for power converters. Control design concepts are also developed. There is also some material taken from the technical literature. With the early emphasis on control of power converters, the students are well prepared when we get to Chapter 8.

The text tries to achieve a balance between mathematical rigor and physical insight. I find the material in Chapters 1 and 2 essential to help with the physical insights. In particular, the discussion of requirements for average power conversion in Section 2.7 can be revisited regularly to drive home the point of how the stator and rotor fields interact within the machine to create torque. These important concepts can help to provide clarity, and boil field-oriented control down to its essence.

An undertaking of this type cannot survive to conclusion without a lot of support. I am fortunate to have a forgiving family, supportive colleagues, and a pedagogical opportunity to capture, all able to provide the right encouragement at the right time. Dean Robert Kozik at Union Graduate College gave me the freedom to teach electric machines the way I think it should be taught. My colleagues at Advanced Energy Conversion have been most generous with their insights over the years, stimulating my own. Drs. Duane Hanselman (University of Maine), Philip Krein (University of Illinois), Dean Patterson (University of Nebraska and Regal Beloit Corporation), Sheppard Salon (Rensselaer Polytechnic Institute), and Yilmaz Sozer (University of Akron) were encouraging about the content. Duane Hanselman was a font of knowledge about the actual publication process, and very patient with my questions. Peter Miller took on the task of editing with resolve and good humor. But, most importantly, Sara, Josh, and Jake gave me the freedom to spend nights and weekends developing the material and writing it all down.

D. A. T.

Schenectady, New York
February 2010

Chapter 1

Preliminaries

1.1 Introduction

This book discusses electric machines, otherwise known as motors and generators. Because any motor can be operated as a generator and vice versa, we adopt the generic term electric machine.

Electric machines are ubiquitous throughout modern society. The vast majority of electric energy is produced using electric machines. On the other end, electric machines represent the largest segment of electricity consumption.

Electric machines are multi-disciplinary by nature, representing the interface between electrical systems and mechanical systems. The conversion between electrical and mechanical energy relies on an intermediate magnetic field. This magnetic field is produced either by electric currents flowing through windings, or (increasingly) by permanent magnets.

It is worth noting that electric machines can also use an intermediate electric field. Many microelectromechanical systems (MEMS) are based on electric fields, not magnetic fields. This makes sense at very small sizes where the energy density in an electric field can be much larger than that of a magnetic field. The use of electric fields is also consistent with microfabrication techniques. For electric machines on the order of a millimeter or larger, the use of magnetic fields makes good sense. This is why we focus on electric machines based on magnetic fields.

There are many classifications that are applied to electric machines: synchronous, induction, asynchronous, commutator, reluctance, dc, ac, etc. Fundamentally, all electric machines are based on the interaction of magnetic fields. Because of the mechanical rotation involved, at least one of the magnetic fields is set up by time-varying current. Accordingly, all electric machines can be fundamentally classified as ac machines.

Some electric machines have an integral mechanical commutator that is responsible for switching the direction of currents in synchronism with the rotation. Thus, even with the application of dc to the terminals of the rotor winding, the switching action of the commutator produces alternating currents. Not to diminish the significance of dc machines, this text focuses on ac machines and their control.

An increasing number of electric machine applications are using ac machines. There are a number of reasons for this, but chief among them is power density. An ac machine is fundamentally more power dense than a dc machine. That is, an ac machine of a given speed and power rating will be smaller than its dc counterpart. In addition, ac machines cover a much wider spread in speed and power levels than dc machines.

As with any subject, we must learn to walk before we can run. The remaining sections of this chapter lay a foundation for our work in subsequent chapters.

1.2 Maxwell's Equations and the Quasistatic Limit

Electric machines are governed by Maxwell's equations and therefore represent our starting point. A differential form of Maxwell's equations are[1]:

$$\nabla \times \vec{E} = -\frac{\partial \vec{B}}{\partial t} \quad ; \tag{1.1}$$

$$\nabla \times \vec{B} = \mu_0 \vec{J} + \mu_0 \frac{\partial \epsilon_0 \vec{E}}{\partial t} \quad ; \tag{1.2}$$

$$\nabla \cdot \epsilon_0 \vec{E} = \rho \quad ; \tag{1.3}$$

$$\nabla \cdot \vec{B} = 0 \quad ; \tag{1.4}$$

$$\nabla \cdot \vec{J} + \frac{\partial \rho}{\partial t} = 0 \quad . \tag{1.5}$$

Here, $\vec{E}$ is the electric field intensity, $\vec{B}$ is the magnetic flux density, $\vec{J}$ is the conduction current density, and ρ is the free charge density. Other formulations might include the electric displacement ($\vec{D}$) and the the magnetic field

---

[1]There are several formulations of Maxwell's equations. The differences among them are generally tied to the treatment of materials, field quantities that are the basis of the formulation, or using the cgs system of units rather than the mks system of units. Handled with care, the formulations describe the same phenomena in a consistent manner.

1.2. Maxwell's Equations and the Quasistatic Limit

intensity ($\vec{H}$). Also, μ_0 is the permeability of free space, ϵ_0 is the permittivity of free space, and t is time.

Equation 1.1 is known as Faraday's law and indicates that time-varying magnetic fields induce an electric field that circulates around the magnetic field. Equation 1.2 is Ampere's law and it tells us that electric currents are sources of magnetic fields. The magnetic field circulates around the electric current. Electric currents can be due either to conduction currents ($\vec{J}$) or displacement currents ($\partial \epsilon_0 \vec{E}/\partial t$). Equation 1.3 is the electric version of Gauss's law, indicating that electric charges are sources of electric fields. Equation 1.4 is the magnetic form of Gauss's law and tells us that magnetic fields must close on themselves; this is sometimes known as being solenoidal. Finally, Eq. 1.5 is a statement of charge conservation; it is not usually included in Maxwell's equations, but it is easily obtained by taking the divergence of Ampere's law.

The corresponding versions of Maxwell's equations in integral form are

$$\oint_C \vec{E} \cdot d\vec{l} = -\frac{d}{dt} \int_S \vec{B} \cdot d\vec{a} \quad ; \tag{1.6}$$

$$\oint_C \vec{B} \cdot d\vec{l} = \mu_0 \int_S \vec{J} \cdot d\vec{a} + \mu_0 \frac{d}{dt} \int_S \epsilon_0 \vec{E} \cdot d\vec{a} \quad ; \tag{1.7}$$

$$\oint_S \vec{E} \cdot d\vec{a} = \int_V \frac{\rho}{\epsilon_0} dV \quad ; \tag{1.8}$$

$$\oint_S \vec{B} \cdot d\vec{a} = 0 \quad ; \tag{1.9}$$

$$\oint_S \vec{J} \cdot d\vec{a} + \frac{d}{dt} \int_V \rho \, dV = 0 \quad . \tag{1.10}$$

In some instances the integral versions of Maxwell's equations are more convenient to use than the differential form. The conversion between the two forms is facilitated by Stoke's law and the divergence theorem from vector calculus.

Regardless of the formulation, Maxwell's equations describe coupling between electric and magnetic fields. The coupling exerts itself through time-varying quantities. That is, time-varying magnetic fields give rise to electric fields (Faraday's law). Time-varying electric fields give rise to magnetic fields (Ampere's law).

For electromechanical phenomena, we can often exploit an effective decoupling between electric and magnetic field evolution. This is not the same as saying that we treat the fields as static. The fields can still vary in time, but we can disregard some of the terms in Maxwell's equations. Under these conditions we consider the fields to be *quasi*static.

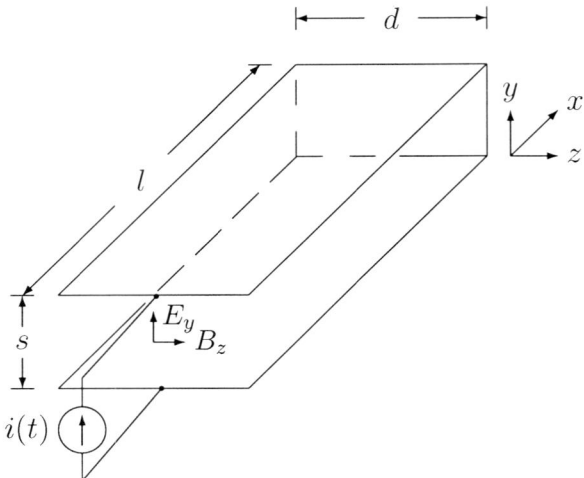

Figure 1.1: A single-turn solenoid excited by a current source.

Since we are primarily concerned with the application of Maxwell's equations to systems dominated by magnetic energy, we will develop the quasistatic limit for magnetic field systems through consideration of Faraday's law and Ampere's law. A parallel development can be undertaken for systems dominated by electric energy, also starting with Faraday's law and Ampere's law.

Consider the system shown in Fig. 1.1 comprised of a single-turn solenoid that is excited by a current source. Our interest is in the field description within the solenoid in the region where $\vec{J} = 0$. To simplify things, we consider only field variations in x and time.

The y component of Ampere's law gives

$$-\frac{\partial B_z}{\partial x} = \mu_0 \epsilon_0 \frac{\partial E_y}{\partial t} \quad . \tag{1.11}$$

The z component of Faraday's law gives

$$\frac{\partial E_y}{\partial x} = -\frac{\partial B_z}{\partial t} \quad . \tag{1.12}$$

Combining these two expressions gives

$$\frac{\partial^2 E_y}{\partial x^2} = \mu_0 \epsilon_0 \frac{\partial^2 E_y}{\partial t^2} = \frac{1}{c^2} \frac{\partial^2 E_y}{\partial t^2} \quad , \tag{1.13}$$

1.2. Maxwell's Equations and the Quasistatic Limit

where $c = 1/\sqrt{\mu_0 \epsilon_0}$ is the speed of light in free space.

Equation 1.13 is a wave equation with a solution of the form

$$E_y(x,t) = E_+(x - ct) + E_-(x + ct) \quad . \tag{1.14}$$

This solution is formed by the superposition of a component traveling in the positive direction and a component traveling in the negative direction. The velocity of each component is the speed of light. A wave equation for $B_z(x,t)$ can be formed in a similar manner, again starting with Eqs. 1.11 and 1.12. It will also have two components, each of which travels at the speed of light.

Electromechanical interactions are usually not appreciably affected by wave phenomena because the velocity of wave propagation is so large. If l is a characteristic dimension, then l/c is the propagation time it takes for a wave to travel from one side of the system to the other. If this time is short compared to the times of interest, we can ignore the wave propagation.

We now explore this in more detail using the system of Fig. 1.1. We apply current source excitation to our solenoid at $x = -l$: $i(t) = I_0 \cos \omega t$. The current source imposes a boundary condition on the magnetic field at $x = -l$ and the short at the origin imposes a boundary condition on the electric field. Formally, our boundary conditions are:

$$B_z(-l,t) = -\frac{\mu_0 i(t)}{d} \quad ; \tag{1.15}$$

$$E_y(0,t) = 0 \quad . \tag{1.16}$$

Solutions to the wave equation (Eq. 1.13) take the form

$$B_z(x,t) = -\frac{\mu_0 I_0 \cos(\omega t) \cos(\omega x/c)}{d \cos(\omega l/c)} \quad ; \tag{1.17}$$

$$E_y(x,t) = -\frac{I_0 \sin(\omega t) \sin(\omega x/c)}{d \epsilon_0 c \cos(\omega l/c)} \quad . \tag{1.18}$$

The product of trigonometric functions in the numerators of Eqs. 1.17 and 1.18 represent traveling waves in the positive and negative x direction, consistent with Eq. 1.14[2]. In addition, Eqs. 1.17 and 1.18 satisfy the boundary conditions given in Eqs. 1.15 and 1.16, respectively.

If we can make the approximation that times of interest $(1/\omega)$ are long compared to wave propagation time (l/c) then we have

$$\frac{\omega l}{c} \ll 1 \quad , \tag{1.19}$$

---

[2] $\sin \alpha \sin \beta = [\cos(\alpha - \beta) - \cos(\alpha + \beta)]/2$
$\cos \alpha \cos \beta = [\cos(\alpha - \beta) + \cos(\alpha + \beta)]/2$

and our field solutions become, approximately,

$$B_z(x,t) \approx -\frac{\mu_0 I_0 \cos(\omega t)}{d} \quad ; \tag{1.20}$$

$$E_y(x,t) \approx -\frac{I_0(\omega x/c)\sin(\omega t)}{d\epsilon_0 c} = -\frac{\mu_0 \omega x I_0 \sin(\omega t)}{d} \quad . \tag{1.21}$$

Note that the magnetic field between the shorted plates has the same distribution as if the fields were static. (An alternative interpretation of $\omega l/c \ll 1$ is that the velocity of interest in the numerator is substantially smaller than the velocity of the wave in the denominator.)

Now if we compute the voltage $v(t)$ at $x = -l$

$$v(t) = -\int_0^s E_y(-l,t)dy \quad , \tag{1.22}$$

we obtain the terminal equation for an inductance:

$$v(t) = L\frac{d}{dt}\left(I_0 \cos \omega t\right) \quad , \tag{1.23}$$

where

$$L = \frac{\mu_0 l s}{d} \quad . \tag{1.24}$$

In the limit of static excitation, we have no electric field since there is no time variation in magnetic field.

The foregoing analysis suggests a reduced formulation of Maxwell's equations under the magnetoquasistatic (MQS) limit in which the displacement current can be neglected. The differential form of this reduced formulation is:

$$\nabla \times \vec{E} = -\frac{\partial \vec{B}}{\partial t} \quad ; \tag{1.25}$$

$$\nabla \times \vec{B} = \mu_0 \vec{J} \quad ; \tag{1.26}$$

$$\nabla \cdot \vec{B} = 0 \quad ; \tag{1.27}$$

$$\nabla \cdot \vec{J} = 0 \quad . \tag{1.28}$$

For completeness, carrying out the parallel development for systems dominated by electric fields, Maxwell's equations under the electroquasistatic (EQS) limit are simplified by ignoring time variation in the magnetic field:

$$\nabla \times \vec{E} = 0 \quad ; \tag{1.29}$$

$$\nabla \times \vec{B} = \mu_0 \vec{J} + \mu_0 \frac{\partial \epsilon_0 \vec{E}}{\partial t} \quad ; \tag{1.30}$$

1.2. Maxwell's Equations and the Quasistatic Limit

$$\nabla \cdot \epsilon_0 \vec{E} = \rho \quad ; \quad (1.31)$$

$$\nabla \cdot \vec{J} + \frac{\partial \rho}{\partial t} = 0 \quad . \quad (1.32)$$

The MQS and EQS limits hold so long as

$$\frac{\omega l}{c} \ll 1 \quad . \quad (1.33)$$

In free space, this condition is easily met for many practical situations. Note that frequency and length are inversely related which correlates well with our practical experience. For example, it would be impractical to try and excite an electric machine that has a characteristic length of 10 m with a radian frequency of 10,000 rad/s.

We are going to use the MQS limit to simplify Maxwell's equations for our study of (magnetic) electric machines. This is consistent with fields that have dynamics, but the wave nature of the electric and magnetic field interactions can be ignored. In effect, we have decoupled the time evolution of $\vec{E}$ and $\vec{B}$.

Before we move into consideration of magnetic circuits, it is worth a brief digression to consider the treatment of magnetic materials and their influence on the relationship between field quantities. So far, the formulation of Maxwell's equations is based exclusively on the magnetic flux density $\vec{B}$ and the electric field intensity $\vec{E}$. In the discussion of magnetic circuits that follows, we are interested not only in the magnetic flux density, but also the magnetic field intensity $\vec{H}$. In general, the magnetic flux density and the magnetic field intensity are related through the constituitive relationship

$$\vec{B} = \mu_0 \vec{H} + \mu_0 \vec{M} \quad , \quad (1.34)$$

where $\vec{M}$ is the magnetization density. For linearly magnetizable materials

$$\vec{M} = \chi_m \vec{H} \quad , \quad (1.35)$$

where χ_m is the magnetic susceptibility. It follows that for linearly magnetizable materials

$$\vec{B} = \mu_0 \left(1 + \chi_m\right) \vec{H} = \mu \vec{H} \quad , \quad (1.36)$$

where μ is the permeability. As discussed in Sec. 1.4, most magnetic materials are nonlinear. However, a linear approximation over some operating range is often possible. This approximation allows for simple analytic descriptions of field quantities.

Consistent with the constituitive relationship between $\vec{B}$, $\vec{H}$, and $\vec{M}$, Ampere's law for MQS systems is frequently written as

$$\nabla \times \vec{H} = \vec{J} \quad (1.37)$$

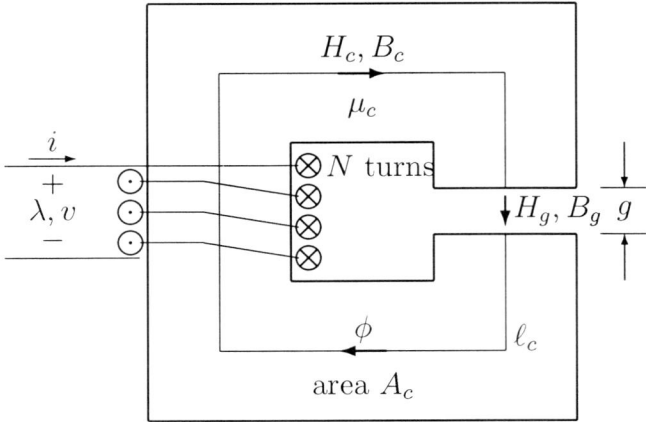

Figure 1.2: A simple magnetic circuit.

in differential form and

$$\oint_C \vec{H} \cdot d\vec{l} = \int_S \vec{J} \cdot d\vec{a} \qquad (1.38)$$

in integral form. As previously discussed, conduction currents are sources of magnetic field. Our discussion of magnetic circuits starts with the version of Ampere's law given by Eq. 1.38.

1.3 Magnetic Circuits

Consider the magnetic structure shown in Fig. 1.2. The structure is comprised of a core with high permeability (but not infinite), an air gap in the core, and a winding. In this type of structure, we are generally interested in relating the terminal quantities to the material properties and dimensions of the structure.

Our assumptions regarding the magnetic circuit are:

1. The core has a permeability of μ_c, where μ_c is large enough to assume that magnetic flux prefers to flow through the core rather than air.

2. The core has a cross-sectional area of A_c.

3. The mean length of the core in the direction traveled by the flux is ℓ_c.

4. The air gap has a cross-sectional area of A_g.

1.3. Magnetic Circuits

5. The length of the air gap is g.

6. The coil has N turns.

7. There is no magnetic saturation in the core.

We analyze the magnetic circuit of Fig. 1.2 by assuming that the magnetic field intensity and magnetic flux density in the core are H_c and B_c, respectively. Similarly, the magnetic field intensity and magnetic flux density in the air gap are H_g and B_g, respectively. Applying Ampere's law to a clockwise contour around the mean path of the magnetic circuit gives

$$H_c \ell_c + H_g g = Ni \quad . \tag{1.39}$$

Applying Gauss's law to a small volume that straddles the magnetic core and the air gap shows the magnetic flux to be

$$\phi = B_c A_c = B_g A_g \quad . \tag{1.40}$$

We connect these two relationships through the material properties:

$$B_c = \mu_c H_c \quad B_g = \mu_0 H_g \quad . \tag{1.41}$$

Using Eqs. 1.41 and 1.40 in Eq. 1.39 gives

$$\phi \left(\frac{\ell_c}{\mu_c A_c} + \frac{g}{\mu_0 A_g} \right) = Ni \quad . \tag{1.42}$$

From this result we see that the same magnetic flux flows through the core and the air gap. This flux is driven by the Ampere-turns applied to the coil, and the proportionality between the flux and the Ampere-turns is tied to the material properties and the geometry. From the way the influence of the core and the air gap combine as a sum, we can recognize parallel behavior between this magnetic circuit and an electric circuit comprised of a source and two series-connected resistors.

The quantity Ni is termed the magnetomotive force (mmf) $\mathcal{F}$, analogous to the electromotive force (voltage) that drives current through an electric circuit. The flux ϕ is analogous to the electric current. The quantity $\ell_c/(\mu_c A_c)$ is termed the reluctance of the core, and denoted by $\mathcal{R}_c$. Similarly, the quantity $g/(\mu_0 A_g)$ is termed the reluctance of the air gap, and denoted by $\mathcal{R}_g$. It follows that the circuit analog of the magnetic circuit of Fig. 1.2 is shown in Fig. 1.3. Table 1.1 summarizes the analogy between magnetic and electric circuits.

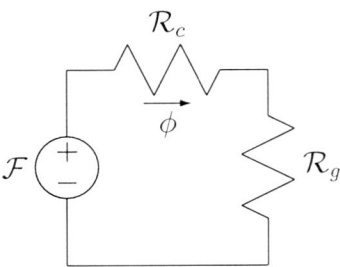

Figure 1.3: The electric analog of the magnetic circuit shown in Fig. 1.2.

Table 1.1: A summary of the analogy between magnetic and electric circuits.

Magnetic		Electric	
Magnetomotive Force	$\mathcal{F} = Ni$	Electromotive Force	v
Flux	ϕ	Current	i
Reluctance	$\mathcal{R}$	Resistance	R
Permeability	μ	Conductivity	σ
Permeance	$\mathcal{P}$	Conductance	G
Flux Density	B	Current Density	J
Magnetic Field Intensity	H	Electric Field Intensity	E

Using the definition of reluctance, the flux flowing through the magnetic circuit of Fig. 1.2 is, from Eq. 1.42,

$$\phi = \frac{Ni}{\mathcal{R}_c + \mathcal{R}_g} \quad, \tag{1.43}$$

so the total flux linking the terminals of the coil is

$$\lambda = \frac{N^2 i}{\mathcal{R}_c + \mathcal{R}_g} \quad, \tag{1.44}$$

making the inductance

$$L = \frac{N^2}{\mathcal{R}_c + \mathcal{R}_g} \quad. \tag{1.45}$$

Hence, we see that inductance is inversely proportional to reluctance and proportional to permeance. This might lead one to conclude that a high inductance can be achieved by making the reluctance small, such as by eliminating the air gap. Energy considerations will reveal this to be of limited

1.3. Magnetic Circuits

value, since the energy density that can be achieved in a high permeability core is substantially smaller than the energy density that can be achieved in an air gap.

It may be somewhat confusing at this point to distinguish between λ and ϕ since they are both given by the integral of flux density over an area. The distinction is subtle but important. Generally the area associated with determining ϕ is the cross-sectional area of a magnetic circuit. On the other hand, the area associated with determining λ is the area of a winding through which magnetic flux passes. Since each turn of a coil typically links ϕ flux, the N turns of the winding link a total flux of $\lambda = N\phi$.

Note that in the limit as $\mu_c \to \infty$, $\mathcal{R}_c \to 0$ the inductance reduces to

$$L = \frac{N^2}{\mathcal{R}_g} = \frac{\mu_0 A_g N^2}{g} \quad . \tag{1.46}$$

For many magnetic materials, a useful understanding of circuit operation is obtained by considering the magnetic material to have infinite permeability. Consideration of large but finite permeability may best be left for more detailed analysis when nonlinear effects are considered.

It is important to remember that the magnetic circuit and its electric analog make a useful modeling tool. However, like all tools it has limitations. To begin, magnetic flux is not easily constrained like electric currents. Electric currents can be constrained to flow within wires. Magnetic flux, on the other hand, is present wherever there are currents in the vicinity. Accordingly, even in a magnetic circuit that has highly permeable materials in it, there will still be some flux that is present in regions where we might not want it. Ignoring it is not going to make it go away. The nonlinearity of magnetic materials often causes our simple magnetic circuit models to break down. This can be compensated to some extent by adopting more complicated magnetic circuit models, but eventually a nonlinear field description may be necessary.

We can apply the approach and reasoning associated with the simple example of Fig. 1.2 to more complicated situations involving irregular shapes by determining the incremental reluctance (or permeance), and integrating over the material region to determine the total reluctance (or permeance). We will need to adopt this approach when, for example, we want to understand the impact of winding slots on the magnetic fields in the air gap of an electric machine. The selection between using the incremental reluctance or permeance depends on the shape of the region relative to the magnetic field lines. If the integration proceeds in the direction of the magnetic field, an incremental reluctance formulation is appropriate. On the other hand, if the integration proceeds in a direction orthogonal to the magnetic field,

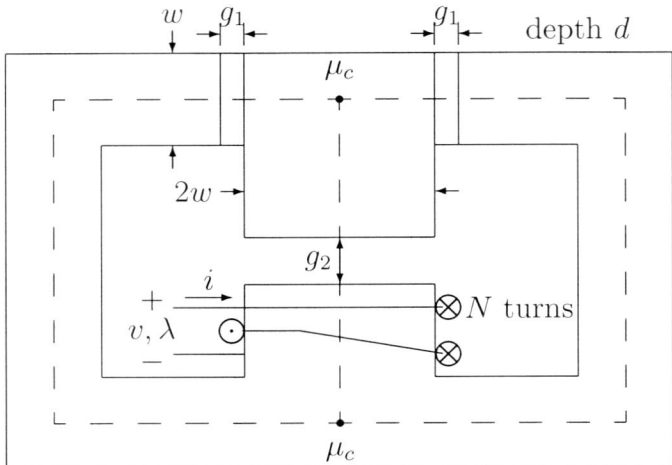

Figure 1.4: A magnetic circuit with two flux pathes.

an incremental permeance formulation is appropriate. We will look at irregular shapes that require incremental permeances in Sec. 4.2. This type of approach is sometimes referred to as flux tube analysis.

Now consider the slightly more complicated magnetic circuit shown in Fig. 1.4. This magnetic circuit has a center leg, the flux through which divides and returns through the outside legs. By virtue of the symmetry, the flux divides evenly. All of the flux is driven by the winding on the center leg. The dashed lines indicate the flux paths that would be used for determining the core reluctances.

Figure 1.5 shows the electric analog of the magnetic structure of Fig. 1.4. The associated reluctances are:

$$\mathcal{R}_{g1} = \frac{g_1}{\mu_0 wd} \quad ; \tag{1.47}$$

$$\mathcal{R}_{g2} = \frac{g_2}{\mu_0 2wd} \quad ; \tag{1.48}$$

$$\mathcal{R}_{\text{outer leg}} = \frac{l_{\text{outer}}}{\mu_0 wd} \quad ; \tag{1.49}$$

$$\mathcal{R}_{\text{center leg}} = \frac{l_{\text{center}}}{\mu_0 2wd} \quad . \tag{1.50}$$

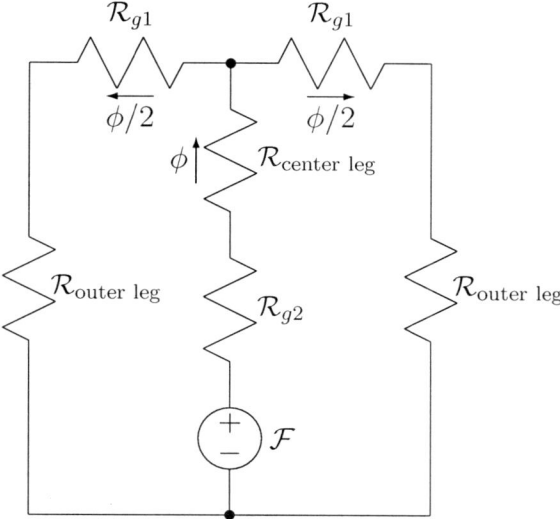

Figure 1.5: The electric analog of the magnetic circuit shown in Fig. 1.4.

The lengths for the core reluctances are defined by the dashed paths in Fig. 1.4.

1.4 Materials Used in Electric Machines

Materials used in electric machines fall into three general categories:

1. Ferromagnetic materials that are used for "conducting" magnetic flux to where it is able to support energy conversion.

2. Conducting materials that are used to support the currents, which create one or more magnetic fields in the machine.

3. Insulating materials that prevent current flow in undesirable locations.

Of the three material types, the ferromagnetic materials generate the most discussion because these materials are nonlinear. In addition, ferromagnetic materials can be classified into two general types: hard and soft. Hard materials are hard to magnetize and hard to demagnetize. Accordingly, hard materials are also known as permanent magnets. Soft materials are easily magnetized and easily demagnetized.

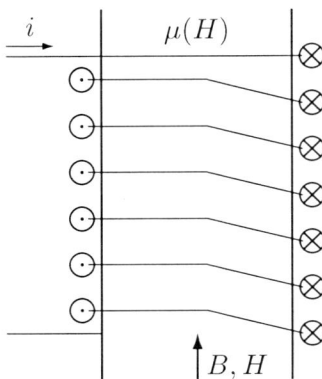

Figure 1.6: A section of ferromagnetic core wrapped with a winding.

1.4.1 Saturation and Hysteresis

Saturation is an issue with all ferromagnetic materials. Figure 1.6 outlines a simple experiment that allows us to investigate saturation, showing a section of ferromagnetic material around which a coil is wound. Initially when we start, the coil has zero $\vec{H}$ and zero $\vec{B}$. As we inject positive current into the winding, $\vec{H}$ and $\vec{B}$ begin increasing. Initially the creation of the $\vec{B}$ field is enhanced by the action of the magnetic domains within the material. These magnetic domains can be viewed as small magnets. Initially the virgin material has the domains in random order so there is no noticeable effect of their presence. As the $\vec{H}$ field is increased, the domains begin swinging into alignment, much as a compass needle seeks to align with a magnetic field.

As the current continues to increase, $\vec{H}$ and $\vec{B}$ continue to build. As more and more magnetic domains come into alignment with the applied field, however, the rate of growth in $\vec{B}$ starts to slow. Eventually, all of the magnetic domains are aligned with the applied field and the material is saturated. Once saturated, additional increases in $\vec{H}$ continue to increase $\vec{B}$, but the material is now effectively acting like air. This description of the initial magnetization of the material is shown in Fig. 1.7 as the trajectory that starts at the origin and moves up into the first quadrant.

After the initial magnetization, if the current in the coil is allowed to collapse, the magnetic domains are no longer being coerced to remain in alignment and they relax. Accordingly, the flux density drops, but it does not drop all the way back to zero. The value to which the flux density collapses in the absence of applied current is known as the residual flux density, denoted by B_r in Fig. 1.7. In order to drive the flux density to

1.4. Materials Used in Electric Machines

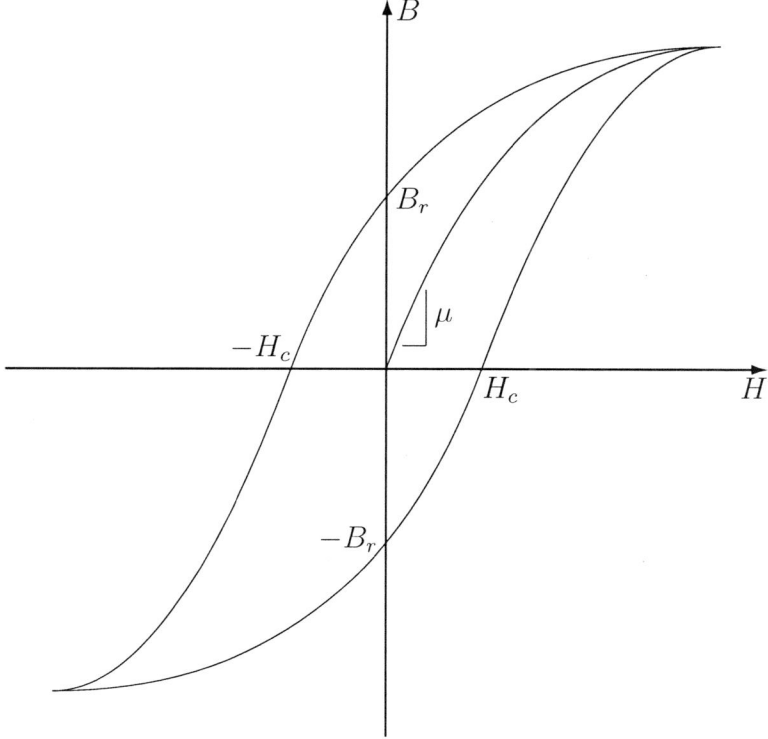

Figure 1.7: A hysteresis loop for a ferromagnetic material.

zero, the magnetic field $\vec{H}$ must be taken negative to coerce the magnetic domains to support zero flux density. This coercive force is denoted as $-H_c$ in Fig. 1.7.

As the magnetic field $\vec{H}$ is caused to alternate, we see the magnetic flux density $\vec{B}$ following, but always lagging behind. This phenomena of lagging behind the excitation is known as hysteresis. The area swept out by the hysteresis loop represents energy that is lost to coercing the magnetic domains to follow the applied field. The hysteresis losses are a function of frequency, and the extent of the excitation $\vec{H}$. Hysteresis losses may also be described as a function of peak flux density $\vec{B}$.

The differences between hard and soft ferromagnetic materials can be seen most easily in a comparison of coercive forces, residual flux densities, and saturation flux densities. Tables 1.2 and 1.3 give important parameters for a number of hard and soft materials, respectively. The data given in Tables 1.2 and 1.3 should be taken as representative numbers. Composition differences and final treatment of the materials will affect their ultimate

Table 1.2: Parameters for a number of hard magnetic materials.

Material	H_c (A/m)	B_r (T)	$BH_{\max}$ (kTA/m)	T_c (C)
Ceramic	222,817	0.41	32,627	450
SmC (bonded)	485,422	0.8	119,366	720
SmC (sintered)	636,620	1.15	250,669	800
AlNiCo	119,366	1.07	71,619	860
NdFeB (bonded)	389,930	0.68	79,578	350
NdFeB (sintered)	636,620	1.5	437,676	310

Table 1.3: Parameters for a number of soft magnetic materials.

Material	H_c (A/m)	B_{sat} (T)
M-19	39.8	1.5
Metglas	2.39	1.5
Nickel Steel	31.8	0.72
Vanadium Permendur	35.0	2.2

performance.

1.4.2 Permanent Magnets

Permanent magnet materials can be formed in two common ways. Bonded permanent magnets are formed by suspending the magnetic material in powder form within a nonmagnetic resin. Because of the space taken up by the resin, bonded magnets give lower performance than sintered magnets which are pressed and sintered to form a solid piece. Sintered magnets offer higher performance at the expense of higher losses. Bonded magnets can made to be flexible, typical for refrigerator magnets.

As we will see, permanent magnet materials operate in the second quadrant of the hysteresis loop. For permanent magnet materials that exhibit significant curvature in the second quadrant (Alnico and ferrite), we must be careful of demagnetization. For materials with an essentially straight demagnetization curve, demagnetization is less of an issue, but full temperature extremes and loadings must still be considered carefully.

Consider Fig. 1.8 which shows the second quadrant of the hysteresis loop

1.4. Materials Used in Electric Machines

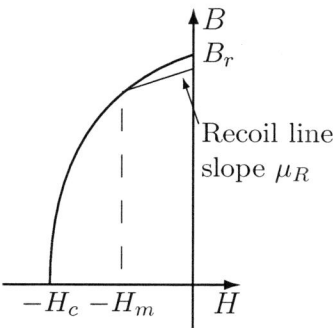

Figure 1.8: The magnetization curve for a permanent magnet showing the recoil line.

for a permanent magnet that has significant curvature. When the operation of the magnetic circuit pushes the magnet down the demagnetization curve, the flux density recovers when the excitation is removed, but not fully to the residual flux density B_r. Instead, the magnet characteristic recoils back to the vertical axis along what is known as the recoil line. The slope of the recoil line is the recoil permeability, which is approximately equal to the slope of the hysteresis loop at $H = 0$. The recoil permeability is usually slightly larger than the permeability of free space, typically $\mu_R = 1.05\mu_0$. Because the recoil permeability is so small, permanent magnets are often treated like air.

Under periodic excitation, operation along the recoil line corresponds to operation around a minor hysteresis loop. Accordingly, losses in the magnet result and will contribute to magnet heating. A careful estimate of magnet losses is necessary to determine the most appropriate means for cooling the magnets.

Figure 1.9 shows a magnetic circuit that contains a permanent magnet and an air gap; we take the core to have infinite permeability. We assume the magnetic field intensity within the magnet is H_m and that within the air gap is H_g. The corresponding magnetic flux densities are B_m and B_g. Ampere's law requires

$$H_m \ell_m + H_g g = 0 \quad , \tag{1.51}$$

giving

$$H_g = -\frac{\ell_m}{g} H_m \quad . \tag{1.52}$$

Applying Gauss's law to the upper section of the core gives

$$B_m = \frac{A_g}{A_m} B_g \quad . \tag{1.53}$$

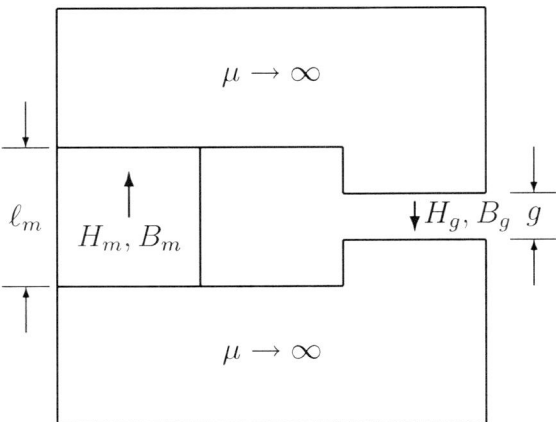

Figure 1.9: A simple magnetic circuit containing a permanent magnet.

Combining Eqs. 1.52 and 1.53 relates the flux density and the magnetic field intensity within the magnet as dictated by the loading imposed by the air gap:

$$B_m = -\frac{\mu_0 A_g \ell_m}{A_m g} H_m \quad . \tag{1.54}$$

The negative sign forces the magnet to operate in its second quadrant of the hysteresis loop with a positive flux density. Equation 1.54 must be satisfied self-consistently with the magnet material properties.

Figure 1.10 shows the magnet $B - H$ characteristic. Superimposed on this characteristic is the load line given by Eq. 1.54. The magnet operates where the two curves intersect. A legitimate question is where to place the operating point of the magnet.

Rearranging Eq. 1.52 and multiplying it by Eq. 1.53 gives

$$B_m H_m = -\frac{A_g g}{A_m \ell_m} \frac{B_g^2}{\mu_0} \quad . \tag{1.55}$$

Recognizing that $A_g g = V_g$ is the volume of the air gap and $A_m \ell_m = V_m$ is the volume of the magnet, we can solve Eq. 1.55 for the magnet volume, giving

$$V_m = -\frac{V_g B_g^2}{\mu_0 B_m H_m} \quad . \tag{1.56}$$

This expression shows that the magnet volume is minimized when the magnet operating point maximizes the product of B_m and H_m. This product represents a hyperbola, with the distance from the origin dictated by the

1.4. Materials Used in Electric Machines

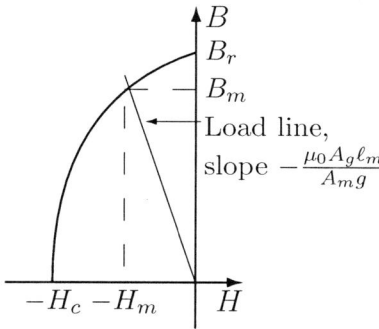

Figure 1.10: Determination of the operating point for a permanent magnet within a magnetic circuit.

energy product $B_m H_m$. The hyperbola that forms a tangent to the magnetization curve represents the maximum energy product and hence the smallest magnet volume. Often in the design of electric machines, there are other considerations that may outweigh minimization of magnet volume. Our result also suggests that it is possible to increase the flux density in the air gap by giving the air gap a smaller cross-sectional area than the magnet. Some machine designs try to "focus the flux" to accomplish this.

1.4.3 Equivalence Between Permanent Magnets and Windings

It is useful to show the equivalence between permanent magnets and conventional windings. Figure 1.11 shows two magnetic circuits that have the same dimensions. One is based on a permanent magnet while the other is based on a winding of N turns. Our model of the permanent magnet is guided by the demagnetization curve shown in Fig. 1.12. In the region of interest, the demagnetization curve can be represented as

$$B_m = \mu_R H_m + B_r \quad , \tag{1.57}$$

where

$$B_r = \mu_R H_c' \quad . \tag{1.58}$$

The quantity H_c' can be thought of as the effective coercive force.

Our analysis proceeds in parallel for the two circuits with the permanent magnet circuit given on the left and the wound circuit given on the right. As with other magnetic circuits, we begin by applying Ampere's law:

$$H_m \ell_m + H_g g = 0 \quad | \quad H_e \ell_m + H_g g = Ni \quad , \tag{1.59}$$

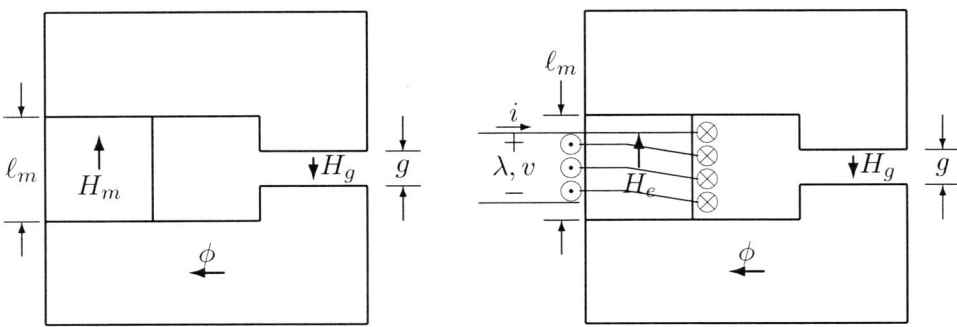

Figure 1.11: Two magnetic circuits. The one on the left is based on a permanent magnet; the one on the right is based on a winding.

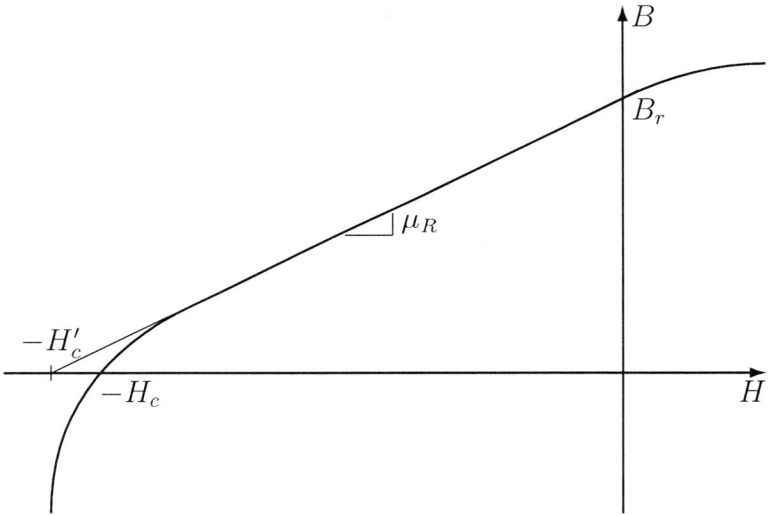

Figure 1.12: The second quadrant of the hysteresis loop for a hard magnetic material.

then Gauss's law:

$$B_m A_m = B_g A_g = \phi \quad | \quad B_e A_m = B_g A_g = \phi \quad , \tag{1.60}$$

then material properties:

$$\begin{aligned} B_g &= \mu_0 H_g \quad | \quad B_g = \mu_0 H_g \\ B_m &= \mu_R H_m + B_r \quad | \quad B_e = \mu_R H_e \end{aligned} \quad . \tag{1.61}$$

1.4. Materials Used in Electric Machines

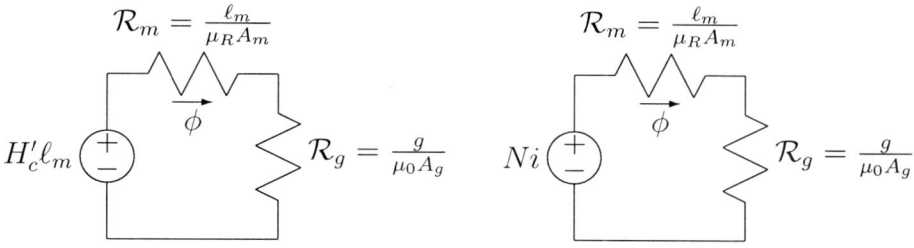

Figure 1.13: The equivalent magnetic circuits of the structures shown in Fig. 1.11.

Using Eqs. 1.60 and 1.61 in Eq. 1.59 gives

$$\frac{B_m - B_r}{\mu_R}\ell_m + \frac{B_g}{\mu_0}g = 0 \quad | \quad \frac{B_e}{\mu_R}\ell_m + \frac{B_g}{\mu_0}g = Ni \quad . \tag{1.62}$$

Rearranging the result for the circuit with the magnet gives

$$\frac{B_m}{\mu_R}\ell_m + \frac{B_g}{\mu_0}g = \frac{B_r}{\mu_R}\ell_m \quad . \tag{1.63}$$

It follows that the two magnetic circuits are equivalent when

$$\frac{B_r}{\mu_R}\ell_m = Ni \quad , \tag{1.64}$$

or

$$H'_c\ell_m = Ni \quad . \tag{1.65}$$

This suggests that each millimeter in thickness of a NdFeB sintered permanent magnet is equivalent to 636.6 Ampere-turns, using the data from Table 1.2. For small motors, it is difficult for current and copper to match the space efficiency of permanent magnets.

Figure 1.13 shows the two magnetic circuits that result from introducing the magnetic flux ϕ into Eq. 1.62. In each circuit, the magnet is treated as a reluctance based on its dimensions and the recoil permeability μ_R. In the circuit describing the magnet, the magnetomotive force is tied to the effective coercive force of the magnet.

In our demonstration of equivalence between a coil and a magnet, we ended up with the magnet being modeled as a source of magnetomotive force in series with a reluctance. In circuit terms, the magnet is modeled as a Thevenin source with an ideal voltage (mmf) source acting behind a source resistance (reluctance). It is also possible to model the magnet as a source of flux in parallel with a permeance, as shown in Fig. 1.14.

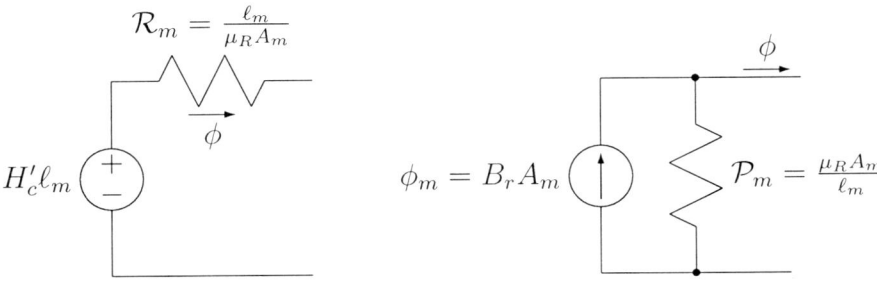

Figure 1.14: The Thevenin and Norton models of a permanent magnet.

1.4.4 Losses

This section has focused almost exclusively on ferromagnetic materials, due mostly to their nonlinear nature. In concluding, it is worth noting that losses occur within the materials of an electric machine, and ultimately it is the power dissipation that usually will dictate the physical size of the machine. Losses within an electric machine include:

Ferromagnetic materials: Losses within the magnetic materials are due to hysteresis as the magnetic domains are coerced into following the applied excitation. Losses also occur within the magnetic materials due to conduction. Soft and hard ferromagnetic materials not only conduct flux, but electric currents as well. Any electric machine has regions that are subjected to time-varying magnetic fields. Accordingly, there are currents induced into the magnetic materials through interaction of the laws of Ampere, Faraday, and Ohm. These losses, often termed eddy current losses, are proportional to the square of the peak flux density and the frequency. Eddy current losses are managed by laminating the magnetic core of the machine, or by using sintered powder metallurgy to effectively decrease the electrical conductivity without substantially impacting the magnetic properties. Permanent magnet materials are prone to these loss mechanisms also. Ignoring magnet losses can lead to disappointing performance because the magnetization will be reduced.

The combination of hysteresis and eddy current losses is generally summarized with curves that give the loss density, usually in W/lb or W/kg, as a function of peak flux density and frequency. Figure 1.15 gives an example of core loss information for M-19, 29 gage nonoriented electrical steel, a common silicon sheet steel for electric machines[3].

---

[3]United States Steel, "Nonoriented sheet steel for magnetic applications," May 1978.

1.4. Materials Used in Electric Machines

Nonoriented steels are generally used in electric machines because the magnetization properties are not oriented in a specific direction. This loss information can be summarized analytically using

$$w = kf^m B^n \quad , \tag{1.66}$$

where w is the core loss density, f is the electrical frequency, B is the peak flux density, and k, m, and n are constants. Typically m and n are between one and two, representing the combination of hysteresis and eddy current losses. The structure of Eq. 1.66 is attributed to Steinmetz.

Conductors: Losses within the magnetic materials are due to the conduction of currents in the face of finite electrical conductivity. The currents will have a tendency to redistribute themselves closer to the edges of the conductors as the frequency is increased. This is a magnetic diffusion phenomena governed by the laws of Ampere, Faraday, and Ohm that is commonly referred to as the skin effect. The skin effect motivates winding electric machines with conductors of a size appropriate for the frequency; additional conducting area is created by paralleling conductors, generally referred to as multiple conductors in hand. Further, the currents in the conductors will have a tendency to redistribute themselves because of local magnetic fields that are created by conductors in close proximity. This is known as the proximity effect.

1.4.5 Eddy Currents and Laminations

The eddy current contribution to the core losses motivates building electric machines out of stacks of laminated sheets of high permeability steel. In addition to the loss considerations, it can be shown that eddy currents act to try and keep time-varying magnetic fields from penetrating a conducting material. Figure 1.16 shows a block of material with electrical conductivity σ and permeability μ upon which a time-varying magnetic field is imposed. The time-varying magnetic field induces an electric field into the conductor due to Faraday's law. A current flows along the electric field because of Ohm's law. The induced current produces a magnetic field by virtue of Ampere's law. The induced field acts to oppose time variations in the applied field, effectively pushing the applied field toward the edges of the material.

Eddy currents are the result of a magnetic diffusion problem that reflects the competition between the rate at which magnetic fields can penetrate a conductor and the rate at which the fields are varying. The rate at which the fields can penetrate a conductor is tied to the conductivity and permeability; in a lumped-parameter circuit this would be described by an L/R

time constant. The rate at which the magnetic fields are varying is tied to the reciprocal of the excitation frequency. Since time variation can be created by relative motion, the diffusion of magnetic fields into a moving conductor depends on material velocity as well[4]. A characteristic parameter of magnetic diffusion problems is the skin depth, given by

$$\delta = \sqrt{\frac{2}{\omega\mu\sigma}} \quad , \quad (1.67)$$

where ω is the radian frequency of the applied field. The skin depth is the distance over which the magnetic field strength drops to $1/e$ of its value at the surface of the material.

To keep eddy currents from preventing the applied field from reaching the middle of a conducting material, the material is built up of thin laminations that are electrically insulated from one another. The idea is illustrated in Fig. 1.17. Breaking the material up into a stack of laminations makes it far more difficult for eddy currents to flow. Effectively, the electrical conductivity of the material has been reduced, thereby increasing the resistance. The skin depth suggests that the lamination thickness should be selected so that it is no thicker than two skin depths. If it is allowed to be thicker, the center of the material will not be used effectively, and eddy currents can produce appreciable heating.

1.5 Mechanical Elements

Since electric machines are an interface between the electrical and mechanical domains, we must be as concerned with the mechanical dynamics as we are with the electrical dynamics. While the electrical dynamics are described through Kirchhoff's laws, the mechanical dynamics are described through Newton's laws.

Table 1.4 summarizes common mechanical elements that influence the mechanical dynamics, in addition to the forces and torques of magnetic origin that will be introduced in Chapter 2. Each of these elements is associated with a force (translational) or torque (rotational) that acts on the mechanical system.

In Chapter 2 we are going to examine the coupling between the electrical and mechanical systems. Because of this coupling, we must be concerned

---

[4]See, for example, H. H. Woodson and J. R. Melcher, *Electromechanical Dynamics, Part II: Fields, Forces, and Motion*, Wiley 1968, Chapter 7, and J. R. Melcher, *Continuum Electromechanics*, MIT Press, 1981, Chapter 6, for a rigorous treatment of magnetic diffusion.

Table 1.4: Common mechanical elements that contribute to mechanical dynamics.

	Translational	Rotational
Mass, Inertia	M	H
Spring	$k(x - X_0)$	$k(\theta - \theta_0)$
Windage	$B(dx/dt)^2$	$B(d\theta/dt)^2$
Viscous Damper	$B(dx/dt)$	$B(d\theta/dt)$

with the self-consistent resolution of dynamic equations that describe both the electrical and mechanical subsystems. In many electric machine systems, the electrical dynamics are sufficiently fast relative to the mechanical dynamics that useful advantage can be taken of a time-scale separation that allows the resolution of the electrical dynamics to assume constant mechanical motion. The mechanical dynamics would consider the electrical system to respond infinitely fast.

1.6 Summary

The objective of this chapter has been to lay a foundation for the work that is to come in subsequent chapters. Beginning with Maxwell's equations, we motivated a useful decoupling between the evolution of the electric and magnetic fields. This simplification does not mean that the fields are static. Rather, it means that we can ignore wave phenomena when the waves travel through our system in times that are very short relative to the characteristic time of interest.

This chapter has also looked at the magnetic materials that are used to build electric machines. Soft magnetic materials are used to direct the flux to where it can be used to convert energy. Hard magnetic materials are used to replace windings with currents. Modern high energy magnet compositions based on rare earth elements can put substantial energy into a volume that is considerably smaller than would be required of coils and current, without the dissipation caused by current flow.

The analogy between electric and magnetic circuits was developed as an alternative to starting from first principles in analyzing a magnetic structure. Like all analogies, care must be taken to keep the analogy within its range of applicability.

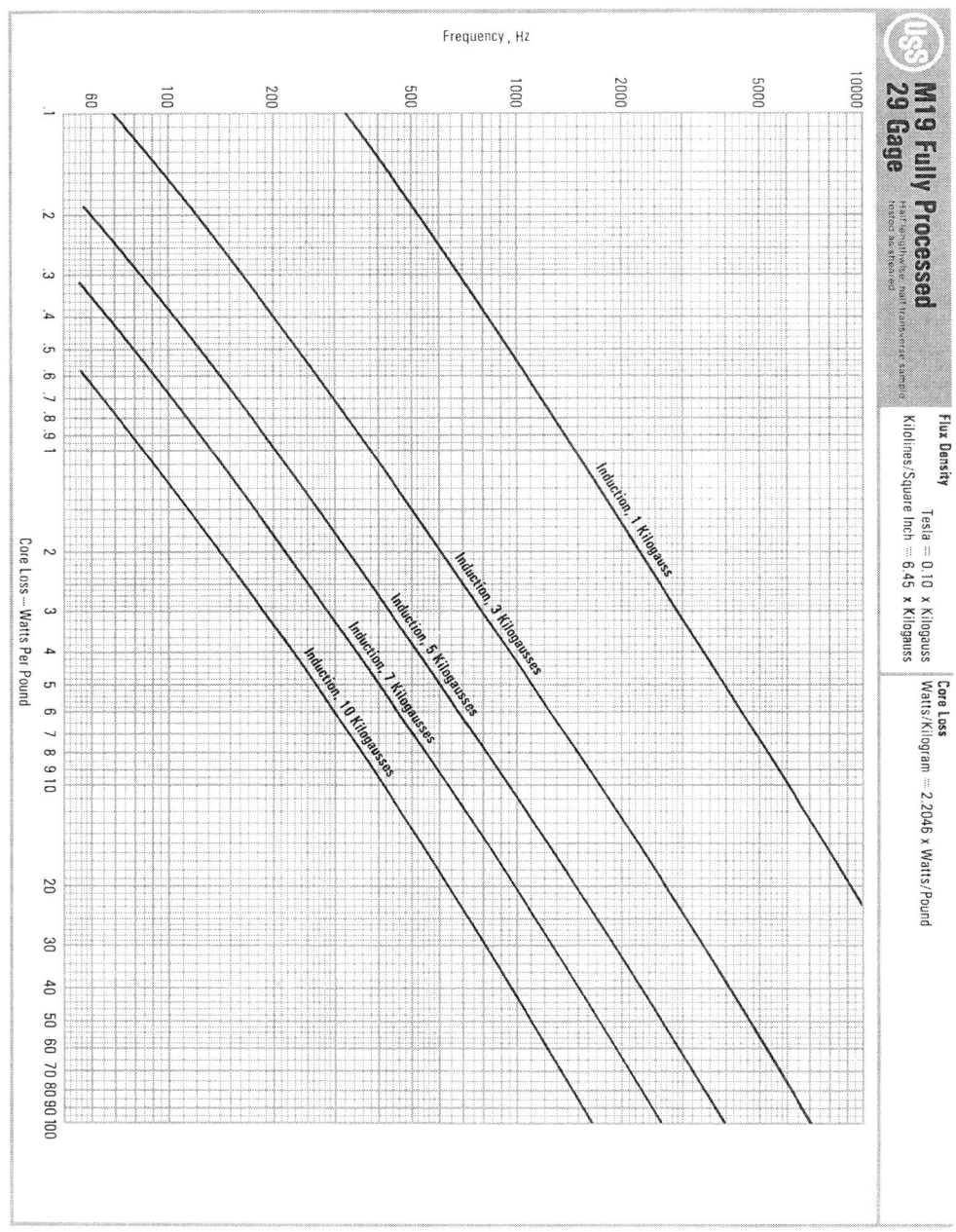

Figure 1.15: Core loss data for M-19 29 gage nonoriented electrical sheet steel.

1.6. Summary

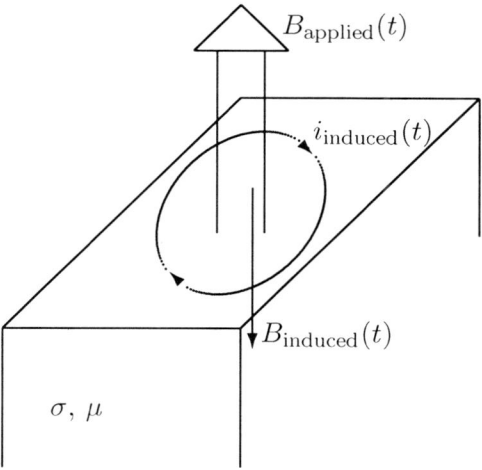

Figure 1.16: A block of conducting material into which eddy currents are induced by a time-varying magnetic field.

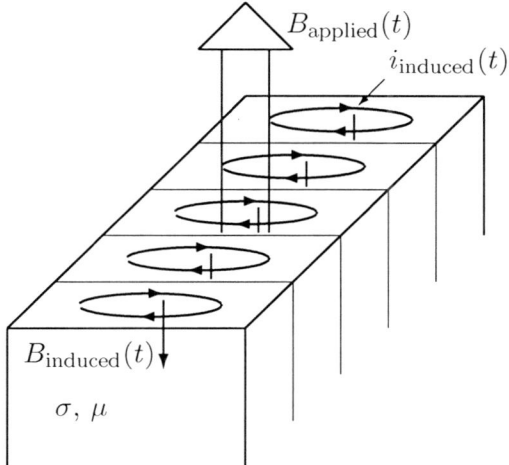

Figure 1.17: The use of laminations to mitigate the creation of eddy currents.

Chapter 2

Principles of Electromechanical Energy Conversion

2.1 Introduction

In this chapter we build on the elements of field theory developed in Chapter 1 as we start to consider how things are made to move. Motion requires one magnetic member (often called the rotor) to be free to move relative to another magnetic member (often called the stator). In addition, magnetic field quantities must depend on the physical position of one or more mechanical components. This requires us to generalize our concept of inductance to include mechanical degrees of freedom.

Consider Fig. 2.1 as it relates to a generic system. On the electrical side, we have a network that must satisfy Kirchhoff's voltage and current laws. Fundamentally, Kirchhoff's laws stem from electromagnetic field theory. (Faraday's law gives rise to Kirchhoff's voltage law, and charge conservation gives rise to Kirchhoff's current law.) On the mechanical side, components must satisfy Newton's laws of motion. Electromechanical coupling is created when the magnetic fields established by currents within the electrical network are able to spontaneously interact with the mechanical components. In this way mechanical motion influences the evolution of voltages and currents, just as the torques created by currents influence the evolution of mechanical displacements.

A consequence of electromechanical coupling is that mechanical degrees of freedom (typically one or more displacements) will appear in our descrip-

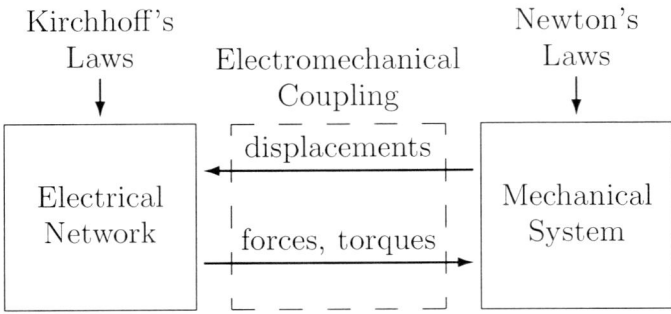

Figure 2.1: A conceptual view of electromechanical coupling.

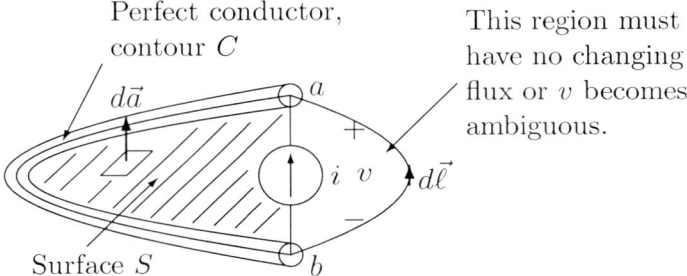

Figure 2.2: A generalized inductor used to explore electromechanical coupling. Some of the inductor can be displaced.

tion of the electrical network. Similarly, a force or torque of magnetic origin will appear in the formulation of Newton's laws for the mechanical components. As we will see, certain conditions must be satisfied in order to produce steady torque and rotation. The developments in this chapter are fundamental to all electric machines.

2.2 Electromechanical Coupling

We can better understand the electromechanical coupling process by considering a generalized form of inductance, in which our inductor is dependent on some mechanical displacement. Our generalized inductor is depicted in Fig. 2.2.

Figure 2.2 shows an inductor as a single turn of a perfect conductor that

2.2. Electromechanical Coupling

defines a contour C and an associated surface S. The inductor is fed by a current source i. The incremental area $d\vec{a}$ associated with surface S has a direction consistent with the counter-clockwise integration path applied to contour C using $d\vec{\ell}$. Outside of the area bounded by the conductor there is no changing flux, or the induced voltage v becomes ambiguous.

Applying Faraday's law to the contour C gives:

$$\oint_C \vec{E} \cdot d\vec{\ell} = \int_a^b \vec{E} \cdot d\vec{\ell} + \int_b^a \vec{E} \cdot d\vec{\ell} = -\frac{d}{dt} \int_S \vec{B} \cdot d\vec{a} \quad . \tag{2.1}$$

Of the two integrals acting on the electric field, the first is zero, because the electric field is zero within the perfect conductor. The second integral of electric field is $-v$. The integral of the flux linkage over the area of the conductor gives the total flux linking our generalized inductor.

Defining the flux linking the inductor to be

$$\lambda = \int_S \vec{B} \cdot d\vec{a} \quad , \tag{2.2}$$

it follows that the induced voltage at the terminals of the inductor is given by

$$v = \frac{d\lambda}{dt} \quad . \tag{2.3}$$

The terminal description of our generalized inductor is related to the field description through the MQS version of Maxwell's equations.

Now, we assume that:

1. System geometry is fixed except for one moveable part whose position can be described through displacement x relative to a fixed reference.

2. The magnetization $\vec{M}$ can be expressed as a function of field quantities alone (and therefore current). That is, this argument does not consider permanent magnets. Given that permanent magnets can be represented by coils with constant current, as we showed in Sec. 1.4, this is not a significant restriction.

It follows that $\lambda = \lambda(i, x)$. Therefore,

$$v = \frac{d\lambda}{dt} = \left.\frac{\partial \lambda}{\partial i}\right|_x \frac{di}{dt} + \left.\frac{\partial \lambda}{\partial x}\right|_i \frac{dx}{dt} \quad . \tag{2.4}$$

The term that depends on the time rate of change in current is referred to as the *transformer voltage*. The term that depends on the time rate of change in displacement is called the *motional voltage*. For linear magnetics $\lambda = L(x)i$ and our voltage expression becomes

$$v = L(x)\frac{di}{dt} + i\frac{dL(x)}{dx}\frac{dx}{dt} \quad . \tag{2.5}$$

We can now generalize to any number of currents and displacements, such that

$$\lambda_k = \lambda_k(i_1, i_2, \ldots, i_k, \ldots, i_N, x_1, x_2, \ldots, x_M) \quad , \tag{2.6}$$

and

$$v_k = \sum_{j=1}^{N} \left.\frac{\partial \lambda_k}{\partial i_j}\right|_x \frac{di_j}{dt} + \sum_{j=1}^{M} \left.\frac{\partial \lambda_k}{\partial x_j}\right|_i \frac{dx_j}{dt} \quad . \tag{2.7}$$

In Sec. 4.4 we will see how important the motional voltages are in describing the behavior of electric machines. The motional voltage in electric machines is commonly referred to as the back emf.

Consider the magnetic circuit shown in Fig. 2.3 which includes a plunger that is constrained to move only in the x direction. This is the magnetic circuit of Fig. 1.4, except the plunger can now move vertically. Analysis of this circuit is straightforward under the following assumptions:

1. The permeability of the plunger and core is infinite.

2. The air gaps g and x are sufficiently small that fringing can be ignored.

3. Leakage flux is negligible. (This is consistent with the assumption of infinite permeability.)

Contour C_2 tells us that the magnetic fields in the air gaps on either side of the plunger must be equal in magnitude and opposite in direction, which we will take to be $\vec{H}_1$ directed from the plunger toward the core. Applying Ampere's law to contour C_1 in the clockwise direction gives

$$H_2 x + H_1 g = Ni \quad . \tag{2.8}$$

Gauss's law applied to the plunger gives

$$-\mu_0 H_2 2wd + 2\mu_0 H_1 wd = 0 \quad \Rightarrow \quad H_1 = H_2 \quad . \tag{2.9}$$

Using this result in Ampere's law gives

$$H_1 = H_2 = \frac{Ni}{x+g} \quad . \tag{2.10}$$

2.2. Electromechanical Coupling

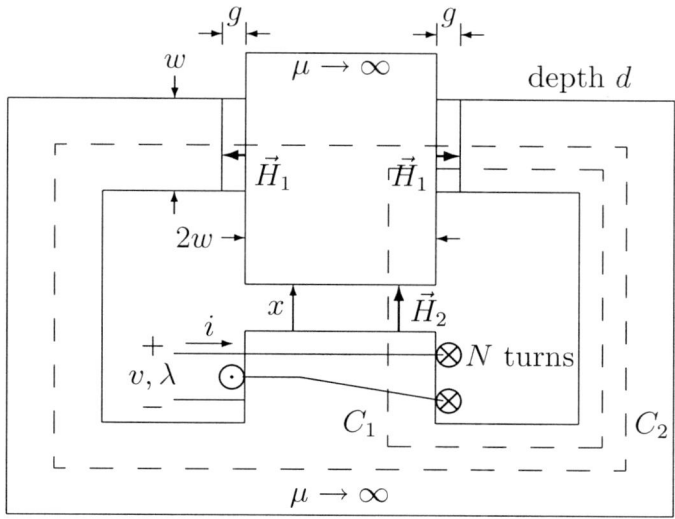

Figure 2.3: A magnetic circuit example that exhibits electromechanical coupling.

The magnetic flux traveling through the center leg of the magnetic circuit is

$$\phi = \int \vec{B}_2 \cdot d\vec{a} = \mu_0 2wd H_2 = \frac{\mu_0 2wd Ni}{x+g} \quad . \tag{2.11}$$

The flux ϕ links each of the N turns in the winding. The total flux linking the winding is, therefore,

$$\lambda = N\phi = \frac{\mu_0 2wd N^2 i}{x+g} \quad . \tag{2.12}$$

From this expression we can identify the inductance as

$$L(x) = \frac{\lambda}{i} = \frac{\mu_0 2wd N^2}{x+g} \quad . \tag{2.13}$$

It follows that the voltage measured at the terminals of the coil is

$$v = \frac{\mu_0 2wd N^2}{x+g} \frac{di}{dt} - \frac{\mu_0 2wd N^2 i}{(x+g)^2} \frac{dx}{dt} \quad , \tag{2.14}$$

showing that the induced voltage is tied to both variations in current and movement of the plunger.

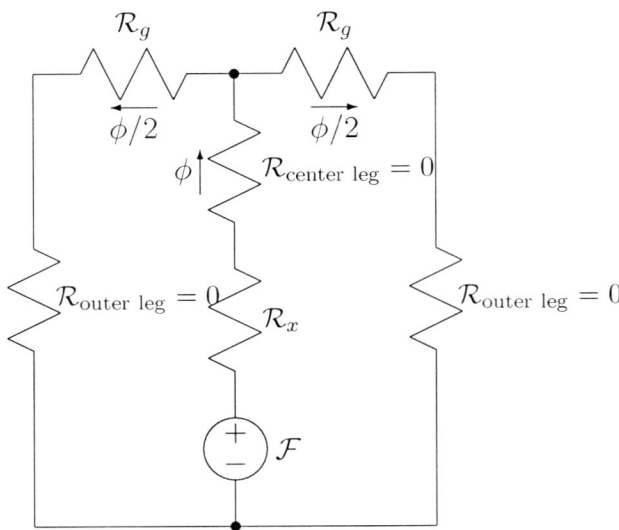

Figure 2.4: The electric analog of the magnetic circuit shown in Fig. 2.3.

An analysis using magnetic circuit concepts would yield the same result. Figure 2.4 summarizes the electric analog of the magnetic circuit structure. Because of the infinite permeability of the core and plunger, the core reluctances are all zero. The reluctances for the air gaps are

$$\mathcal{R}_g = \frac{g}{\mu_0 wd} \quad ; \tag{2.15}$$

$$\mathcal{R}_x = \frac{x}{\mu_0 2wd} \quad , \tag{2.16}$$

and the flux through the center leg is

$$\phi = \frac{Ni}{\frac{g}{\mu_0 2wd} + \frac{x}{\mu_0 2wd}} \quad , \tag{2.17}$$

giving the flux linkage

$$\lambda = N\phi = \frac{\mu_0 2wdN^2 i}{x + g} \quad , \tag{2.18}$$

the same result we obtained from first principles.

Consider the simple electric machine shown in Fig. 2.5, which has a rotor that can rotate within a stator. The rotor and stator each carry a winding that is embedded in slots within the core. The rotor is mounted on bearings

2.2. Electromechanical Coupling

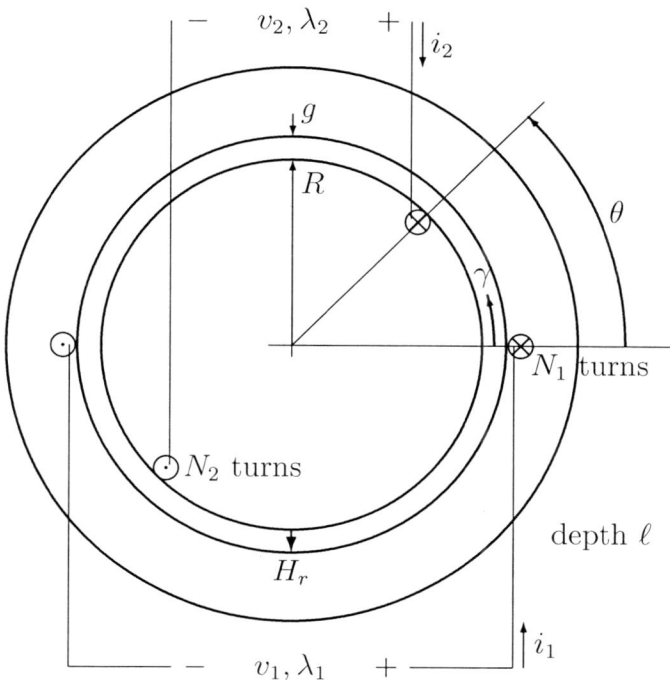

Figure 2.5: A primitive electric machine that exhibits electromechanical coupling.

so that it is constrained to move only in the θ direction. Similar to the last example, analysis of this machine is straightforward under the following assumptions:

1. The permeability of the stator and rotor cores is infinite.

2. The air gap g is sufficiently small that magnetic field fringing can be ignored at the ends of the machine.

3. The slots are sufficiently small that the fields are not perturbed.

We take the magnetic field to be positive when directed radially outward.

Application of Ampere's law to the simple machine of Fig. 2.5 is used to determine the magnetic field within each of the four regions of the air gap. Two contours are used to determine the radial magnetic field, each contour consisting of a semicircle. The first contour starts at the center of the machine, goes radially outward crossing the air gap between $\gamma = 0$ and $\gamma = \theta$, then circles around through the stator core (often referred to as the back iron), crossing the air gap again between $\gamma = \pi$ and $\gamma = \pi + \theta$. The second contour is similar to the first, but it crosses the air gap first between $\gamma = \theta$ and $\gamma = \pi$; the second crossing is between $\gamma = \pi + \theta$ and $\gamma = 2\pi$.

Symmetry is used to conclude that the diametrically-opposed magnetic fields must be equal in magnitude and of opposite sign, consistent with Gauss's law[1]. The air gap magnetic fields are:

$$0 < \gamma \leq \theta : \quad H_r = \frac{N_1 i_1 - N_2 i_2}{2g} \quad ; \quad (2.19)$$

$$\theta < \gamma \leq \pi : \quad H_r = \frac{N_1 i_1 + N_2 i_2}{2g} \quad ; \quad (2.20)$$

$$\pi < \gamma \leq \pi + \theta : \quad H_r = \frac{-N_1 i_1 + N_2 i_2}{2g} \quad ; \quad (2.21)$$

$$\pi + \theta < \gamma \leq 2\pi : \quad H_r = \frac{-N_1 i_1 - N_2 i_2}{2g} \quad . \quad (2.22)$$

The flux linking winding 1 is found by integrating along the surface of the rotor. We can do this because this is the same flux passing through the plane of the winding as required by Gauss's law. The flux linking winding 1 is given by

$$\lambda_1 = N_1 \int_0^\pi \mu_0 H_r \ell R d\gamma = L_{11} i_1 + L_{12} i_2 \quad , \quad (2.23)$$

and the flux linking winding 2 is given by

$$\lambda_2 = N_2 \int_\theta^{\pi+\theta} \mu_0 H_r \ell R d\gamma = L_{21} i_1 + L_{22} i_2 \quad , \quad (2.24)$$

where the inductances are

$$L_{11} = \frac{\mu_0 \pi N_1^2 \ell R}{2g} \quad , \quad (2.25)$$

$$L_{22} = \frac{\mu_0 \pi N_2^2 \ell R}{2g} \quad , \quad (2.26)$$

$$L_{12} = L_{21} = \begin{array}{l} \frac{\mu_0 \pi N_1 N_2 \ell R}{2g} \left(1 - \frac{2\theta}{\pi}\right) \text{ for } 0 < \theta <= \pi \quad , \quad (2.27) \\ \frac{\mu_0 \pi N_1 N_2 \ell R}{2g} \left(1 + \frac{2\theta}{\pi}\right) \text{ for } -\pi < \theta <= 0 \quad . \quad (2.28) \end{array}$$

---

[1]If the symmetry in the field structure is not clear, apply Gauss's law to the surface of the rotor. Factoring according to the geometry terms will require diametrically opposed fields to be equal and opposite in order to have Gauss's law satisfied at all rotor positions.

2.3. Electromagnetic Forces

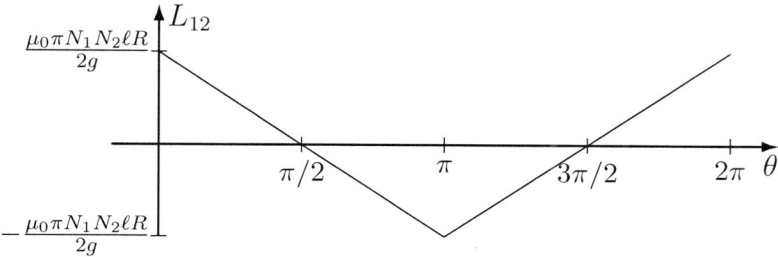

Figure 2.6: The mutual inductance of the simple machine in Fig. 2.5.

The inductances L_{12} and L_{21} are known as mutual inductances because they represent the coupling between windings 1 and 2. Inductance L_{12} represents the flux linking winding 1 due to currents in winding 2. Similarly, inductance L_{21} represents the flux linking winding 2 due to currents in winding 1. A plot of the mutual inductance is given in Fig. 2.6. Note that the mutual inductance is maximum when coils 1 and 2 take on the same physical orientation. It is also significant that the mutual inductance is zero when the coils are orthogonal. Real electric machines would be wound to reduce the space harmonic components of the mutual inductance so that torque ripple is minimized. Chapter 3 addresses shaping the air gap magnetic field through proper design of the windings. Chapter 4 discusses the implications of space harmonics, and their interaction with time harmonics in the current waveform.

In both of these examples, the dependence of electrical parameters on mechanical displacement has been made explicit. Since mechanical displacement influences the electrical dynamics, we should expect that the electrical dynamics will impact the mechanical dynamics. Section 2.4 develops this aspect of electromechanical coupling.

2.3 Electromagnetic Forces

Often the starting point for the description of electromagnetic forces is the Lorentz force equation:

$$d\vec{f}^e = \vec{J} \times \vec{B} \quad . \tag{2.29}$$

This expression is a force density (force per unit volume), given the dimensions of the current density and the magnetic flux density. For a conductor placed in a magnetic field, integration over the conductor volume yields the force to be

$$\vec{f}^e = l\vec{i} \times \vec{B} \quad , \tag{2.30}$$

where l is the length of the conductor that is exposed to the flux density. This assumes that the cross-sectional area of the conductor is uniform, and the $\vec{B}$ field to which the conductor is exposed is uniform. Otherwise, Eq. 2.29 must be integrated over the volume taken up by the conductor.

In many electric machines, working with Eq. 2.30 is cumbersome, because the current may not be known precisely, or, more likely, the $\vec{B}$ field is difficult to quantify in the vicinity of the conductors. In the next section, we will develop a powerful approach for determining forces and torques using conservation of energy. However, before we leave this discussion of the Lorentz force, it is worthwhile to note that the force density can be expressed in terms of the fields only, because we know how the current density is related to the magnetic field. Incorporating Ampere's law (Eq. 1.26) into the Lorentz force density gives

$$d\vec{f}^e = \left(\nabla \times \vec{H}\right) \times \vec{B} \quad , \tag{2.31}$$

or, for linearly magnetizable materials,

$$d\vec{f}^e = \mu \left(\nabla \times \vec{H}\right) \times \vec{H} \quad . \tag{2.32}$$

Using a vector identity[2], Eq. 2.32 can be written as

$$d\vec{f}^e = \mu \left\{ \left(\vec{H} \cdot \nabla\right) \vec{H} - \frac{1}{2} \nabla \left(\vec{H} \cdot \vec{H}\right) \right\} \quad . \tag{2.33}$$

There are three components to this expression. There are some manipulations that are easier if they are carried out on a component-by-component basis. This is most easily handled using index notation[3].

Assume a right-hand Cartesian coordinate system based on x_1, x_2, and x_3. The component of a vector in the direction of an axis carries the subscript of that axis, so the component of a force in the x_1 direction would be f_1. In general we describe the force by f_m, where m could be 1, 2, or 3. When we use the differential operator $\partial/\partial x_n$, we mean $\partial/\partial x_1$, $\partial/\partial x_2$, or $\partial/\partial x_3$. When an index is repeated in a single term, we sum over all three values of the index. For example,

$$\frac{\partial H_n}{\partial x_n} = \frac{\partial H_1}{\partial x_1} + \frac{\partial H_2}{\partial x_2} + \frac{\partial H_3}{\partial x_3} = \nabla \cdot \vec{H} \quad , \tag{2.34}$$

and

$$H_n \frac{\partial}{\partial x_n} = H_1 \frac{\partial}{\partial x_1} + H_2 \frac{\partial}{\partial x_2} + H_3 \frac{\partial}{\partial x_3} = \vec{H} \cdot \nabla \quad . \tag{2.35}$$

---

[2] $(\nabla \times \vec{A}) \times \vec{A} = (\vec{A} \cdot \nabla)\vec{A} - \frac{1}{2}\nabla(\vec{A} \cdot \vec{A})$

[3] For a discussion of index notation and its application to tensors, see, for example, A. I. Borisenko and I. E. Tarapov (translated by R. A. Silverman), *Vector and Tensor Analysis with Applications*, Dover, 1979.

2.3. Electromagnetic Forces

The practice of summing over all components is known as the summation convention. On the other hand, $\partial H_m/\partial x_n$ represents any one of the nine possible derivatives of components of $\vec{H}$ with respect to coordinates.

The Kronecker delta function δ_{mn} is defined as

$$\delta_{mn} = \begin{cases} 1 & m = n \\ 0 & m \neq n \end{cases} . \tag{2.36}$$

The Kronecker delta has the property

$$\delta_{mn} H_n = H_m , \tag{2.37}$$

and

$$\delta_{mn} \frac{\partial}{\partial x_n} = \frac{\partial}{\partial x_m} . \tag{2.38}$$

Using Eqs. 2.34 and 2.35 we can write the mth component of Eq. 2.33 as

$$df_m = \mu \left\{ H_n \frac{\partial H_m}{\partial x_n} - \frac{1}{2} \frac{\partial}{\partial x_m} (H_k H_k) \right\} . \tag{2.39}$$

Using the property of the Kronecker delta, this can be rewritten as

$$df_m = \mu \left\{ H_n \frac{\partial H_m}{\partial x_n} - \frac{1}{2} \delta_{mn} \frac{\partial}{\partial x_n} (H_k H_k) \right\} . \tag{2.40}$$

From the chain rule,

$$\frac{\partial}{\partial x_n} (H_n H_m) = H_n \frac{\partial H_m}{\partial x_n} + H_m \frac{\partial H_n}{\partial x_n} , \tag{2.41}$$

so Eq. 2.40 becomes

$$df_m = \mu \left\{ \frac{\partial}{\partial x_n} (H_n H_m) - \frac{1}{2} \delta_{mn} \frac{\partial}{\partial x_n} (H_k H_k) - H_m \frac{\partial H_n}{\partial x_n} \right\} . \tag{2.42}$$

We recognize the last term on the right as

$$H_m \left(\nabla \cdot \mu \vec{H} \right) = H_m \left(\nabla \cdot \vec{B} \right) = 0 , \tag{2.43}$$

so Eq. 2.42 reduces to

$$df_m = \frac{\partial T_{mn}}{\partial x_n} , \tag{2.44}$$

where

$$T_{mn} = \mu \left\{ (H_n H_m) - \frac{1}{2} \delta_{mn} (H_k H_k) \right\} \tag{2.45}$$

is referred to as the Maxwell stress tensor.

If we know $\vec{H}$ in a region of space, we can calculate the components of the stress tensor T_{mn}. We only need to calculate at most six terms since $T_{mn} = T_{nm}$. Differentiation of Eq. 2.45 as indicated in Eq. 2.44 gives the mth component of $\vec{J} \times \vec{B}$. We must use the total $\vec{H}$ to obtain the correct answer. That is, we must consider not only the portion of $\vec{H}$ associated with the applied field, but also the portion of $\vec{H}$ that is created by the free current density. It is worth noting that since permanent magnets can be represented as equivalent free currents, Eqs. 2.44 and 2.45 hold for bodies that contain either current or permanent magnets.

Equation 2.44 gives us a force density. To determine the total force acting on an object, we must integrate the force density over an appropriate volume. It follows that

$$f^e = \int_V df_m dV = \int_V \frac{\partial T_{mn}}{\partial x_n} dV \quad . \tag{2.46}$$

When we define the vector $\vec{A}$ as

$$\vec{A} = T_{m1}\vec{i}_1 + T_{m2}\vec{i}_2 + T_{m3}\vec{i}_3 \tag{2.47}$$

we can write Eq. 2.46 as

$$f^e = \int_V \frac{\partial A_n}{\partial x_n} dV = \int_V \left(\nabla \cdot \vec{A} \right) dV \quad . \tag{2.48}$$

We can now use the divergence theorem to write the force as the integral over a closed surface:

$$f^e = \oint_S \vec{A} \cdot \vec{n} da = \oint_S A_n n_n da \quad , \tag{2.49}$$

where n_n is the nth component of the normal vector of the outward-directed unit vector $\vec{n}$ normal to the surface S. The surface S encloses the volume V. Substitution from Eq. 2.47 gives

$$f^e = \oint_S T_{mn} n_n da \quad . \tag{2.50}$$

This tells us we can find the total force of magnetic origin on the matter in volume V by knowing only the fields on the surface S of the volume. This suggests we can find the force acting on the rotor of an electric machine by knowing only the fields in the air gap. A subtle but valuable aspect of working with the stress tensor is that we are free to choose any closed surface, so long as it encloses the object on which we seek the force. This allows us to choose a surface that is particularly convenient.

The stress tensor of Eq. 2.45 does not take magnetostriction into account. Magnetostriction is stress caused by gradients in the permeability.

2.3. Electromagnetic Forces

Other stress tensor formulations are able to accomplish this, but are beyond the scope of this discussion[4]. In electric machines, the effect of magnetostriction is generally small compared to the forces created by the interaction of magnetic fields.

The development of the Maxwell stress tensor was based on working with a Cartesian coordinate system. This was particularly convenient with the index notation since the spatial derivative in each direction was of exactly the same form. In other coordinate systems, such as cylindrical or spherical, the Maxwell stress tensor of Eq. 2.45 still applies, but one must be careful to interpret the spatial derivatives of Eq. 2.46 properly based on direction. For example, in the cylindrical coordinate system where $\vec{H}(r, \theta, z)$, the radial force density would be given by

$$df_r = \frac{\partial T_{rr}}{\partial r} + \frac{1}{r}\frac{\partial T_{r\theta}}{\partial \theta} + \frac{\partial T_{rz}}{\partial z} \quad , \tag{2.51}$$

where

$$T_{rr} = \frac{\mu}{2}\left(H_r^2 - H_\theta^2 - H_z^2\right) \quad , \tag{2.52}$$

$$T_{r\theta} = \mu H_r H_\theta \quad , \tag{2.53}$$

$$T_{rz} = \mu H_r H_z \quad , \tag{2.54}$$

consistent with Eq. 2.45.

We have seen that the Maxwell stress tensor is a symmetric ordered array of nine functions of space and time. The usual way to write this array is in matrix form as

$$T_{mn} = \begin{bmatrix} T_{11} & T_{12} & T_{13} \\ T_{21} & T_{22} & T_{23} \\ T_{31} & T_{32} & T_{33} \end{bmatrix} \quad , \tag{2.55}$$

where the first index marks the row and the second marks the column. The row index specifies the force component, and the column specifies the direction of the vector normal to the surface. The symmetry is about the major diagonal.

The integrand of our force expression (Eq. 2.50) has the dimension of force per unit area. Consistent with the summation convention with a repeated index, $T_{mn}n_n$ is the mth component of a vector that is referred to as the traction,

$$\tau_m = T_{mn}n_n = T_{m1}n_1 + T_{m2}n_2 + T_{m3}n_3 \quad . \tag{2.56}$$

---

[4]Other stress tensor formulations, including those that consider magnetostriction, are discussed in, for example: H. H. Woodson and J. R. Melcher, *Electromechanical Dynamics, Part II: Fields, Forces, and Motion*, Wiley 1968, Chapter 8; J. A. Stratton, *Electromagnetic Theory*, McGraw-Hill, 1941, Chapter 2; and J. R. Melcher, *Continuum Electromechanics*, MIT Press, 1981, Chapter 3.

The traction $\vec{\tau}$ is the vector force per unit area applied to a surface of arbitrary orientation, and the component T_{mn} of the stress tensor can be physically interpreted as the mth component of the traction applied to a surface with a normal vector in the n direction.

For a traditional radial field machine that is described in cylindrical coordinates, it is instructive to envision a cylindrical surface that encloses the rotor. It is particularly convenient to have the outside boundary of the surface fall in the cylindrical annulus of the air gap, perhaps just inside of the cylinder formed by the stator bore itself but much longer so we can take the fields acting on the ends of the surface to be zero. There are three important components to the traction:

1. The tangential component in the θ direction is responsible for creating torque on the rotor. This component is

$$\tau_\theta = T_{\theta r} = \mu_0 H_\theta H_r \quad , \tag{2.57}$$

where μ_0 is appropriate because it is assumed we are determining the traction on the rotor through integration over a surface in the air gap. There is only one term in the expression because the normal of our cylindrical surface is in the radial direction. The structure of this term will be a useful starting point when we discuss some elements of machine design in Chapter 5.

2. The tangential component in the z direction is responsible for trying to keep the rotor magnetically centered within the stator bore. This component is

$$\tau_z = T_{zr} = \mu_0 H_z H_r \quad . \tag{2.58}$$

3. The normal component in the r direction can be viewed as an attraction force between the rotor and stator. This component is

$$\tau_r = T_{rr} = \frac{\mu_0}{2} \left(H_r^2 - H_\theta^2 - H_z^2 \right) \quad . \tag{2.59}$$

Since the θ and z components together form the fields tangential to the rotor, we see that the radial force depends on the difference in magnitudes (squared) between the normal and tangential fields. By virtue of the different expressions for the tangential and radial traction, there is some opportunity to simultaneously control both torque and radial force. This can be valuable in providing insight into the creation and control of vibrations.

2.4. Energy, Force, and Torque 43

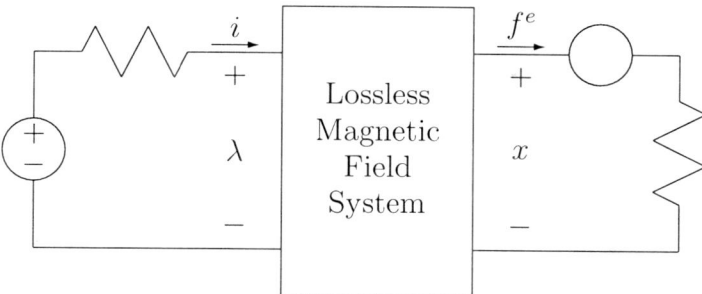

Figure 2.7: A lossless magnetic system and its interface to an electrical and mechanical system.

2.4 Energy, Force, and Torque

Figure 2.7 shows a generic lossless electromechanical system. Any losses in a physical system can still be modeled, but for purposes of this discussion, they are removed from the magnetic field system. Through conservation of energy we are able to develop expressions for the force (or torque in a rotary system) of magnetic origin. This approach is sometimes known as the method of virtual work.

On the electrical side of the system, we assume that the flux linkage is a function of current and mechanical position. That is, $\lambda = \lambda(i, x)$. In this formulation, the current i and the displacement x dictate the state of the system. Alternatively, we could choose a formulation that describes the current as a function of flux linkage and displacement. That is, $i = i(\lambda, x)$ where λ and x dictate the state of the system.

On the mechanical side of the system, we assume that $f^e = f^e(i, x)$ or $f^e = f^e(\lambda, x)$. Note that the functions are different but should give the same values of f^e for a consistent set of i, λ, and x.

Take the total energy in the magnetic field to be W_m. Conservation of energy applied to our lossless system gives

$$\frac{dW_m}{dt} = i\frac{d\lambda}{dt} - f^e\frac{dx}{dt} \quad , \tag{2.60}$$

where $d\lambda/dt$ is the voltage given by Faraday's law at the terminals of our lossless system, and the sign of the force f^e is based on positive f^e acting to increase x. The quantity $i\, d\lambda/dt$ is the electrical power input, and $f^e\, dx/dt$ is the mechanical power output, typically through a rotating shaft. (Here we refer to electrical power in and mechanical power out because of our reference

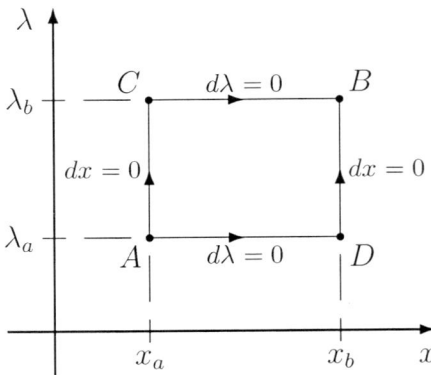

Figure 2.8: Two integration paths through the energy state space.

sign conventions given in Fig. 2.7. There is nothing in our argument that prevents these quantities from taking on any particular sign.)

The structure of Eq. 2.60 shows that the energy is a function of flux linkage and displacement. That is,

$$dW_m = i d\lambda - f^e dx \quad , \tag{2.61}$$

implying $W_m = W_m(\lambda, x)$. Equation 2.61 shows that the change in magnetic energy can be computed in a straightforward manner as

$$W_m(\lambda_b, x_b) - W_m(\lambda_a, x_a) = \int_{\lambda_a}^{\lambda_b} i(\lambda, x_a) d\lambda - \int_{x_a}^{x_b} f^e(\lambda_b, x) dx \quad , \tag{2.62}$$

or

$$W_m(\lambda_b, x_b) - W_m(\lambda_a, x_a) = -\int_{x_a}^{x_b} f^e(\lambda_a, x) dx + \int_{\lambda_a}^{\lambda_b} i(\lambda, x_b) d\lambda \quad . \tag{2.63}$$

The differences between Eqs. 2.63 and 2.62 are subtle and can be understood with reference to Fig. 2.8. State changes in a lossless system are independent of integration path. Accordingly, we can develop the energy difference between points A and B by integrating along any path of interest. Equation 2.63 corresponds to integration along the vertical path from A to C first, then integration along the horizontal path from C to B second. Equation 2.62 corresponds to integration along the horizontal path from A to D first, then integration along the vertical path from D to B second.

Since we define electrical terminal pairs that account for the excitation of all electric or magnetic fields in the system, when the electrical terminal variables are zero ($\lambda_a = 0$) we can say that there is no force of electric

2.4. Energy, Force, and Torque

origin. This makes the integration of Eq. 2.63 easier to compute, since the integration over x yields zero because $f^e = 0$. It must be emphasized that this formulation runs into trouble in the presence of permanent magnets. However, because of our ability to model permanent magnets as equivalent windings, we can resolve our formulation to include permanent magnets. This will be done at the end of this section.

Since $W_m = W_m(\lambda, x)$, incremental changes in W_m are determined using the chain rule:

$$dW_m = \left.\frac{\partial W_m}{\partial \lambda}\right|_x d\lambda + \left.\frac{\partial W_m}{\partial x}\right|_\lambda dx \quad . \tag{2.64}$$

Comparing this with Eq. 2.61 requires us to conclude that

$$i = \left.\frac{\partial W_m}{\partial \lambda}\right|_x \quad , \tag{2.65}$$

and

$$f^e = -\left.\frac{\partial W_m}{\partial x}\right|_\lambda \quad . \tag{2.66}$$

This can be generalized to any number of electrical and mechanical terminal pairs. For N electrical terminal pairs and M mechanical terminal pairs

$$dW_m = \sum_{j=1}^{N} \left.\frac{\partial W_m}{\partial \lambda_j}\right|_x d\lambda_j + \sum_{j=1}^{M} \left.\frac{\partial W_m}{\partial x_j}\right|_\lambda dx_j \quad , \tag{2.67}$$

from which we conclude

$$i_j = \left.\frac{\partial W_m}{\partial \lambda_j}\right|_x \quad , \tag{2.68}$$

and

$$f^e_j = -\left.\frac{\partial W_m}{\partial x_j}\right|_\lambda \quad . \tag{2.69}$$

The problem with this formulation is that flux is hard to measure, and in practice it is more involved to control flux than current. For this reason, we modify our formulation to make use of current rather than flux linkage.

From Eq. 2.61 we have,

$$dW_m = id\lambda - f^e dx \quad . \tag{2.70}$$

If we note that

$$id\lambda = d(\lambda i) - id\lambda \quad , \tag{2.71}$$

Eq. 2.70 can be rewritten as

$$d(\lambda i - W_m) = \lambda di + f^e dx \quad . \tag{2.72}$$

We define the coenergy to be

$$W'_m = \lambda i - W_m \quad . \tag{2.73}$$

Proceeding in the same manner as we did for the energy formulation, we conclude that for N electrical terminal pairs and M mechanical terminal pairs

$$dW'_m = \sum_{j=1}^{N} \left. \frac{\partial W'_m}{\partial i_j} \right|_x di_j + \sum_{j=1}^{M} \left. \frac{\partial W'_m}{\partial x_j} \right|_\lambda dx_j \quad , \tag{2.74}$$

from which we conclude

$$\lambda_j = \left. \frac{\partial W'_m}{\partial i_j} \right|_x \quad , \tag{2.75}$$

and

$$f^e_j = \left. \frac{\partial W'_m}{\partial x_j} \right|_i \quad . \tag{2.76}$$

The relationship between energy and coenergy can be better understood with reference to Fig. 2.9. In Fig. 2.9, the relationship between flux linkage and current is shown for a particular value of displacement. This curve of $\lambda = \lambda(i, x)$ serves to divide the rectangular region defined by λi into its constituent pieces of energy and coenergy. Since $dW_m = id\lambda$ for $dx = 0$, the total energy up to flux linkage λ_0 and current i_0 is given by

$$W_m = \int_0^{\lambda_0} i d\lambda' \quad . \tag{2.77}$$

This corresponds to the area above the $\lambda = \lambda(i, x)$ curve. On the other hand, $dW'_m = \lambda di$ for $dx = 0$, and the total coenergy up to flux linkage λ_0 and current i_0 is given by

$$W'_m = \int_0^{i_0} \lambda di' \quad , \tag{2.78}$$

corresponding to the area below the $\lambda = \lambda(i, x)$ curve. Consistent with Eq. 2.73,

$$\lambda i = W'_m + W_m \quad . \tag{2.79}$$

Note that for a given λi, increases in coenergy create decreases in energy and vice versa. This explains why

$$f^e = \left. \frac{\partial W'_m}{\partial x} \right|_i = - \left. \frac{\partial W_m}{\partial x} \right|_\lambda \quad . \tag{2.80}$$

It is important to note that energy is associated with flux linkage and coenergy is associated with current. The appropriate quantity must be held constant during differentiation if the correct force is to be calculated.

2.4. Energy, Force, and Torque

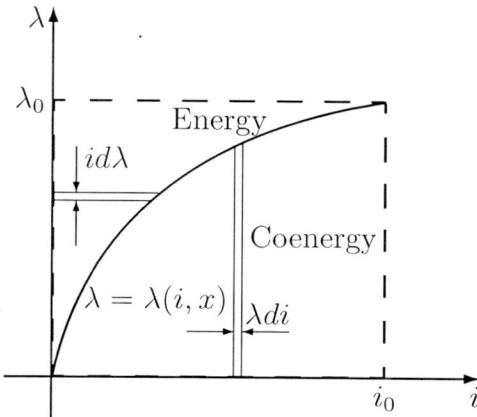

Figure 2.9: The relationship between energy and coenergy.

Early electrical engineering classes tend to condition us to think that the energy stored in an inductor is $1/2Li^2$. In fact, this is the coenergy. For a linear inductor the energy and coenergy are numerically equal, but force calculations can be wrong. For a linear inductor

$$W_m = \frac{1}{2}\frac{\lambda^2}{L(x)} \quad \Rightarrow \quad f^e = \frac{\lambda^2}{2L(x)^2}\frac{dL(x)}{dx} \quad , \tag{2.81}$$

and

$$W'_m = \frac{1}{2}L(x)i^2 \quad \Rightarrow \quad f^e = \frac{i^2}{2}\frac{dL(x)}{dx} \quad , \tag{2.82}$$

which evaluate to the same force for $\lambda = L(x)i$. However, starting the force calculation taking $W_m = 1/2L(x)i^2$ will lead to a force of the wrong sign if current is not first converted into flux linkage. Things will be even worse for nonlinear magnetics.

The calculation of coenergy is performed in a similar manner to that suggested by Eq. 2.63. Consider a system that has two electrical terminal pairs and one mechanical terminal pair such that $\lambda_1 = \lambda_1(i_1, i_2, \theta)$ and $\lambda_2 = \lambda_2(i_1, i_2, \theta)$ where

$$\lambda_1 = L_{11}i_1 + L_{12}i_2 \quad ; \tag{2.83}$$

$$\lambda_2 = L_{21}i_1 + L_{22}i_2 \quad . \tag{2.84}$$

One or more of the inductances will be a function of θ for a useful electric machine. Since rotary motion is involved, coenergy will give us the torque of magnetic origin, not the force.

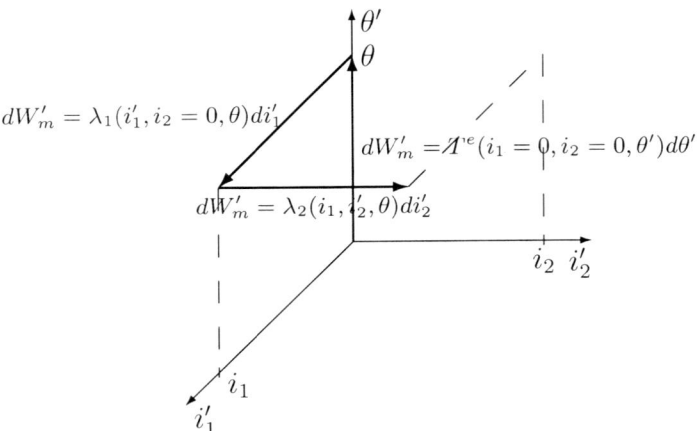

Figure 2.10: An integration path through the state space for a system with two electrical terminal pairs and one mechanical terminal pair.

Our determination of the coenergy involves integration of the incremental coenergy through the state space. The incremental coenergy is

$$dW'_m = \lambda_1 di_1 + \lambda_2 di_2 + T^e d\theta \tag{2.85}$$

based on first principles. Consistent with our earlier discussion, our integration path will include three segments. Because $T^e(i_1 = 0, i_2 = 0, \theta) = 0$, we choose to integrate over displacement first. Consistent with the integration path shown in Fig. 2.10, we then integrate over i_1 before integrating over i_2.

Formally, our coenergy calculation, consistent with Fig. 2.10 is

$$\begin{aligned} W'_m &= \int_0^\theta T^e(i_1 = 0, i_2 = 0, \theta') d\theta' + \\ &\quad \int_0^{i_1} \lambda_1(i_1 = i'_1, i_2 = 0, \theta) di'_1 + \\ &\quad \int_0^{i_2} \lambda_2(i_1, i_2 = i'_2, \theta) di'_2 \quad . \end{aligned} \tag{2.86}$$

Using Eqs. 2.83 and 2.84, Eq. 2.86 becomes

$$W'_m = 0 + \int_0^{i_1} L_{11} i'_1 di'_1 + \int_0^{i_2} (L_{21} i_1 + L_{22} i'_2) di'_2 \quad , \tag{2.87}$$

which reduces to

$$W'_m = \frac{1}{2} L_{11} i_1^2 + L_{21} i_1 i_2 + \frac{1}{2} L_{22} i_2^2 \quad . \tag{2.88}$$

2.4. Energy, Force, and Torque

Note that integration over i_2 before i_1 would give the cross term as $L_{12}i_1i_2$. Since $W'_m(i_1, i_2, \theta)$ is invariant of integration path, we conclude that $L_{12} = L_{21}$. The electromagnetic torque is given by

$$T^e = \left.\frac{\partial W'_m}{\partial \theta}\right|_i = \frac{1}{2}i_1^2 \frac{\partial L_{11}}{\partial \theta} + i_1 i_2 \frac{\partial L_{21}}{\partial \theta} + \frac{1}{2}i_2^2 \frac{\partial L_{22}}{\partial \theta} \quad . \tag{2.89}$$

Our development of magnetic circuit models in Chapter 1 showed how to develop the flux linkage as a function of current in a magnetic circuit. In some situations, such as, for example, idealized distributed windings, it may be easier to determine how flux linkage depends on current through an indirect method based on energy or coenergy. In such a situation, the structure of the model is identified as in Eqs. 2.83 and 2.84. From this structure, we know the coenergy takes the form given in Eq. 2.88. We can then force parity between the desired form of the coenergy (Eq. 2.88) and the coenergy based on the underlying fields. We will use coenergy equivalence to determine the inductance associated with winding slots in Sec. 4.3.2.

The coenergy can be computed directly from the fields that underlie the terminal relationships. The flux linkage is tied to magnetic flux density, just as the current is tied to the magnetic field intensity. The connection between flux linkage and flux density is an area. The connection between current and magnetic field intensity is a length. It follows that the coenergy density is

$$w'_m = \int \vec{B} \cdot d\vec{H} \quad , \tag{2.90}$$

where the dimensions of w'_m are energy per unit volume. In a linear magnetic material, the coenergy density is given by

$$w'_m = \int \mu \vec{H} \cdot d\vec{H} = \frac{1}{2}\mu H^2 \quad . \tag{2.91}$$

It follows that the total coenergy can be determined by integrating the coenergy density over the volume that stores the coenergy. That is,

$$W'_m = \int w'_m \, dV \quad . \tag{2.92}$$

The region containing the coenergy is found where the magnetic field intensity is not zero. The energy contained in the field can be found by integrating the energy density over the volume, where the energy density is given by

$$w_m = \int \vec{H} \cdot d\vec{B} \quad . \tag{2.93}$$

Our energy and coenergy formulations were based upon zero force or torque with zero current. With permanent magnets, we know that there can be magnetic field even without current. However, we can still use our coenergy formulation to determine forces and torques in the presence of permanent magnets. In Section 1.4 we saw how permanent magnets can be modeled with an equivalent winding. This equivalence gives us a way to quickly determine the coenergy of a system that is based on a permanent magnet. For example, the coenergy of the magnetic circuits of Fig. 1.13 must be the same, since they are designed to be equivalent. For the circuit based on the winding, the coenergy is

$$W'_m = \frac{(Ni)^2}{2\left(\frac{\ell_m}{\mu_R A_m} + \frac{g}{\mu_0 A_g}\right)} \quad . \tag{2.94}$$

Since we have equivalence when $Ni = H'_c \ell_m$, the coenergy for the magnet system is

$$W'_m = \frac{(H'_c \ell_m)^2}{2\left(\frac{\ell_m}{\mu_R A_m} + \frac{g}{\mu_0 A_g}\right)} \quad . \tag{2.95}$$

Now that we have the coenergy, we can calculate forces and torques in the usual way, when one or more of the reluctances depend on position. With permanent magnets, one must be careful to distinguish between energy and coenergy. This distinction is straightforward recognizing that H_m is proportional to i and B_m is proportional to λ.

2.5 Reciprocity

Reciprocity is a requirement for a conservative (lossless) system. There are times when using reciprocity allows us to gain additional insights into the electromechanical coupling process. First consider the energy for a system with one electrical terminal pair and one mechanical terminal pair, such that $W_m = W_m(\lambda, x)$. By virtue of how the partial derivative is defined,

$$\frac{\partial^2 W_m}{\partial x \partial \lambda} = \frac{\partial^2 W_m}{\partial \lambda \partial x} \quad , \tag{2.96}$$

but we know that

$$i(\lambda, x) = \frac{\partial W_m}{\partial \lambda} \quad , \tag{2.97}$$

and

$$f^e(\lambda, x) = -\frac{\partial W_m}{\partial x} \quad . \tag{2.98}$$

It follows that

$$\frac{\partial i(\lambda, x)}{\partial x} = -\frac{\partial f^e(\lambda, x)}{\partial \lambda} \quad . \tag{2.99}$$

Similar arguments hold for coenergy. That is,

$$\frac{\partial^2 W'_m}{\partial x \partial i} = \frac{\partial^2 W'_m}{\partial i \partial x} \quad , \tag{2.100}$$

but we know that

$$\lambda(i, x) = \frac{\partial W'_m}{\partial i} \quad , \tag{2.101}$$

and

$$f^e(i, x) = \frac{\partial W'_m}{\partial x} \quad . \tag{2.102}$$

It follows that

$$\frac{\partial \lambda(i, x)}{\partial x} = \frac{\partial f^e(i, x)}{\partial i} \quad . \tag{2.103}$$

These reciprocity relationships allow us to infer how, for example, flux linkage changes with position if we know how force changes with current.

2.6 Energy Conversion Cycles

Our determination of forces and torques of magnetic origin was based on conservation of energy, which said:

[electrical energy in] + [mechanical energy in] =
[change in stored energy] .

This statement was the genesis of Eq. 2.60. For a complete cycle of operation in the periodic steady state, we should expect the net change in stored energy to be zero. That is, we should expect the energy to finish the cycle where it started. It follows that for a complete cycle of operation

$$\begin{bmatrix} \text{net electrical energy} \\ \text{input for one cycle} \end{bmatrix} + \begin{bmatrix} \text{net mechanical energy} \\ \text{input for one cycle} \end{bmatrix} = 0 \quad . \tag{2.104}$$

Consider a simple electric machine that has a single stator winding. The machine is designed to make the magnetic circuit of the stator winding vary with rotor position, so that $\lambda = \lambda(i, \theta)$. This effect is often created by using pole structures that exhibit saliency on the rotor, and sometimes also on the stator. Figure 2.11 shows $\lambda(i, \theta)$ for four different rotor positions. Rotor position θ_1 corresponds to the position of minimum inductance. Position θ_2 corresponds to the position of maximum inductance. The curvature suggests magnetic saturation, the degree of which depends on both rotor position and current magnitude. There is no winding on the rotor, so the machine only

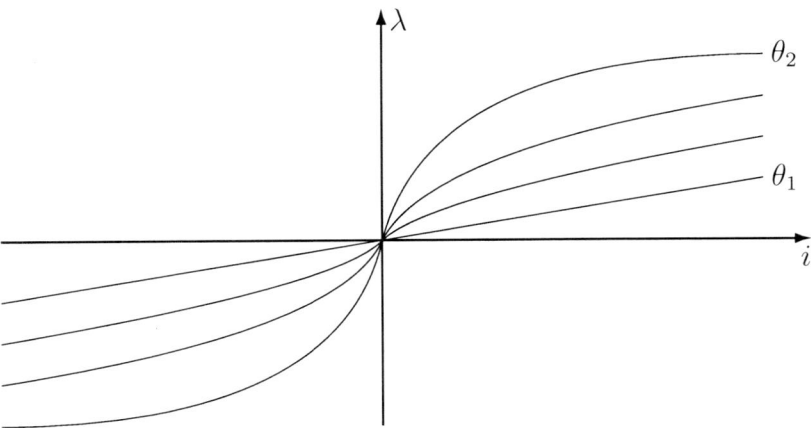

Figure 2.11: Flux linkage as a function of current and rotor position for a simple single phase machine.

produces reluctance torque. This type of machine is known as a variable reluctance machine, and is the subject of Chapter 13.

Take our simple machine to be excited so that positive current is put into the phase winding at position $\theta = \theta_1$, the current is held constant until the rotor moves to position $\theta = \theta_2$, and the current is removed from the phase winding while $\theta = \theta_2$. Using the relationship among flux linkage, current, and rotor position of Fig. 2.11, we can identify the electrical energy put into the machine, the electrical energy removed from the machine, and the energy converted from electrical to mechanical form during one cycle of operation.

Referring to Fig. 2.12, as current is put into the stator winding at position θ_1, the flux builds with the current along path A. Electrical energy continues to flow into the winding as the rotor moves from θ_1 to θ_2 along path B. The total electrical energy input is the shaded area above the combination of paths A and B. This is a simplification of what really happens, because the rotor is not going to stop while the current is fed to the machine, and we know from the finite inductance that the current cannot build instantaneously in the phase winding.

With the rotor at θ_2, current is pulled out of the phase winding, with the flux linkage collapsing with current along path C. Because both flux linkage and current are decreasing, the area above path C corresponds to electrical energy removed from the phase winding. Consistent with Eq. 2.104, the area enclosed by paths A, B, and C is the energy that is converted from electrical to mechanical form. When we traverse the energy conversion

2.6. Energy Conversion Cycles

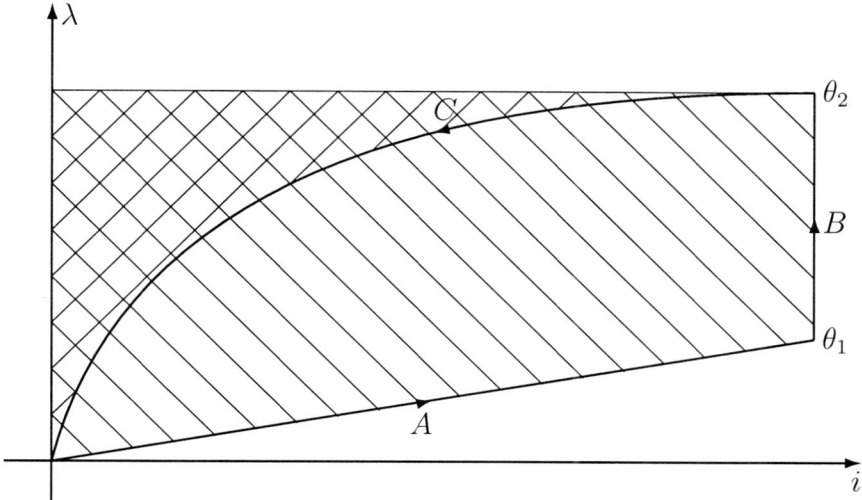

Figure 2.12: An energy conversion cycle for the simple machine of Fig. 2.11. The shaded area above paths A and B corresponds to electrical energy put into the machine. The shaded area above path C corresponds to electrical energy extracted from the terminals.

cycle in the counter-clockwise direction, electrical energy is converted to mechanical energy consistent with motor operation. When we traverse the energy conversion cycle in the clockwise direction, the net electrical energy input is negative consistent with generator operation.

The average torque produced by this energy conversion cycle is the energy converted to mechanical form divided by the angle of rotation corresponding to one cycle. The $\lambda(i,\theta)$ information given in Fig. 2.11 does not convey the periodicity of the simple machine. That is, one revolution of the rotor shaft of the machine could result in 2, 4, 6, or any even number of electrical cycles. Accordingly, the energy conversion cycle shown in Fig. 2.12 could correspond to the energy converted in π, $\pi/2$, or $\pi/3$ radians for 2, 4, or 6 electrical cycles per revolution, respectively.

The energy above path C of Fig. 2.12 represents energy that is put into the machine, only to be removed later. This recirculation of energy is a characteristic of many electric machines, and is generally undesirable because it represents the movement of energy that never gets converted. Since the energy never gets converted, it contributes loss without contributing to accomplishing work. Accordingly, the recirculated energy should be minimized to the extent possible while meeting other performance criteria.

The recirculated energy can be interpreted as creating the magnetic field within the machine that supports the energy conversion process. Accordingly, this recirculated energy is fundamental to the conversion process and it cannot be eliminated without rendering the machine useless. The recirculated energy imposes an excitation penalty on those machines that rely on it to convert energy. Because of the recirculated energy, the power supply to the machine must support higher power flow for a given amount of energy conversion. This could be interpreted as the machine presenting the supply with a power factor that is less than one since the machine appears to require more electrical power than is converted to mechanical form.

The shape of the energy conversion cycle gives us an understanding of the energy conversion process and the interrelationship between the machine characteristics and its control. For example, suppose that the flux linkage were a linear function of current at rotor position θ_2. Our energy conversion cycle would show us that the energy converted into mechanical form goes down, because the area swept out has been reduced. Apparently the presence of magnetic saturation actually allowed us to convert a greater percentage of the energy input into mechanical form than if our machine were magnetically linear. A control objective would be to maximize the swept area of the energy conversion cycle for a given amount of recirculated energy.

The comment about magnetic saturation and energy conversion deserves additional discussion because there is a distinction to be drawn between the value of saturation to the machine and to its supply. Figure 2.12 shows that the machine supply benefits from saturation because a greater percentage of the input energy is converted into mechanical energy. The machine, however, sees its potential output decreased because of saturation.

A machine that does not saturate would be able to convert significantly more energy for a given amount of current. Figure 2.13 shows two energy conversion cycles. The first energy conversion cycle is defined by the points $OABO$ for a machine that exhibits saturation. The second energy conversion cycle is defined by the points $OACO$ that would result from a machine that is magnetically linear. The swept area of the magnetically linear energy conversion cycle is substantially larger. Thus, we conclude that the machine pays a substantial performance penalty because of magnetic saturation, while the machine supply benefits from saturation. We see from energy conversion cycle $OACO$ that for the linear machine less than one half of the energy input is converted to mechanical output. For a heavily saturated machine, it is possible to convert 90% of the energy input to mechanical output.

While saturation limits the performance of the electric machine, most electric machines do experience saturation and are often driven at least par-

2.7. Conditions for Average Power Production

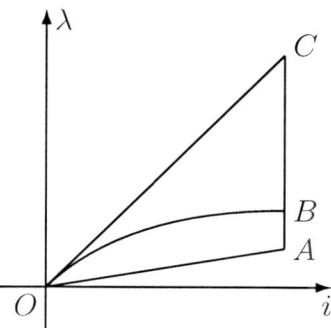

Figure 2.13: A comparison between energy cycles for a saturated machine and a magnetically linear machine.

tially into saturation. Because the materials used within the machine exhibit saturation, a machine that is sized to remain magnetically linear will be larger than a machine that is pushed into saturation. Magnetic saturation serves to limit the energy conversion potential, but the presence of saturation does not end the energy conversion process. Only when the machine is so heavily saturated that no additional area can be swept out is there no point in pushing the machine any further. However, the incremental area swept goes down with increasing current for a saturated machine while winding losses continue to increase quadratically so there will be an upper limit on practical winding current.

2.7 Conditions for Average Power Production

An objective of electric machines is to convert energy between electrical and mechanical forms. We have seen that this is accomplished when the electrical dynamics couple into the mechanical dynamics. This coupling, however, only supports steady energy conversion when certain relationships hold between the electrical and mechanical frequencies.

Consider the electric machine of Fig. 2.14. This is the same machine given in Fig. 2.5 with some changes in labels. While the stator and rotor windings are shown as concentrated coils, we will consider them to be distributed in slots in such a way that they approximate a sinusoidal distribution. We will ignore the presence of the slots and their influence on the magnetic field distribution, so that we can describe the stator and rotor flux linkages as

$$\lambda_s = L_s i_s + L_{sr}(\theta) i_r \quad (2.105)$$

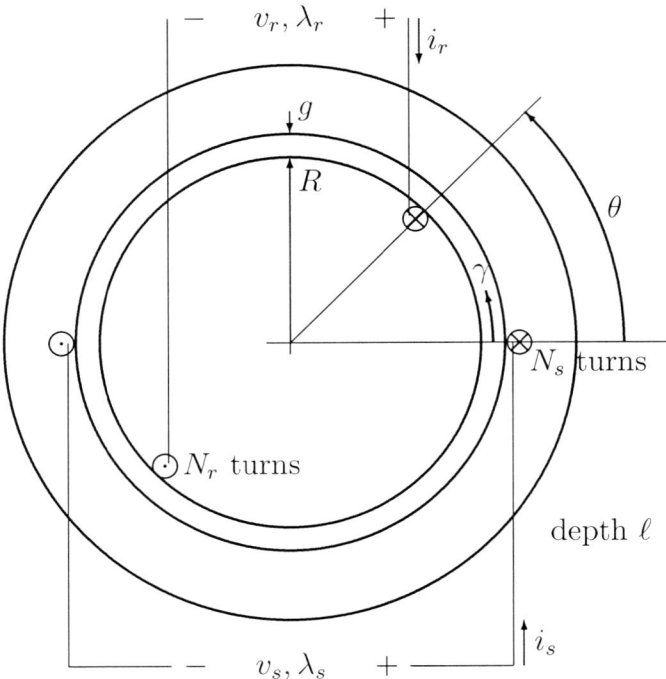

Figure 2.14: A simple electric machine with a single stator winding and a single rotor winding.

and
$$\lambda_r = L_{rs}(\theta)i_s + L_r i_r \quad , \tag{2.106}$$
respectively. It follows from a coenergy analysis that the torque of magnetic origin is
$$T^e = i_s i_r \frac{dL_{sr}(\theta)}{d\theta} \quad . \tag{2.107}$$

By inspection of Fig. 2.14 and physical reasoning, we conclude that
$$L_{sr}(\theta) = L_{rs}(\theta) = M_1 \cos\theta + M_3 \cos 3\theta + M_5 \cos 5\theta \ldots \quad . \tag{2.108}$$

Generally, windings are designed to emphasize the fundamental component and to deemphasize the harmonics. We will discuss how this is done when we focus on windings in Chapter 3. For now, we will take
$$L_{sr}(\theta) = L_{rs}(\theta) = M \cos\theta \quad . \tag{2.109}$$

2.7. Conditions for Average Power Production

Using Eq. 2.109, the torque expression becomes

$$T^e = -Mi_s i_r \sin\theta \quad . \tag{2.110}$$

For constant currents in the windings, the torque tries to align the two windings. We can now examine the conditions under which this machine converts average power, since average power is the only way to accomplish useful work on a sustained basis.

We take the stator current to be

$$i_s = I_s \sin\omega_s t \quad , \tag{2.111}$$

the rotor current to be

$$i_r = I_r \sin\omega_r t \quad , \tag{2.112}$$

and the rotor position to be

$$\theta = \omega_m t + \beta \quad . \tag{2.113}$$

Putting these expressions into Eq. 2.110 gives

$$T^e = -MI_s I_r \sin(\omega_s t)\sin(\omega_r t)\sin(\omega_m t + \beta) \quad . \tag{2.114}$$

By using two trigonometric identities[5], the torque expression becomes

$$\begin{aligned}T^e = \quad -\tfrac{MI_sI_r}{4} \quad &\{\sin\left((\omega_s+\omega_r+\omega_m)\,t+\beta\right)+ \\ &\sin\left((\omega_s+\omega_r-\omega_m)\,t-\beta\right)- \\ &\sin\left((\omega_s-\omega_r+\omega_m)\,t+\beta\right)- \\ &\sin\left((\omega_s-\omega_r-\omega_m)\,t-\beta\right)\} \quad .\end{aligned} \tag{2.115}$$

In order for average power production, we must have a time argument in one of the sin() terms go to zero. This requires the mechanical speed to be one of the four variants of

$$\omega_m = \pm\omega_s \pm \omega_r \quad . \tag{2.116}$$

We conclude from this result that average power production requires a specific relationship among rotational speed, the frequency of stator currents, and the frequency of rotor currents. Again, electromechanical coupling requires that electrical and mechanical frequencies are related.

Since the mechanical speed can only be dictated by one combination of ω_s and ω_r at a time, three of the sin() terms in Eq. 2.115 will be contributing

---

[5] $\sin\alpha\sin\beta = [\cos(\alpha-\beta)-\cos(\alpha+\beta)]/2$
$\sin\alpha\cos\beta = [\sin(\alpha+\beta)+\sin(\alpha-\beta)]/2$

to instantaneous torque but not average torque. These time-varying terms are considered torque ripple, and it is a rare application that would benefit from torque ripple. To be sure, there is a strong desire in most applications to eliminate torque ripple. We will see below how it is possible to eliminate this torque ripple. Before we do that, it is instructive to explore the origin of the relationships in Eq. 2.116 from a more physical perspective.

Referring to Fig. 2.14, the radial magnetic field in the air gap due to the stator currents with a sinusoidally distributed winding is

$$H_s = \frac{N_s i_s}{2g} \sin \gamma \quad . \tag{2.117}$$

This can be verified directly by using Ampere's law. When we apply a time varying current to the winding, the air gap magnetic field becomes

$$H_s = \frac{N_s I_s}{2g} \sin(\omega_s t) \sin(\gamma) \quad . \tag{2.118}$$

Again using a trigonometric identity, this can be rewritten as

$$H_s = \frac{N_s I_s}{4g} [\cos(\omega_s t - \gamma) - \cos(\omega_s t + \gamma)] \quad . \tag{2.119}$$

We can recognize the cos() terms as representing traveling waves, one in the positive γ direction, the other in the negative γ direction. The velocity of each traveling wave is ω_s.

We can go through similar reasoning with the air gap magnetic field created by the rotor winding. That is, the air gap magnetic field created by the rotor will consist of positive and negative traveling waves, each with velocity ω_r relative to the rotor. To get the total velocity of the rotor field, we must superimpose the rotational speed of the rotor ω_m.

In order to create average torque, physical reasoning tells us that one component of the stator magnetic field must be traveling at the same speed as one component of the rotor magnetic field. From this we conclude that

$$\pm \omega_s = \omega_m \pm \omega_r \quad . \tag{2.120}$$

Rearranging this gives Eq. 2.116. The torque ripple is created by the combinations of magnetic fields that do not contribute to average torque. This suggests that we need to reduce the number of field components if we are to reduce the torque ripple.

The electric machine of Fig. 2.15 is like the one of Fig. 2.14 except a second winding has been added to both the stator and rotor. The winding quantities now carry a double subscript, the first letter indicating stator or

2.7. Conditions for Average Power Production

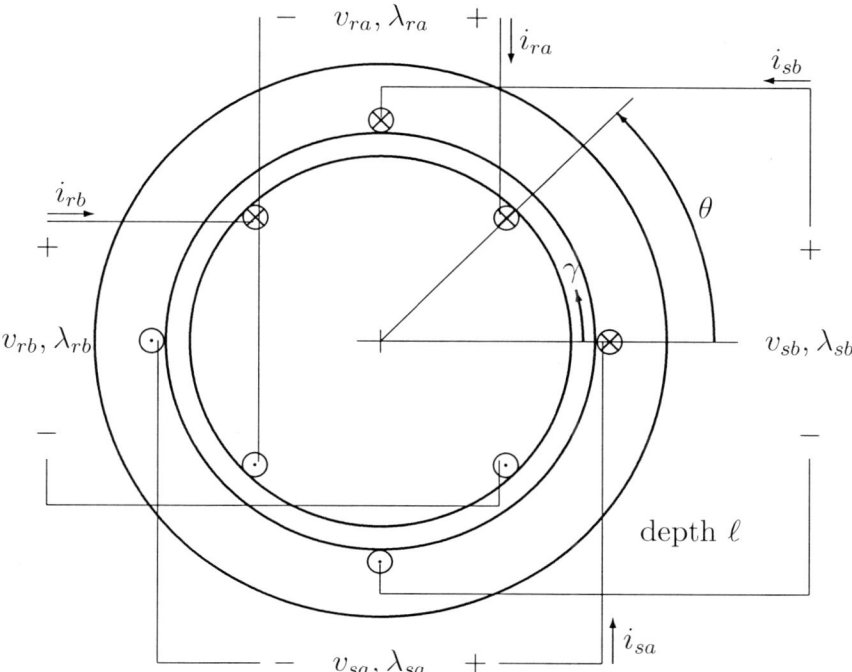

Figure 2.15: An electric machine with two stator windings and two rotor windings.

rotor, and the second letter indicating winding a or b. Since windings a and b on the stator are orthogonal to one another, currents in winding a do not create flux in winding b and vice versa. The same is true with the rotor windings. In general, however, both windings on the rotor will couple to both windings on the stator. Physical reasoning suggests the following flux linkage relationships:

$$\lambda_{sa} = L_s i_{sa} + M i_{ra} \cos\theta - M i_{rb} \sin\theta \quad ; \quad (2.121)$$

$$\lambda_{sb} = L_s i_{sb} + M i_{ra} \sin\theta + M i_{rb} \cos\theta \quad ; \quad (2.122)$$

$$\lambda_{ra} = M i_{sa} \cos\theta + M i_{sb} \sin\theta + L_r i_{ra} \quad ; \quad (2.123)$$

$$\lambda_{rb} = -M i_{sa} \sin\theta + M i_{sb} \cos\theta + L_r i_{rb} \quad . \quad (2.124)$$

Using Eqs. 2.121 through 2.124 in a coenergy formulation gives the coenergy to be

$$W'_m = \frac{1}{2}\left[L_s\left(i_{sa}^2 + i_{sb}^2\right) + L_r\left(i_{ra}^2 + i_{rb}^2\right)\right] +$$

$$M\left[(i_{sa}i_{ra} + i_{sb}i_{rb})\cos\theta + (i_{sb}i_{ra} - i_{sa}i_{rb})\sin\theta\right] \quad, \quad (2.125)$$

from which the torque is determined to be

$$T^e = M\left[(i_{sb}i_{ra} - i_{sa}i_{rb})\cos\theta - (i_{sa}i_{ra} + i_{sb}i_{rb})\sin\theta\right] \quad. \quad (2.126)$$

Similar to before, we take the winding currents to be

$$i_{sa} = I_s \cos\omega_s t \quad; \quad (2.127)$$

$$i_{sb} = I_s \sin\omega_s t \quad; \quad (2.128)$$

$$i_{ra} = I_r \cos\omega_r t \quad; \quad (2.129)$$

$$i_{rb} = I_r \sin\omega_r t \quad. \quad (2.130)$$

The rotor position is taken to be

$$\theta = \omega_m t + \beta \quad. \quad (2.131)$$

Substituting these quantities into Eq. 2.126, and reducing with trigonometric identities gives the torque to be

$$T^e = M I_s I_r \sin\left[(\omega_s - \omega_r - \omega_m)t + \beta\right] \quad. \quad (2.132)$$

We have average torque, and consequently average power, when

$$\omega_m = \omega_s - \omega_r \quad. \quad (2.133)$$

It is apparent that the additional windings supporting currents properly phased have eliminated the time-varying torques! Phasing the stator and rotor currents differently would give a different relationship coupling rotor speed with stator and rotor frequencies; all four combinations embodied in Eq. 2.116 are possible.

To see how the additional windings eliminated the ripple torques, consider the field produced by the stator windings. Similar to our prior reasoning, the radial magnetic field in the air gap due to the stator currents with sinusoidally distributed windings is

$$H_s = \frac{N_s}{2g}\left[i_{sa}\sin\gamma - i_{sb}\cos\gamma\right] \quad. \quad (2.134)$$

Putting our balanced currents into the windings gives

$$H_s = \frac{N_s I_s}{2g}\left[\cos(\omega_s t)\sin(\gamma) - \sin(\omega_s t)\cos(\gamma)\right] \quad. \quad (2.135)$$

Each term in brackets within Eq. 2.135 has two traveling wave components, one in the positive direction and one in the negative direction, just as we had with the single winding. We can also observe that the field reduces through a trigonometric identity[6] to

$$H_s = -\frac{N_s I_s}{2g} \sin(\omega_s t - \gamma) \quad . \tag{2.136}$$

The resultant stator magnetic field travels only in the positive γ direction with velocity ω_s. The field components traveling in the negative direction have canceled. It can be shown through direct substitution that exchanging the winding currents will cause the stator magnetic field to travel in the negative γ direction.

From this example, we see that spatially displaced windings that are fed with temporally displaced currents create magnetic fields that travel in only one direction. In this example we showed how this is accomplished for two windings. Typically it is accomplished with three windings that are symmetrically displaced in space by $2\pi/3$ electrical radians when fed by currents that are symmetrically displaced in time by $2\pi/3$ electrical radians. Because of the intended phase displacement of the currents, these windings are often called phase windings.

2.8 Summary

This chapter has focused on the electromechanical coupling process. First we examined the influence of mechanical displacement on the electrical dynamics, typically forcing flux linkage to be not only a function of current but also of position. Turning things around, we then discussed how forces of electric origin are created based on conservation of energy. While forces can be described using energy, we found that coenergy is actually much more convenient, as well as practical.

The discussion of energy and coenergy naturally lead into a discussion of energy conversion cycles, and the insight they can project into how best to control a machine to maximize energy conversion for a given amount of current. Energy conversion cycles also identify recirculated energy and suggest that magnetic saturation may reduce the recirculated energy relative to the energy converted. Magnetic saturation, however, limits the energy that can be converted by the machine.

We finished off the chapter by looking at the conditions under which an electric machine can produce average torque and power. The conclusion was that average torque can be produced with single windings on the stator and rotor, however, there will be significant time-varying torque superimposed

---

[6] $\sin(\alpha \pm \beta) = \sin\alpha\cos\beta \pm \sin\beta\cos\alpha$

on the average torque. Moving to multiple windings displaced in space and feeding them with currents that are displaced in time is an effective method for eliminating magnetic field components that contribute to time-varying torques. This feature of multiple windings is why this type of machine is preferred for applications that require smooth torque.

Chapter 3

Windings

3.1 Introduction

Electric machines require windings to support the electromechanical energy conversion process. While one or more permanent magnets can be used to replace some windings, the machine will still require one or more windings. Permanent magnets are used to set up a nominally constant field, but the energy conversion process in the face of motion requires fields that alternate and this can only be facilitated with windings that carry alternating currents.

Fundamentally, a winding is used to create a magnetic field in the air gap of an electric machine. It is generally an objective of the winding design to control the spatial distribution of the magnetic field to develop the required torque subject to other constraints such as cost, torque ripple, etc. From the discussion provided in Chapter 4, it should be possible to better understand how the winding structure is related to torque ripple, the terminal characteristics of the electric machine, and the complexity of making the machine.

This chapter introduces some of the issues associated with the design of windings. We begin with some terminology that will draw on concepts that have already been developed, putting them in terms commonly associated with electric machines and windings. As we proceed, additional concepts will be introduced, including some that are more historical in nature, since with computers, it is possible to succinctly represent windings and manipulate the fields they create.

3.2 Phases, Poles, Slots, and Pitches

We saw in Sec. 2.7 that temporally displaced currents fed to spatially displaced windings can create magnetic fields that travel in a single and known direction. This is a valuable tool in eliminating ripple torques. In the example of the machine given in Fig. 2.15, two windings were used on the stator and rotor, and it was shown how the stator and rotor magnetic fields traveled in a single direction by virtue of the phase relationship between the currents in the windings. Because we had two windings that are orthogonal in space, our currents were phased to be orthogonal in time.

A *phase* is taken to be a group of turns and/or windings that are intended to be excited by the same voltage or current. When a machine has multiple phases, symmetric displacement of the windings in space is an important objective. The currents fed to the phases should also be symmetrically displaced in time. Phase windings can be connected in series or parallel. Most industrial motors allow for connection either way to support multiple voltages. The machine of Fig. 2.15 has two phase windings on the stator and two phase windings on the rotor.

Phase windings are designed to create a desired number of magnetic *poles*, and the phase windings are generally placed in *slots* within the magnetic structure. For a conventional machine, the slots for the stator windings are on the inside periphery of the stator core; the slots for the rotor windings are on the outside periphery of the rotor core for a conventional motor. The slots are generally punched into the laminations as they are stamped.

Magnetic poles always travel in pairs, so for every north pole there is a south pole. The number of poles in an electric machine is a parameter decided in the design of the machine. The number of poles describes a conversion between electrical and mechanical angular measure. Consider the machine in Fig. 3.1 that has a single winding that forms two magnetic poles. The magnetic field pattern goes through one full cycle in one full revolution.

Compare this to the machine in Fig. 3.2 that has four magnetic poles. Now the field pattern goes through two full cycles in each revolution around the air gap. Accordingly, the magnetic field pattern is

$$H_s = \frac{N_s i_s}{2g} \sin(2\theta) \quad . \tag{3.1}$$

Exciting this winding with current $i_s = I_s \sin \omega_s t$ gives

$$H_s = \frac{N_s I_s}{2g} \sin(\omega_s t) \sin(2\theta) \quad . \tag{3.2}$$

3.2. Phases, Poles, Slots, and Pitches

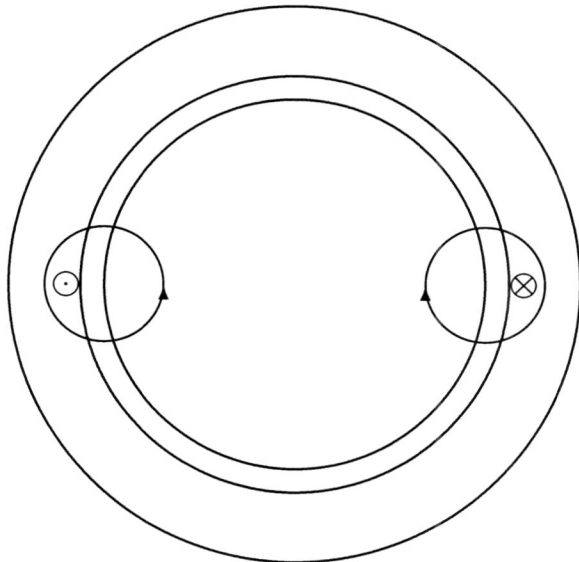

Figure 3.1: A simple electric machine with a single stator winding and two magnetic poles.

Using a trigonometric identity[1], this can be rewritten as

$$H_s = \frac{N_s I_s}{4g} \left[\cos(\omega_s t - 2\theta) - \cos(\omega_s t + 2\theta)\right] \quad . \tag{3.3}$$

This magnetic field has two traveling wave components, one in each direction. The velocity of each wave component is $\omega_s/2$. Generalizing, a machine with N_p magnetic poles will have traveling waves established by windings with velocity $2\omega_s/N_p$.

The velocity of the traveling waves has a direct impact on the mechanical speeds associated with average power conversion. Since the velocity of the stator field drops with increasing pole count, the mechanical speed associated with average power conversion also drops. Because the spatial rate of change in the magnetic field increases with pole count, torque production goes up with pole count. However, while the spatial rate of change increases, the associated change in field magnitude tends to drop, so the increase in torque with pole count is not as dramatic as one might hope.

The *pitch* is the angular extent of the coils used to form a phase winding. A full-pitch coil implies the start and return coil sides are 180° electrical degrees apart. A fractional pitch coil does not extend 180° electrical degrees.

---
[1] $\sin \alpha \sin \beta = [\cos(\alpha - \beta) - \cos(\alpha + \beta)]/2$

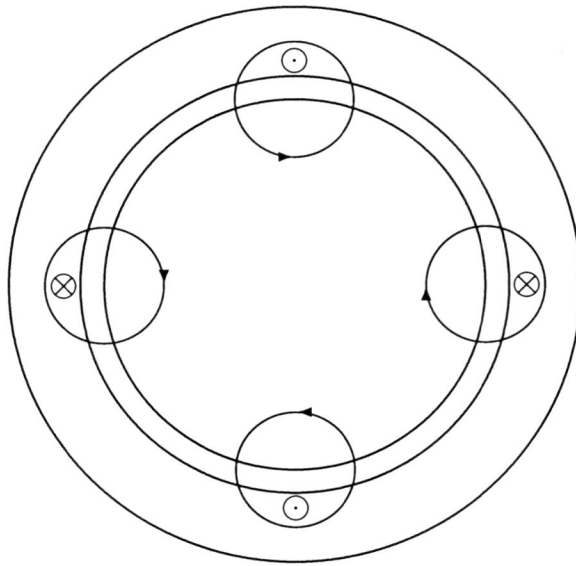

Figure 3.2: A simple electric machine with a single stator winding and four magnetic poles.

The windings in the machines of Figs. 3.1 and 3.2 are full-pitch. As we shall see, the number of slots per pole can be an integer, or fractional. The choice of integral or fractional slot winding will have substantial impact on the spatial harmonics produced by the winding.

3.3 Types of Windings

There are many names given to the windings in electric machines. Some adjectives are intended to indicate how coils share a slot. A *single-layer* winding has only one coil side in each slot, while a *double-layer* winding has two coil sides in each slot. Double-layer windings are generally more common than single-layer windings. As you can imagine, a double-layer winding offers greater flexibility.

A *distributed winding* is comprised of a number of coils that are distributed over a number of slots. This type of winding is used to force the air gap mmf to approximate a sinusoid or a trapezoid. On the other hand, a *concentrated windings* makes use of single coils, probably in an effort to minimize the number of slots needed. Concentrated windings are sometimes wound around a single tooth, as in variable-reluctance machines and some

3.3. Types of Windings

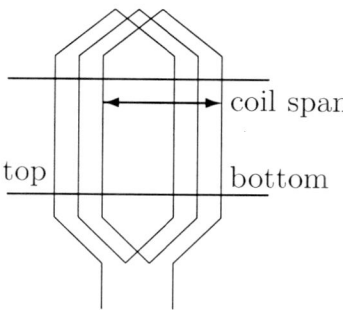

Figure 3.3: The structure of a lap winding.

permanent magnet machines. Other uses of the modifier term *concentrated* are simply to distinguish the winding from a distributed winding.

In double-layer windings, it is usual to use a *lap winding* technique to put the coils into the slots. Figure 3.3 shows the structure of a lap winding, in which the end turns overlap, or lap, one another. In the figure, the vertical lines represent the winding conductors traveling the axial length of the machine; angled lines represent the treatment of the end turns. In a double-layer winding, one side of each coil would occupy the top of the winding slot, with the other side occupying the bottom of the slot. It is also possible to form single-layer lap windings. This ordered treatment of the windings not only helps to maintain balance among each coil, but it facilitates uniform treatment of the end turns. Lap windings have constant end-turn lengths among the coils. Lap windings can be formed by bars in large machines, or form-wound coils in smaller machines. An alternative to the lap winding is the *concentric winding*, shown in Fig. 3.4. In the concentric winding the end turns do not overlap, giving less build up in the end-turn region, and potentially making the slots easier to populate with automatic machinery. However, each coil has a different resistance and makes a different contribution to the field.

Figure 3.5 shows a more physical depiction of how the end turns are treated in a lap winding. Each coil has one coil side occupying the top of the slot; the other coil side occupies the bottom of the slot. Based on the treatment of the end turns, one can envision how the coils are made. The coil can be wound on a mandrel, resulting in an oval coil. After winding, the two long straight sections that will go into the slot can be held. Pulling each straight section in opposite directions will create the coil span. Alternative winding approaches can avoid the "knuckle" where the wire bundle transitions from the outer radius (the bottom coil side) to the inner radius

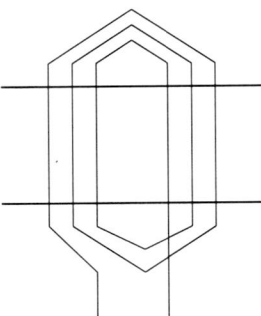

Figure 3.4: The structure of a concentric winding.

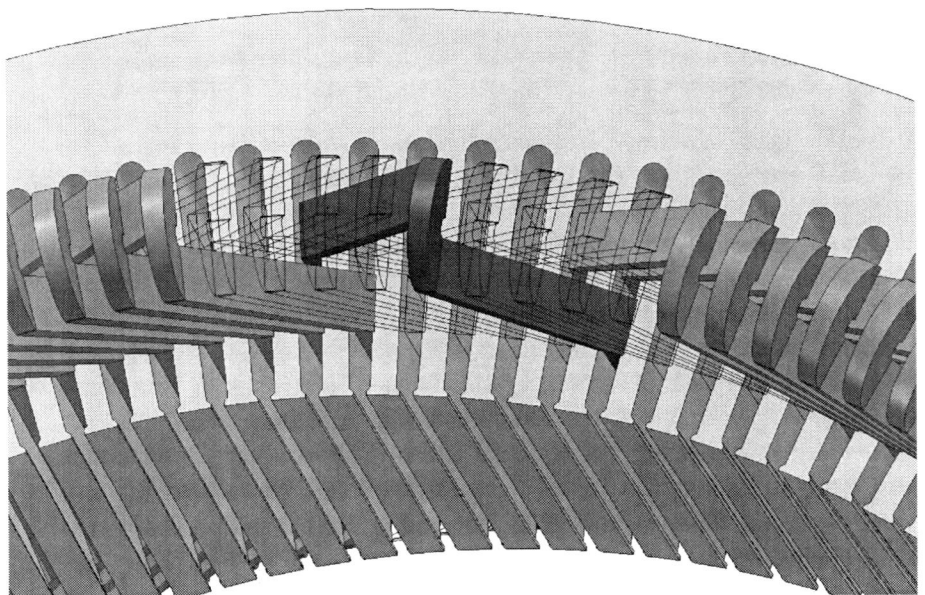

Figure 3.5: The solid model of a stator that uses lap windings.

(the top coil side). The important point is that all coils treat the end turns the same way so that the end turns end up in orderly fashion. After the machine is wound, the end turns would be laced, and perhaps braced to handle short-circuit forces.

The *wave winding* is an alternative to the lap winding, but is seldom used because it is more difficult to insert the windings into the slots. The concept of a wave winding is shown in Fig. 3.6.

We take the number of slots used for the winding to be N_s. The number

3.3. Types of Windings

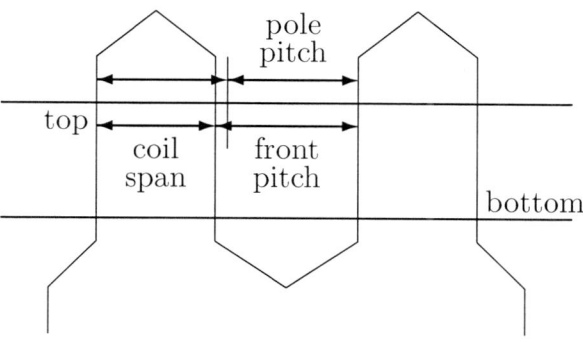

Figure 3.6: The structure of a wave winding.

of magnetic poles is taken to be N_p, and the number of phases is N_ϕ. The number of slots per pole is

$$N_{sp} = \frac{N_s}{N_p} \quad . \tag{3.4}$$

Another important number is the number of slots per pole per phase:

$$N_{sp\phi} = \frac{N_s}{N_p N_\phi} \quad . \tag{3.5}$$

If $N_{sp\phi}$ is an integer, the machine will have an integral slot winding; otherwise the machine will have a fractional slot winding. The angular pitch between slots is

$$\theta_s = \frac{2\pi}{N_s} \quad . \tag{3.6}$$

The angular pitch between slots in electrical measure is

$$\theta_{se} = \frac{\pi N_p}{N_s} \quad . \tag{3.7}$$

Integral slot windings, as the name implies, have an integer number of slots per pole per phase. This requires that the number of slots is divisible by both the number of poles and the number of phases. In integral slot windings, the windings are generally distributed to reduce the spatial harmonics in the air gap mmf. A distributed winding requires $N_{sp\phi} \geq 2$ to provide sufficient room for multiple coils within a phase for each pole.

The alternative to an integral slot winding is the *fractional slot winding*. In this winding the number of slots is not a multiple of the number of magnetic poles. This type of winding makes use of staggered coils to reduce the harmonic content of the air gap mmf. The coils are naturally staggered because the pole boundaries fall between winding slots, making the coils inherently fractional pitch. This technique is commonly used in permanent magnet machines with any number of poles. It is also used in other machines when the number of poles get too large to accommodate an integral slot winding. A fractional slot winding can provide acceptable performance in many cases with $N_{sp\phi} \leq 2$.

In the discussion of integral slot and fractional slot windings that follows, the following assumptions are made to focus the discussion, not to impose fundamental limitations:

1. The machine has three phases: $N_\phi = 3$. Our discussion can be extended to an arbitrary number of phases, but since most ac machines have three phases, we will work with that number.

2. All slots are filled. This requires the number of slots to be a multiple of $N_\phi = 3$.

3. The winding is a double-layer winding, so each slot contains two coil sides. This implies that there are N_s coils in the machine.

4. Winding balance is a requirement. Only pole and slot combinations are considered that result in the phase windings being spatially displaced by $2\pi/N_\phi$ electrical radians.

5. All coils have the same span and the same number of turns. This implies all coils have the same resistance and inductance.

These assumptions are consistent with trying to build a machine that is electrically balanced and provides reasonable performance. This is generally sound practice, giving a machine that exhibits less reason to be noisy or inherently produce a lot of torque ripple.

3.4 Integral Slot Double-layer Windings

The basic building block of the integral slot winding is a pair of coils symmetrically displaced over adjacent magnetic poles, as shown in Fig. 3.7. The coils in the winding of Fig. 3.7 are short-pitched windings since they do not extend for π electrical radians. Electrical angle δ is the angle of the first coil side; the mate for this coil is at electrical angle $\pi - \delta$. The angle between the coil sides must be an integer number of slot pitches such that $\pi - 2\delta = m\theta_{se}$ where m is an integer. The angle δ must be related to the number of slots, but it is not restricted to an integer relationship. The *span* of a winding is

3.4. Integral Slot Double-layer Windings

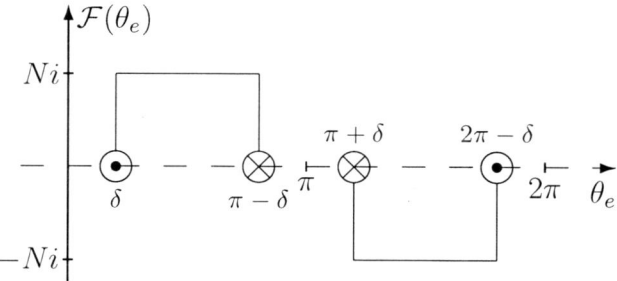

Figure 3.7: The basic structure used to build up an integral slot winding using two coils that have pitch $\pi - 2\delta$.

the number of slots between the coil sides. That is, if one side of the winding is in slot k and the return side is in slot $k+2$, the span is two slots.

Observe that the mmf distribution of these coils possesses both odd- and half-wave symmetry[2]. Accordingly, we can write the mmf distribution of the coil pair as

$$\mathcal{F}(\theta_e) = \sum_{\substack{n=-\infty \\ n \text{ odd}}}^{\infty} f_n \exp(jn\theta_e) \quad , \tag{3.8}$$

where

$$f_n = -j\frac{1}{\pi} \int_0^\pi f(\theta_e) \sin(n\theta_e) \, d\theta_e \quad , \tag{3.9}$$

and θ_e is the electrical angle such that each pole covers π radians. Using the mmf distribution of Fig. 3.7 gives

$$f_n = -j\frac{1}{\pi} \int_\delta^{\pi-\delta} f(\theta_e) \sin(n\theta_e) \, d\theta_e \quad ; \tag{3.10}$$

$$f_n = -j\frac{2Ni}{n\pi} \cos(n\delta) \quad n \text{ odd} \quad . \tag{3.11}$$

The act of short-pitching the winding gives an opportunity to control the spatial harmonic content of the winding mmf through the $\cos n\delta$ term. In addition, the end turns become shorter. Because this term is due to the pitch of the coils, it is often referred to as the pitch factor:

$$k_p = \cos(n\delta) \quad . \tag{3.12}$$

---

[2]See Appendix B for a discussion of Fourier series and related signal concepts such as symmetry, ac and dc components, and their relationship to other waveform attributes.

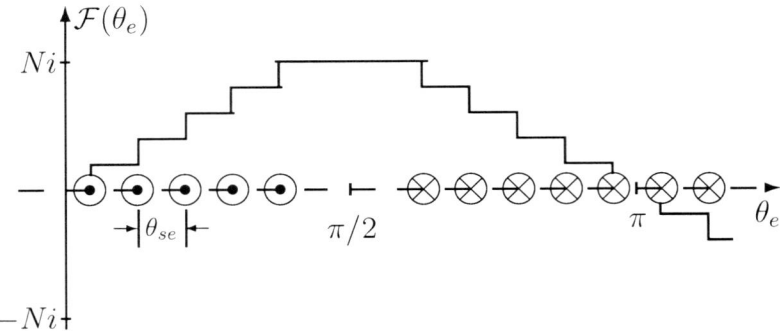

Figure 3.8: One pole of a five-coil phase belt, where coils are displaced from one another by θ_{se}.

Now we take a number of our primitive windings and assemble them together into a group of windings, or *phase belt*. Figure 3.8 shows a phase belt comprised of five short-pitched coils forming one pole of an integral slot winding. Because the coils are spread out, this configuration is also known as a *distributed winding*. In general we assume that there are q coils in the phase belt. Note that $q = N_{sp\phi}$ if balance among the phase windings is to be maintained while keeping all slots filled in the double-layer winding.

In Fig. 3.8, there are three ways of visualizing the winding structure:

1. As five identical short-pitched coils displaced from one another by the electrical angle θ_{se}. The middle coil retains odd symmetry, but the symmetry of the other four coils is masked by the spatial phase shift.

2. As five short-pitched coils, each with a different pitch but possessing odd symmetry. The end turns of each coil will be of different length.

3. As ten full-pitched coils displaced from one another. The coil sides sit over adjacent poles.

The first two ways of looking at the winding structure are the more productive since the end turns are shorter and the mmf distribution is the same for all three. We will consider the winding of Fig. 3.8 as five identical coils with spatial displacement, consistent with our assumption that all coils are identical. Considering the winding to be comprised of five nested short-pitched coils opens up the possibility of changing the number of turns per coil to optimize the mmf distribution with a concentric winding, but this is beyond the scope of our discussion.

3.4. Integral Slot Double-layer Windings

The phase belt shown in Fig. 3.8 is for a specific coil span and coil placement within the slots. With five coils in the phase belt, we know that $N_{sp\phi} = 5$ so there are 15 slots per pole. From this we conclude that each coil spans 9 slots so that the phase belt occupies the five winding slots at each end of the pole. The windings for the other two phases would have similar structure but be displaced by 10 slots from the phase belt shown. Other coil arrangements are possible, created by coils with a different coil span. These other arrangements might have two coil sides from the same phase occupying the same slot.

For a phase belt comprised of q coils, we begin by noting that the total number of Ampere-turns in the winding is Ni, so the number of Ampere-turns in each coil is Ni/q. We also note that the span of each coil is k slots where $k \leq N_s/N_p$, making the coil angle $k\theta_{se}$ so that the pitch angle must satisfy

$$\delta = \frac{\pi - k\theta_{se}}{2} \quad . \tag{3.13}$$

Note that for $k = 2q$ and $N_\phi = 3$ we have $\delta = \pi/6$, automatically eliminating all harmonics that are a multiple of 3, often referred to as the triple-n or triplen harmonics in the double-layer winding.

Coil placement within the phase belt is based on maintaining odd symmetry. The analytic representation of the mmf produced by the mth coil is given by Eqs. 3.8 and 3.11 with appropriate adjustment of the Ampere-turns and the addition of a term representing the spatial shift:

$$\mathcal{F}_m(\theta_e) = \sum_{\substack{n=-\infty \\ n \text{ odd}}}^{\infty} \frac{f_n}{q} \exp(jn\gamma_m) \exp(-jn\theta_e) \quad , \tag{3.14}$$

where

$$f_n = -j\frac{2Ni}{n\pi} \cos(n\delta) \quad . \tag{3.15}$$

It follows that the mmf of the total winding is given by the superposition of the mmf from each of the q coils:

$$\mathcal{F}(\theta_e) = \sum_{\substack{n=-\infty \\ n \text{ odd}}}^{\infty} \frac{f_n}{q} \exp(-jn\theta_e) \left\{ \sum_{m=1}^{q} \exp(jn\gamma_m) \right\} \quad . \tag{3.16}$$

The term in curly braces represents the phasor addition of the mmf space vectors from each coil, as shown in Fig. 3.9 for the winding of Fig. 3.8. This term becomes the distribution factor for the winding, denoted by k_d.

Because of the symmetry with which the coils are placed relative to zero phase shift, our sum of complex exponentials reduces to the sum of circular

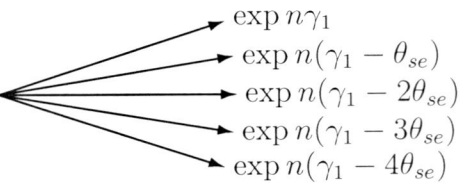

Figure 3.9: The mmf space vectors associated with the coils in Fig. 3.8.

cosines:

$$\sum_{m=1}^{q} \exp\left(\jmath n \gamma_m\right) = \sum_{m=1}^{q} \cos\left(n \gamma_m\right) \quad . \tag{3.17}$$

This applies whether q is even or odd, and regardless of coil span k. For the case of the winding shown in Fig. 3.8, this series reduces to an important result.

For the winding of Fig. 3.8, we denote spatial phase shift of the first coil by γ_1. The phase shifts of the other coils are tracked relative to γ_1. The angle γ_1 is tied to θ_{se}, being an integer multiple of θ_{se} for q odd and an odd multiple of $\theta_{se}/2$ for q even. In general we have

$$\gamma_1 = \frac{q-1}{2} \theta_{se} \quad , \tag{3.18}$$

and

$$\gamma_{m+1} = \gamma_m - \theta_{se} ; \quad m = 1, \ldots, q-1 \quad . \tag{3.19}$$

Expanding the series of Eq. 3.17 using the definition of γ_m given by Eq. 3.19 gives

$$\sum_{m=1}^{q} \cos\left(n \gamma_m\right) = \cos\left(n\left(\gamma_1\right)\right) + \cos\left(n\left(\gamma_1 - \theta_{se}\right)\right) + \cos\left(n\left(\gamma_1 - 2\theta_{se}\right)\right) + \cdots + \cos\left(n\left(\gamma_1 - (q-1)\theta_{se}\right)\right) \quad . \tag{3.20}$$

Making use of a trigonometric identity for the terms involving the difference of two angles[3] results in

$$\sum_{m=1}^{q} \cos\left(n \gamma_m\right) = \cos\left(n \gamma_1\right) \{1 + \cos\left(n \theta_{se}\right) + \cos\left(n 2 \theta_{se}\right) + \cdots + \cos\left(n(q-1)\theta_{se}\right)\} + \sin\left(n \gamma_1\right) \{\sin\left(n \theta_{se}\right) + \sin\left(n 2 \theta_{se}\right) + \cdots + \sin\left(n(q-1)\theta_{se}\right)\} \quad . \tag{3.21}$$

---

[3] $\cos(\alpha \pm \beta) = \cos\alpha\cos\beta \mp \sin\alpha\sin\beta$

3.4. Integral Slot Double-layer Windings

The two series within curly braces in Eq. 3.21 are found in mathematical tables[4] to be

$$\{1 + \cos(n\theta_{se}) + \cos(n2\theta_{se}) + \cdots + \cos(n(q-1)\theta_{se})\} = \sin\left(\frac{qn\theta_{se}}{2}\right) \frac{\cos\left(\frac{(q-1)n\theta_{se}}{2}\right)}{\sin\left(\frac{n\theta_{se}}{2}\right)} \quad (3.22)$$

and

$$\{\sin(n\theta_{se}) + \sin(n2\theta_{se}) + \cdots + \sin(n(q-1)\theta_{se})\} = \sin\left(\frac{qn\theta_{se}}{2}\right) \frac{\sin\left(\frac{(q-1)n\theta_{se}}{2}\right)}{\sin\left(\frac{n\theta_{se}}{2}\right)}, \quad (3.23)$$

respectively. Inserting these results into Eq. 3.21 gives

$$\sum_{m=1}^{q} \cos(n\gamma_m) = \frac{\sin\left(\frac{qn\theta_{se}}{2}\right)}{\sin\left(\frac{n\theta_{se}}{2}\right)} \times \left\{\cos n\gamma_1 \cos\left(\frac{(q-1)n\theta_{se}}{2}\right) + \sin n\gamma_1 \sin\left(\frac{(q-1)n\theta_{se}}{2}\right)\right\}. \quad (3.24)$$

Noting that

$$\left\{\cos n\gamma_1 \cos\left(\frac{(q-1)n\theta_{se}}{2}\right) + \sin n\gamma_1 \sin\left(\frac{(q-1)n\theta_{se}}{2}\right)\right\} = \cos\left(n\left(\gamma_1 - \frac{(q-1)\theta_{se}}{2}\right)\right), \quad (3.25)$$

and using the definition of $\gamma_1 = (q-1)\theta_{se}/2$ from Eq. 3.18 we have (finally!)

$$\sum_{m=1}^{q} \cos(n\gamma_m) = \frac{\sin\left(\frac{qn\theta_{se}}{2}\right)}{\sin\left(\frac{n\theta_{se}}{2}\right)}. \quad (3.26)$$

Using this in Eq. 3.16 gives the resultant mmf from the distributed winding to be

$$\mathcal{F}(\theta_e) = \sum_{\substack{n=-\infty \\ n \text{ odd}}}^{\infty} f_n \left\{\frac{\sin\left(\frac{qn\theta_{se}}{2}\right)}{q\sin\left(\frac{n\theta_{se}}{2}\right)}\right\} \exp(-jn\theta_e). \quad (3.27)$$

---

[4]H. B. Dwight, *Tables of Integrals and Other Mathematical Data*, 4th ed., MacMillan, 1961.

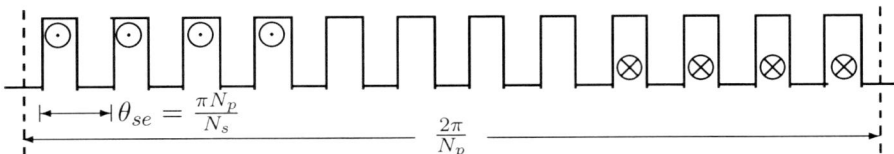

Figure 3.10: The coil arrangement for one phase of a machine with 48 slots and 4 poles. Each coil spans 8 slots. Only one pole is shown.

The term in curly braces represents the effect of distributing multiple coils, and is therefore called the distribution factor:

$$k_d = \frac{\sin\left(\frac{qn\theta_{se}}{2}\right)}{q\sin\left(\frac{n\theta_{se}}{2}\right)} \quad . \tag{3.28}$$

This expression for the distribution factor applies to a phase belt with the structure shown in Fig. 3.9. It also applies to phase belts in which the coil span forces the phase belt to extend beyond one pole. In this case, coil sides from adjacent poles end up sharing a slot.

Consider the application of the distribution factor to a three-phase machine with four poles and 48 slots. First, consider the winding to be comprised of four coils, with each coil spanning 8 slots and $Ni = 4$. This coil distribution is shown in Fig. 3.10. Figure 3.11 shows the mmf distribution over one electrical cycle (one pole pair) based on Eq. 3.27, since this coil distribution is consistent with Fig. 3.8, only with a different number of coils.

Now consider the winding to be comprised of four coils, with each coil spanning 10 slots and $Ni = 4$. This coil distribution is shown in Fig. 3.12. It will be observed that the coil arrangement is still consistent with Fig. 3.8, but because of the coil span, the total extent of the phase belt is more than a pole. Because of this, some of the Ampere turns that are contributing to mmf over each pole are generated by coil sides that belong to another phase belt within the same phase. For this winding we have

$$\delta = \frac{1}{2}\left\{\pi - 10\frac{\pi N_p}{N_s}\right\} = \frac{\pi}{12} \quad , \tag{3.29}$$

consistent with Eq. 3.13. Due to the overlap of the phase belts, we should now expect two large steps in the mmf distribution and two small steps in the mmf distribution. Figure 3.13 shows the mmf distribution over one electrical cycle (one pole pair) based on Eq. 3.27. We see that the mmf distribution is consistent with what we should have expected based on Fig. 3.12, with the line of odd symmetry based between the two slots that are fully occupied by the phase current.

3.4. Integral Slot Double-layer Windings

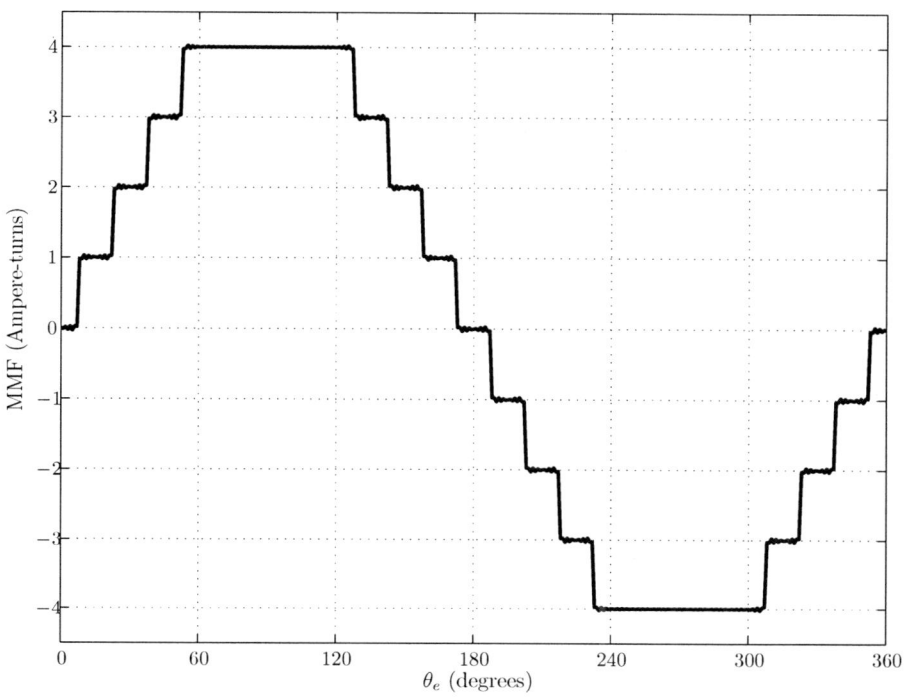

Figure 3.11: The mmf distribution given by Eq. 3.27 for $N_s = 48$, $N_p = 4$, $N_\phi = 3$, and $Ni = 4$ for coils that span 8 slots.

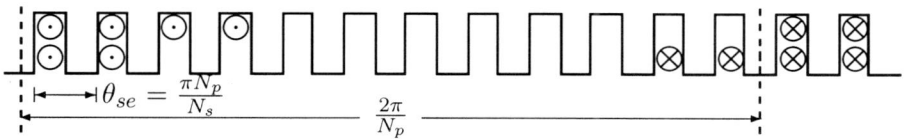

Figure 3.12: The coil arrangement for one phase of a machine with 48 slots and 4 poles. Each coil spans 10 slots. Because of the coil span, the phase belts for one phase overlap by two slots.

Table 3.1 gives a listing of the first fifteen odd harmonic amplitudes normalized to the fundamental for coil spans of 8 and 10 slots. The amplitude of the fundamental for the mmf distribution with the 8 slot coil span is 4.2239; the amplitude of the fundamental for the mmf distribution with the 10 slot coil span is 4.7111. While the distribution based on 10 slot coils has a higher fundamental amplitude, the total harmonic distortion (THD) is also substantially higher[5]. The waveform of Fig. 3.11 has a THD of 9.27%, while

---
[5]See Appendix B for a definition and discussion of Total Harmonic Distortion.

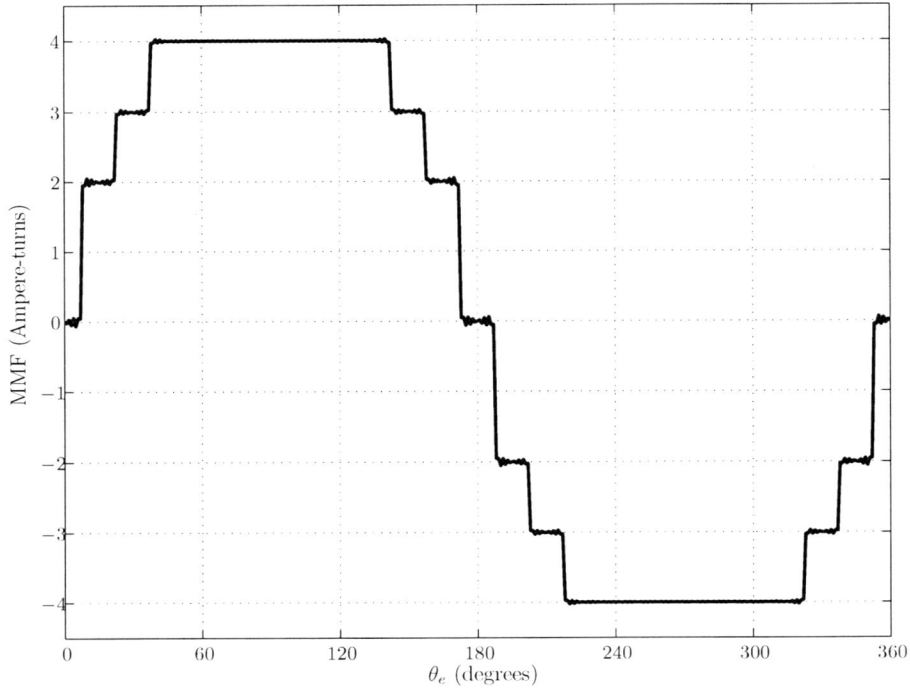

Figure 3.13: The mmf distribution given by Eq. 3.27 for $N_s = 48$, $N_p = 4$, $N_\phi = 3$, and $Ni = 4$ for coils that span 10 slots. The mmf distribution now has large and small steps, consistent with the current placement shown in Fig. 3.12.

the waveform of Fig. 3.13 has a THD of 18.95%. The relatively low THD in Fig. 3.11 can be traced to the elimination of the triplen harmonics.

Before leaving these two example winding distributions, it is worth examining the details of placing the coils into the slots. For the winding of Fig. 3.10, the coil sides for which current is coming out of the page all occupy the bottom of the slot. This is also true for the windings of Fig. 3.12, even though the coil span starts to cause the coil sides of adjacent phase belts to overlap. Within a single phase belt, the coil sides for which current is going into the page all occupy the top of the slot. (The slot bottom is conventionally taken to be adjacent to the back iron.) With both of these winding structures, it is straightforward to see that the four coils are of identical shape and will nest together naturally as the machine is built.

Table 3.1: The normalized harmonic amplitudes for the mmf distribution for $N_s = 48$, $N_p = 4$, and $N_\phi = 3$ with coil spans of 8 and 10 slots.

Harmonic	8 Slot Amplitude	10 Slot Amplitude
1	1	1
3	0.0000	0.1665
5	-0.0429	0.0115
7	0.0235	0.0063
9	0.0000	0.0230
11	-0.0120	0.0120
13	0.0101	-0.0101
15	0.0000	-0.0138
17	-0.0097	-0.0026
19	0.0113	-0.0030
21	0.0000	-0.0238
23	-0.0435	-0.0435
25	-0.0400	-0.0400
27	0.0000	-0.0185
29	0.0074	-0.0020
31	-0.0053	-0.0014

3.5 Fractional Slot Double-layer Windings

The basic building block of the fractional slot winding is a single coil placed within two slots that are separated by an electrical angle that depends on the coil span, as shown in Fig. 3.14. Unlike the integral slot winding, the combination of slot numbers, pole numbers, and phases may give an odd number of coils in each phase winding. In this case, the air gap mmf distribution can have a dc offset which could have undesired consequences on the opposing member, particularly in the case of an induction machine.

Based on the even symmetry of the mmf distribution, we have

$$\mathcal{F}(\theta_e) = \sum_{n=-\infty}^{\infty} f_n \exp\left(\jmath n \theta_e\right) \quad , \tag{3.30}$$

where

$$f_n = \frac{Ni}{\pi} \int_0^\delta \cos n\theta_e \, d\theta_e \quad . \tag{3.31}$$

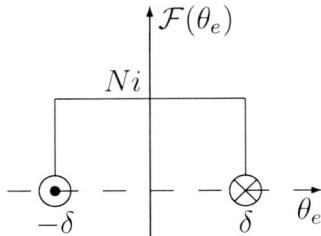

Figure 3.14: The basic winding structure for a fractional slot winding.

Carrying out the intended integration gives

$$f_n = \frac{Ni}{n\pi} \sin n\delta \quad . \tag{3.32}$$

Note that $f_0 = Ni\delta/\pi$, which indicates the presence of a dc offset in the mmf as noted above. In addition, we also note that there is no half-wave symmetry to the waveform (unless $\delta = \pi/2$), suggesting that the mmf distribution is richer in harmonic content than the integral slot winding of Fig. 3.7.

In a fractional slot winding, the coil span is generally chosen to be as close to one pole pitch as possible, but generally not larger. The coil span, however, must be at least one slot, leading us to conclude that the coil span in number of slots is

$$S = \max\left(\text{fix}\left(\frac{N_s}{N_p}\right), 1\right) \quad , \tag{3.33}$$

where $\text{fix}(\cdot)$ returns the integer portion of the argument. We need for S to be an integer so that the coils fit neatly into slots. Coil span could be chosen to be less than $N_{sp}(= N_s/N_p)$ and may be chosen due to considerations of end turns or inductance harmonics.

We are interested only in combinations of poles and slots that enable balance among the phases. In particular, the spatial shift of one phase relative to its neighbors must be $120°$ electrical. If we cannot get this phase shift, we reject the combination of poles and slots. We also constrain all coils to have the same span, and all phases to have the same number of coils.

The symmetric displacement of the phases requires identification of a phase offset K_o, denoted in slots, that is used to shift the winding pattern between phases. That is,

$$K_o \theta_{se} = 120° + q360° \quad , \tag{3.34}$$

3.5. Fractional Slot Double-layer Windings

where we allow for the possibility that the appropriate offset may require more than one electrical cycle (pole pair). Using the definition of θ_{se} from Eq. 3.7, we have

$$K_o \frac{N_p}{N_s} 180° = 120° + q360° \quad , \tag{3.35}$$

where q is an integer. This reduces to

$$K_o = \frac{N_s}{N_p}\left(\frac{2}{3} + 2q\right) \quad . \tag{3.36}$$

With a known phase offset, we are assured of developing windings that are balanced. We can now proceed with winding layout.

The winding layout process is one of determining the coils that have the greatest electrical alignment. Otherwise the voltages induced in these coils will have a tendency to cancel, which limits the performance of the machine. As we proceed around the periphery of the machine, coils of span S are sequentially displaced by angle θ_{se}. That is, we end up with a set of N_s space vectors, each displaced by θ_{se}. The objective of winding layout is to determine the coils that most closely align, recognizing that the space vector for any coil can be flipped by interchanging the direction of current flow in the coil sides.

Figure 3.15 shows the space vectors for a three-phase, four pole machine with 15 slots. The number associated with each space vector is the slot number in which the coil starts. Figure 3.16 shows the reconfigured space vectors, where a prime on the number indicates that the polarity has been reversed to get better alignment with the coil that starts in slot 1. From Fig. 3.16 it becomes straightforward to identify the five coils that should be associated with phase a.

It is worth noting in Fig. 3.15 that the coil that starts in slot 11 has 120° electrical displacement from coil 1, identifying the phase offset of 10 slots. We also note that the coil that starts in slot 6 is displaced from slot 1 by 240°, consistent with a 10 slot offset from phase b. Extending our alignment process to phases b and c identifies the five coils associated with each phase. This is shown in Fig. 3.17. In the coils associated with each phase winding, there are three coils in their original orientation and two coils where the use of the coil sides is interchanged to change the angle of the coil by 180° electrical. Table 3.2 summarizes the windings for the four pole machine with 15 slots. Each slot holds two coil sides, with some slots holding coil sides from only one phase with other slots holding coil sides from two phases.

If we look at the spatial distribution of, for example, the phase a winding relative to the underlying pole structure as shown in Fig. 3.18, we see that there is one coil substantially covering each pole. In addition, there is one

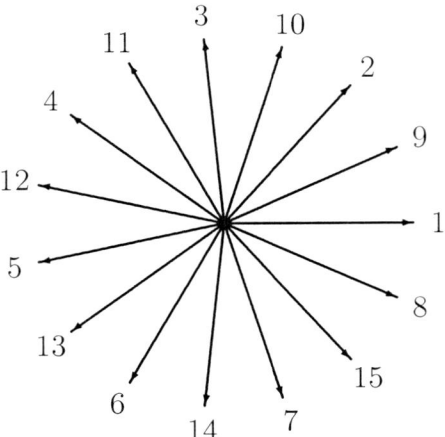

Figure 3.15: The 15 space vectors associated with coils placed in each slot of a three-phase machine with 15 slots and 4 poles.

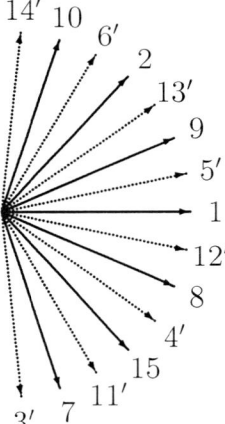

Figure 3.16: The 15 space vectors associated with coils placed in each slot of a three-phase machine with 15 slots and 4 poles, where the polarity of coils has been changed as necessary to get maximum alignment with coil 1.

pole that sees a second coil. Further, no coils align with the underlying pole structure in exactly the same way. This unbalanced use of the winding coils relative to the pole structure is characteristic of a fractional slot winding. If

3.5. Fractional Slot Double-layer Windings

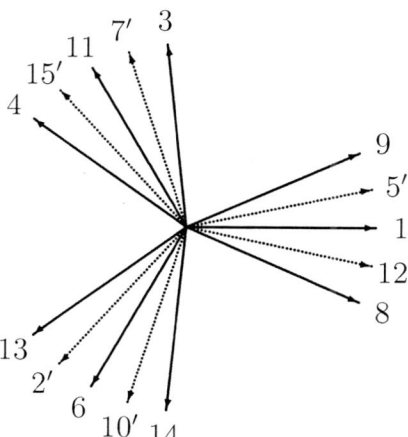

Figure 3.17: The 15 space vectors associated with coils placed in each slot of a three-phase machine with 15 slots and 4 poles, where each coil has been associated with one of the three phases. The group centered on coil 1 goes with phase a, the group centered on coil 11 goes with phase b, and the group centered on coil 6 goes with phase c.

we were to increase the number of slots in the machine to 24, there would now be 8 coils per phase and each pole would be acted on by two coils. The coils acting on each pole would likely not have exactly the same electrical angle. The different electrical angles associated with the phase coils allows for cancellation of some spatial harmonics. We will see this in greater detail when we look at the voltage induced in the phase windings in Sec. 4.4.

In general, the development of the fractional slot winding involves the following steps for phase a:

1. Determine the number of coils per phase:

$$N_{c\phi} = \frac{N_s}{N_\phi} \qquad (3.37)$$

   since each slot holds two coil sides (in a double-layer winding) there is one coil for each slot.

2. Determine the coil span according to Eq. 3.33.

3. Determine the phase offset using Eq. 3.36.

Table 3.2: A summary of the winding pattern for a three-phase machine with four poles and 15 winding slots. S stands for the start side of a coil, while F stands for the finish side of a coil.

Slot	Phase a		Phase b		Phase c	
1	S			↑	F	↓
2	↓			↑	F	F
3	↓		S	S	↑	
4	F		↓	S	↑	
5		F	↓	↓	S	
6		↑	F	↓	S	
7		↑	F	F	↓	
8	S	S	↑		↓	
9	↓	S	↑		F	
10	↓	↓	S			F
11	F	↓	S			↑
12	F	F	↓			↑
13	↑		↓		S	S
14	↑		F		↓	S
15	S			F	↓	↓

4. Define coil 1 to start in slot 1 with electrical angle 0°.

5. Place one coil in each slot with electrical angle $(k-1)\theta_{se}$, where k is the slot number.

6. Adjust the electrical angle of each coil so that it falls on the interval $[-90°, 90°]$, by interchanging the direction of current in the coil as necessary.

7. Select the $N_{c\phi}$ coils that have the angles closest to zero to form the phase a winding.

The windings for the other phases are identical to the phase a winding except they are progressively offset by K_o slots.

3.6 Slot and Skew Winding Factors

To this point coils have been modeled to provide step changes in mmf. This is not what we would expect given that the slot openings have finite width.

3.6. *Slot and Skew Winding Factors* 85

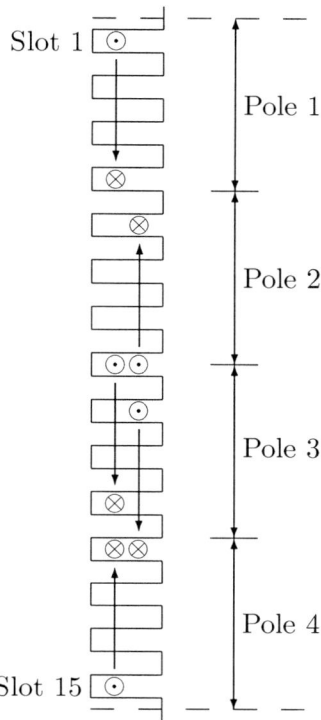

Figure 3.18: The winding layout for phase a of a three-phase machine with 15 slots and four poles in relation to the underlying pole structure.

Accordingly, the distribution shown in Fig. 3.7 should really be modeled as shown in Fig. 3.19, where the angle χ is the angular width of the slot opening. If the slot opening is w_{sl} at radius R, we have

$$\chi = \frac{w_{sl} N_p}{2R} \tag{3.38}$$

in electrical measure.

We should expect the linear transitions in the mmf to translate into reduced higher-order harmonics in the Fourier series representation. For the integral slot winding, recomputing the Fourier coefficient of the mmf distribution, taking into account the finite slot opening, gives

$$f_n = -j\frac{Ni}{\pi} \left\{ \int_{\delta-\chi/2}^{\delta+\chi/2} \left[\frac{1}{\chi}\left(\theta_e - \delta + \frac{\chi}{2}\right)\right] \sin(n\theta_e)\,d\theta_e + \int_{\delta+\chi/2}^{\pi-\delta-\chi/2} \sin(n\theta_e)\,d\theta_e + \right.$$

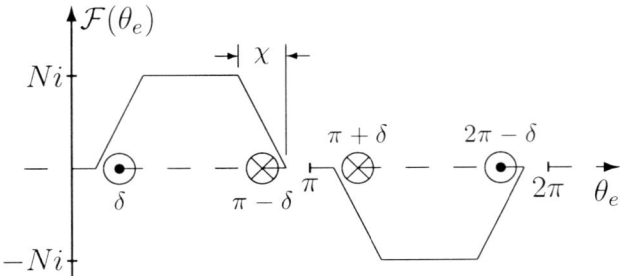

Figure 3.19: The air gap mmf distribution for an integral slot winding, taking into consideration the finite slot width χ, where χ is centered on the coil sides.

$$\int_{\pi-\delta-\chi/2}^{\pi-\delta+\chi/2} \left[\frac{1}{\chi}\left(-\theta_e + \pi - \delta + \frac{\chi}{2}\right)\right] \sin(n\theta_e)\, d\theta_e \biggr\} \, . \quad (3.39)$$

Carrying out the indicated integrations and reducing gives

$$f_n = -j\frac{2Ni}{n\pi}\cos(n\delta)\frac{\sin(n\chi/2)}{(n\chi/2)} \, . \quad (3.40)$$

The additional term $\sin(n\chi/2)/(n\chi/2)$ acts as a low pass filter and as a modulation function; it is shown in Fig. 3.20. That is, the effect of finite slot width is represented by a single term that represents low pass filtering by a sampling function. This term is often referred to as the slot factor:

$$k_{sl} = \frac{\sin(n\chi/2)}{(n\chi/2)} \, . \quad (3.41)$$

If we add a finite slot width to our sample four pole machine, the mmf distribution changes from that shown in Fig. 3.11 to that shown in Fig. 3.21. Note that the transitions in the mmf are more gradual. The application of Eq. 3.40 to the machine used as the basis of Table 3.3 gives a listing of the first fifteen odd harmonic amplitudes normalized to the fundamental. The data from Table 3.1 has been included for easy comparison of amplitudes.

It is also common to introduce skew into a machine. Skewing offsets each lamination by a small angular displacement relative to its neighbors. The effect is one of twisting the lamination stack over a total angle $\alpha\theta_{se} = \alpha\pi N_p/N_s$ in electrical measure, where α is the amount of skew normalized to one slot pitch. Figure 3.22 depicts the introduction of skew into one of the coils of Fig. 3.7. Figure 3.23 shows a solid model of a wound stator that incorporates skew.

3.6. Slot and Skew Winding Factors

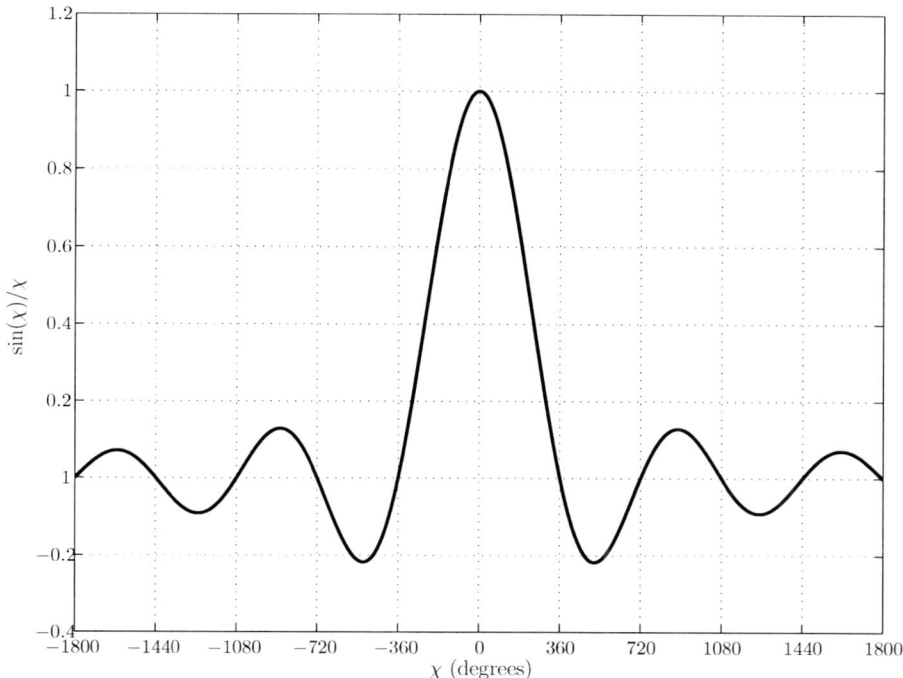

Figure 3.20: The filter function produced by finite slot openings.

If the angular position of the short-pitch winding of Fig. 3.7 is taken as $-\alpha/2$ at one end of the stack and $\alpha/2$ at the other end of the stack, the Fourier coefficient of the radial mmf distribution as a function of axial position along the stack is given by

$$f_n(z) = -j\frac{2Ni}{n\pi}\frac{1}{L_s}\cos(n\delta)\exp\left[-j\frac{n\alpha}{2}\left(1-\frac{2z}{L_s}\right)\right]dz \quad , \tag{3.42}$$

where L_s is the axial length of the stack, and the exponential term represents the spatial shifting created by the skew at axial position z. The total influence of the skew is found by summing the mmf distribution over the length of the machine:

$$f_n = \int_0^{L_s} f_n(z) \quad . \tag{3.43}$$

Substituting Eq. 3.42 gives

$$f_n = -j\frac{2Ni}{n\pi}\frac{1}{L_s}\cos(n\delta)\left\{\int_0^{L_s}\exp\left[-j\frac{n\alpha}{2}\left(1-\frac{2z}{L_s}\right)\right]dz\right\} \quad . \tag{3.44}$$

Carrying out the indicated integration and making use of the relationship

Table 3.3: The normalized harmonic amplitudes for the mmf distribution for $N_s = 48$, $N_p = 4$, $N_\phi = 3$, and $\chi = 0.3\theta_{se}$. The middle column gives the harmonic amplitudes considering only the distribution factor; the third column includes the slot factor.

Harmonic	k_d Only	k_d and k_{sl}
1	1	1
3	0.0000	0.0000
5	-0.0429	-0.0426
7	0.0235	0.0232
9	0.0000	0.0000
11	-0.0120	-0.0116
13	0.0101	0.0097
15	0.0000	0.0000
17	-0.0097	-0.0090
19	0.0113	0.0103
21	0.0000	0.0000
23	-0.0435	-0.0378
25	-0.0400	-0.0339
27	0.0000	0.0000
29	0.0074	0.0059
31	-0.0053	-0.0041

between complex exponentials and the circular sine function[6] gives

$$f_n = -j\frac{2Ni}{n\pi}\cos(n\delta)\,\frac{\sin(n\alpha/2)}{(n\alpha/2)} \quad . \tag{3.45}$$

The effect of skew manifests itself exactly like the finite slot opening, leading to the skew factor:

$$k_\alpha = \frac{\sin(n\alpha/2)}{(n\alpha/2)} \quad . \tag{3.46}$$

Because the effects of finite slot opening and skew impact the mmf distribution of each coil in the same way, the slot and skew factors become multiplicative factors for the mmf distribution of the composite winding, regardless of whether the winding pattern is integral slot, fractional slot, full-pitch, or fractional-pitch.

Table 3.4 and Fig. 3.24 update the mmf distribution taking skew into consideration. Figure 3.24 shows that the skew completely dominates the

---
[6] $\sin\alpha = [\exp(j\alpha) - \exp(-j\alpha)]/(j2)$.

3.7. Winding Connections

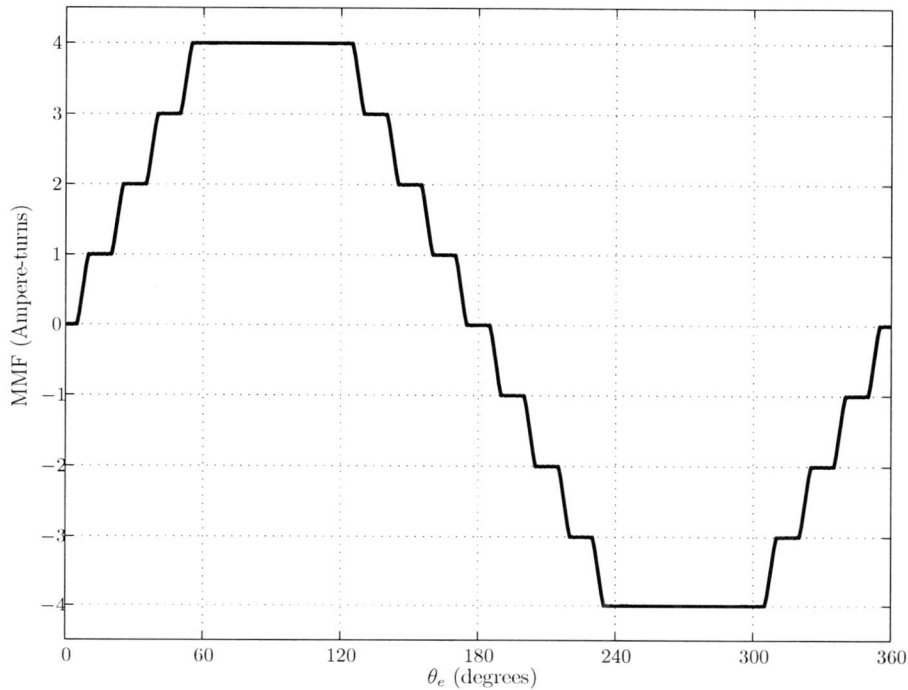

Figure 3.21: The mmf distribution given by Eq. 3.40 for $N_s = 48$, $N_p = 4$, $N_\phi = 3$, and $Ni = 4$, including the influence of $\chi = 0.3\theta_{se}$. The coils span 8 slots.

effect of slotting shown in Fig. 3.21. In addition, Fig. 3.24 shows that skewing is able to make a substantial contribution to shaping the air gap mmf distribution.

3.7 Winding Connections

At some point in the design of an electric machine, the windings must be made compatible with the power source. One way to do this is to adjust the number of turns in the coils used to form the phase windings. However, this does not allow for accommodating multiple supply voltages. Most industrial machines are compatible with two supply voltages.

The coils that comprise the phase windings can be connected in series or parallel. Series connections tend toward higher supply voltages while parallel connections tend toward lower supply voltages. The decision of which approach to take is not, however, simply a matter of juggling coil

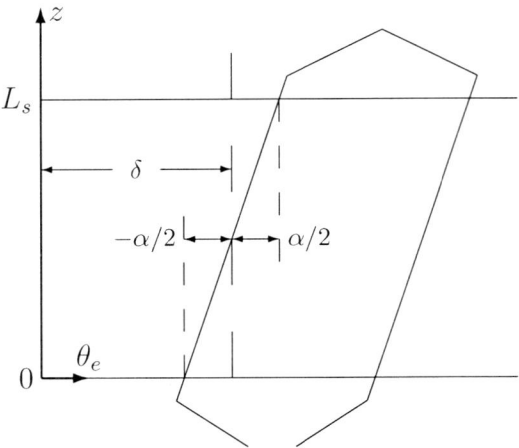

Figure 3.22: One coil showing the introduction of skew angle α.

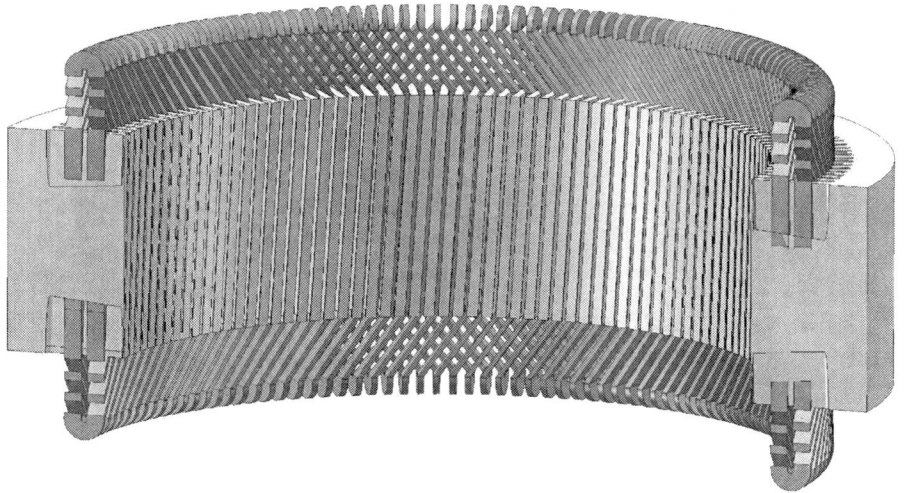

Figure 3.23: A solid model of a wound stator showing the skew. The skew angle is one slot pitch.

connections with the number of turns in each coil.

Coils within a phase winding should only be considered compatible with a parallel connection if they have the same spatial relationship relative to the pole structure. Otherwise, the voltages induced in the coils will be out of phase and large circulating currents will result. Consider the integral slot

3.7. Winding Connections

Table 3.4: The normalized harmonic amplitudes for the mmf distribution for $N_s = 48$, $N_p = 4$, $N_\phi = 3$, $\chi = 0.3\theta_{se}$, and $\alpha = \theta_{se}$ for coils that span 8 slots. The second column gives the harmonic amplitudes with just the distribution factor; the third column includes the distribution and slot factors only; and the fourth column includes distribution, slotting, and skew.

Harmonic	k_d Only	k_d and k_{sl}	k_d, k_{sl}, and k_α
1	1	1	1
3	0.0000	0.0000	0.0000
5	-0.0429	-0.0426	-0.0398
7	0.0235	0.0232	0.0202
9	0.0000	0.0000	0.0000
11	-0.0120	-0.0116	-0.0080
13	0.0101	0.0097	0.0057
15	0.0000	0.0000	0.0000
17	-0.0097	-0.0090	-0.0032
19	0.0113	0.0103	0.0025
21	0.0000	0.0000	0.0000
23	-0.0435	-0.0378	-0.0016
25	-0.0400	-0.0339	0.0014
27	0.0000	0.0000	0.0000
29	0.0074	0.0059	-0.0009
31	-0.0053	-0.0041	0.0008

winding of Fig. 3.8 that has five identical coils per pole. While the coils are identical in that they have the same number of turns and the same span, they are spatially displaced from one another relative to the underlying pole structure. Machine performance would suffer badly if the coils within a pole are connected in parallel. Accordingly, at a minimum the five coils associated with each pole should be connected in series. Because of the integral slot nature of the winding, it is possible to connect the windings for each pole in parallel with the windings for other poles.

The flexibility for connecting the coils within the phase windings is substantially reduced for fractional slot windings because it is not possible for all coils to have the same spatial relationship relative to the pole structure. It may be possible to connect some subset of phase coils in parallel, but it will not be possible to connect all of them in parallel.

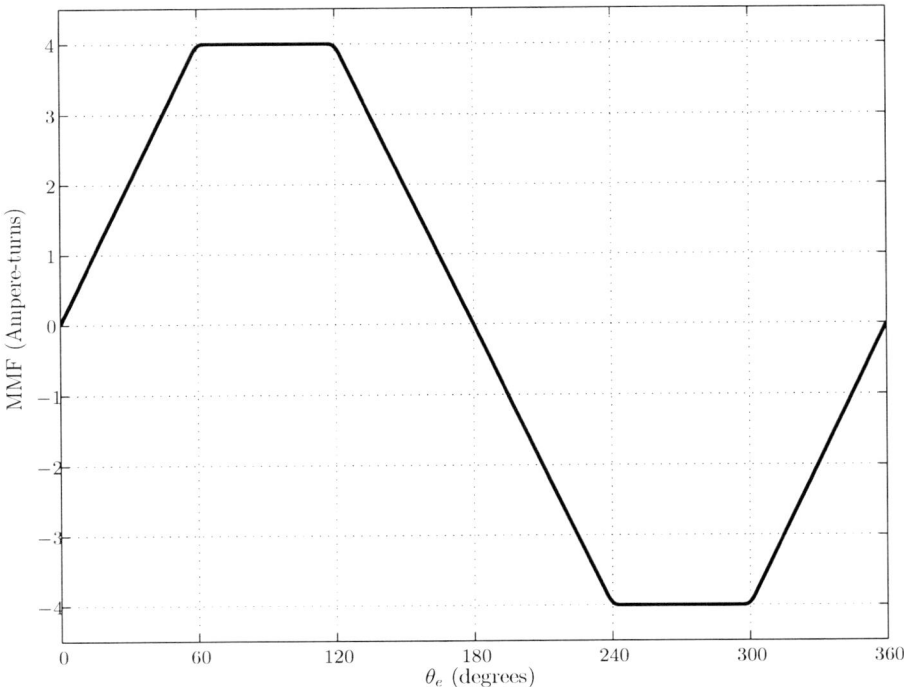

Figure 3.24: The mmf distribution given by Eq. 3.27 for $N_s = 48$, $N_p = 4$, $N_\phi = 3$, and $Ni = 4$, including the influence of $\chi = 0.3\theta_{se}$ and one slot of skew. The coils span 8 slots.

3.8 Summary

This chapter has examined how windings are developed for ac machines. Separate consideration was given to windings with an integral number of slots per pole and windings with a fractional number of slots per pole.

Integral slot windings are generally formed by short-pitched coils that are spatially distributed to approximate a sinusoidal mmf distribution. Fractional slot windings make use of harmonic cancellation to reduce the harmonic content in the mmf distribution. However, it was shown that for fractional slot windings with an odd number of coils, an mmf offset is introduced into the distribution that can be problematic for induction machines.

The issues of finite slot width and skew each affect integral slot windings and fractional slot windings in the same manner. Both finite slot width and skew effectively pass the mmf distribution through a low pass filter, reducing the magnitude of the higher-order space harmonics by softening the transitions through each coil side.

Chapter 4

Inductances, Back Emfs, and Space Harmonics

4.1 Introduction

This chapter builds on the winding structures discussed in Chapter 3 and integrates them into the energy conversion process. We begin with consideration of how to model an air gap that is faced by a slotted structure, such as created by the slots used to hold the phase windings. Once we know how to model the air gap, we can determine the inductance of each coil in the phase windings to build up the total phase inductance.

There are a number of components that make up the total phase inductance. The most important component is the air gap inductance since that is most directly responsible for energy transfer between the stator and rotor. However, there are also mutual inductances, slot inductances, and leakage inductances that also need to be considered. Inductance is often an undesired characteristic of an electric machine. The inductance represents energy storage, but the purpose of the machine is to convert energy, not store it. The inductance does, however, help with smoothing phase currents when the machine is excited through a power electronic inverter.

An objective of this chapter is to show that energy conversion depends heavily on the interaction between back emf and phase current. The back emf profile, as we will see, is integrally related to the winding structure. The interaction between the space harmonics of the back emf and the temporal harmonics in the currents applied to the phase windings gives rise to time-varying torques that are undesirable. We will analyze these interactions to see that the torque harmonics occur at specific frequencies, suggesting that

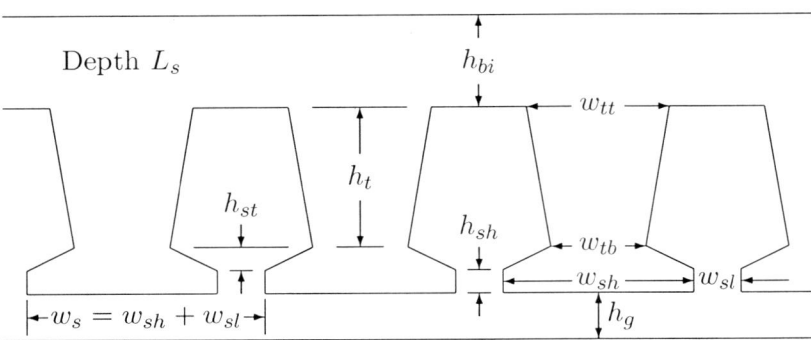

Figure 4.1: The assumed geometry for modeling the air gap when it is bordered by winding slots.

there are ways to compensate for these harmonics.

4.2 Air Gap Modeling

In the vast majority of electric machines the windings are placed in slots within the magnetic structure. The presence of the slots makes the air gap appear larger than its actual physical dimension. Traditionally, a compensation factor known as the Carter coefficient is applied to the actual air gap dimension to account for the slot openings.

Using the dimensions defined in Fig. 4.1 the Carter coefficient is

$$k_C = \frac{w_s}{w_s - \frac{4h_g}{\pi}\left[\frac{w_{sl}}{2h_g}\tan^{-1}\left(\frac{w_{sl}}{2h_g}\right) - \ln\sqrt{1+\left(\frac{w_{sl}}{2h_g}\right)^2}\right]} \quad , \qquad (4.1)$$

so that the modeled air gap length is

$$h'_g = k_C h_g \quad . \qquad (4.2)$$

The Carter coefficient is based on conformal mapping of the magnetic fields in regularly spaced slots that have parallel sides and are infinitely deep.

Another approach to correcting for the winding slots is to use the method of flux tubes. In flux tube analysis it is assumed that all flux lines follow either straight lines or circular arc segments. These segments are pieced

4.2. Air Gap Modeling

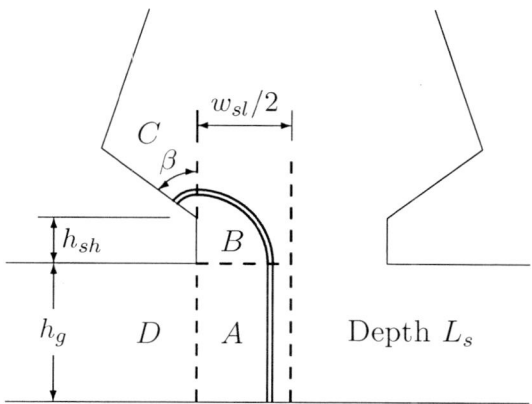

Figure 4.2: An example of how flux tubes can be used to model the region at the entrance to the winding slots.

together to develop an approximate picture of the magnetic field within the region of interest. The permeance associated with each region is then computed, and the permeances are combined into an equivalent permeance. For the slot geometry of Fig. 4.1 a reasonable choice of flux tubes is shown in Fig. 4.2 for the case where $w_{sl}/2 > h_{sh}$. The use of circular arcs is motivated by flux lines entering and leaving high permeability materials at right angles.

In the rectangular region under the pole face, the permeance is simply

$$\mathcal{P}_D = \frac{\mu_0 w_{sh} L_s}{h_g} \quad . \tag{4.3}$$

The incremental permeance that begins in region A and terminates on the vertical side of the pole shoe is given by

$$d\mathcal{P}_{AB} = \frac{\mu_0 L_s dr}{\frac{\pi}{2} r + h_g} \quad , \tag{4.4}$$

where r is measured from the corner of the pole shoe adjacent to the air gap. The total permeance in this region is

$$\mathcal{P}_{AB} = \int_0^{h_{sh}} \frac{\mu_0 L_s dr}{\frac{\pi}{2} r + h_g} \quad . \tag{4.5}$$

Performing the indicated integration gives

$$\mathcal{P}_{AB} = \frac{2}{\pi} \mu_0 L_s \ln\left[\frac{\pi h_{sh}}{2 h_g} + 1\right] \quad . \tag{4.6}$$

For the flux lines that originate in region A and terminate on the back side of the pole shoe the incremental permeance is given by

$$d\mathcal{P}_{ABC} = \frac{\mu_0 L_s dr}{\beta r + \frac{\pi}{2}(r + h_{sh}) + h_g} \quad , \tag{4.7}$$

where r is measured from the top corner of the pole shoe and β is the angle between the back of the pole shoe and the vertical:

$$\beta = \tan^{-1} \frac{2h_{st}}{w_{sh} - w_{tb}} \quad . \tag{4.8}$$

The total permeance for this region is

$$\mathcal{P}_{ABC} = \int_0^{w_{sl}/2 - h_{sh}} \frac{\mu_0 L_s dr}{\beta r + \frac{\pi}{2}(r + h_{sh}) + h_g} \quad . \tag{4.9}$$

Carrying out the indicated integration gives

$$\mathcal{P}_{ABC} = \frac{\mu_0 L_s}{\frac{\pi}{2} + \beta} \ln \left[\frac{\left(\frac{\pi}{2} + \beta\right)\left(\frac{w_{sl}}{2} - h_{sh}\right)}{\frac{\pi}{2} h_{sh} + h_g} + 1 \right] \quad . \tag{4.10}$$

The total air gap permeance associated with the single tooth is

$$\mathcal{P}_{\text{tooth}} = 2\mathcal{P}_{AB} + 2\mathcal{P}_{ABC} + \mathcal{P}_D \quad , \tag{4.11}$$

where the factors of 2 are necessary to account for the flux tubes on each side of the tooth. A gap correction factor for the slot permeance based on flux tube analysis can be determined by setting

$$\mathcal{P}_{\text{tooth}} = \frac{\mu_0 L_s w_s}{k_G h_g} \quad , \tag{4.12}$$

where k_G is the correction factor. Compiling the results of the individual permeances and solving for the correction factor gives

$$k_G = \frac{w_s}{\frac{4h_g}{\pi} \ln\left[\frac{\pi h_{sh}}{2h_g} + 1\right] + \frac{2h_g}{\frac{\pi}{2} + \beta} \ln\left[\frac{\left(\frac{\pi}{2} + \beta\right)\left(\frac{w_{sl}}{2} - h_{sh}\right)}{\frac{\pi}{2} h_{sh} + h_g} + 1\right] + w_{sh}} \quad . \tag{4.13}$$

Both k_C and k_G are reasonable approximations of the impact of winding slots in the case where the flux densities are low enough to prevent significant saturation in the teeth. If there is saturation, more of the flux will get pushed into the slot. Some of this flux may actually be inclined to travel to the bottom of the slot rather than going through the tooth. Figure 4.3 shows a comparison of k_C and k_G for different slot widths as a function of h_g/w_s. The factor k_G is consistently larger than k_C, but the trends are the same.

4.3. Winding Inductances

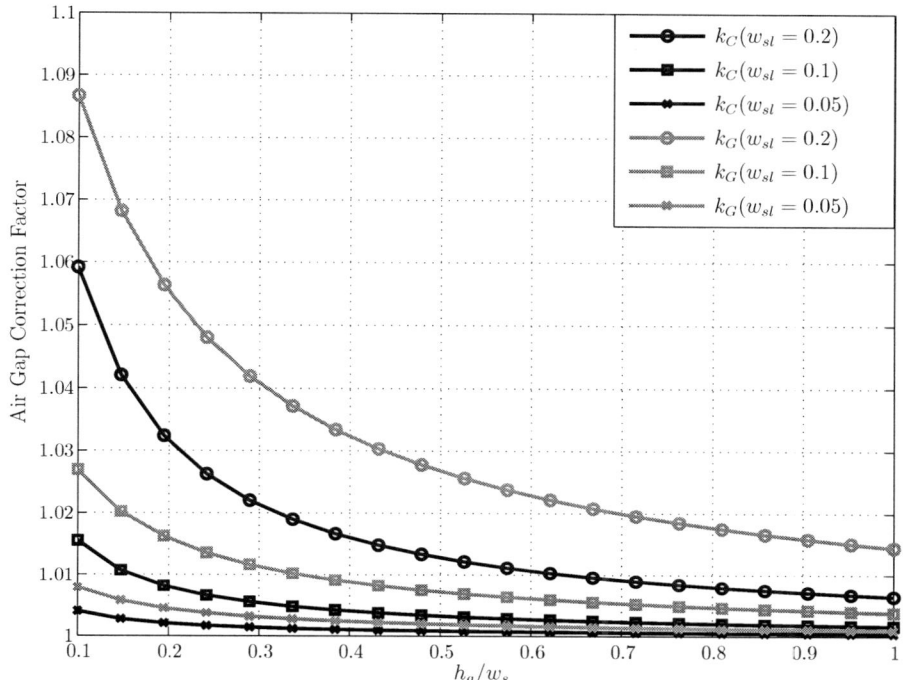

Figure 4.3: A comparison of k_C and k_G for different slot widths as a function of h_g/w_s.

Our model of the air gap has been based on slotting on only one side of the air gap. If there is slotting on both sides of the air gap, the model would need to be extended to include two correction factors, one for each side of the air gap. In this case the air gap can be modeled as two permeances (or reluctances) in series, one for each side of the air gap.

The use of flux tube analysis is a way to build a magnetic model in situations where the geometry does not allow an analytic solution and where finite element analysis is difficult to justify. Another example of flux tube analysis is given in Sec. 11.4 when it is applied to estimating the cogging torque caused by the attraction of rotor-mounted permanent magnets to stator teeth.

4.3 Winding Inductances

There are a number of components to the winding inductances:

1. Self inductance due to the air gap flux.
2. Self inductance due to slot leakage flux.
3. Self inductance due to end turn flux.
4. Self inductance due to zig-zag flux.
5. Mutual inductance due to air gap flux.
6. Mutual inductance due to slot leakage flux.
7. Mutual inductance due to end turn flux.
8. Mutual inductance due to zig-zag flux.

Of these components, the components tied to air gap flux are the most important, since machine performance is most tightly connected to the fields in the air gap. This section focuses on the air gap and slot inductance components. The end turn and zig-zag inductances get quite involved, and are best left to books that are focused on machine design[1].

4.3.1 Air Gap Inductances

In a well-designed machine, the air gap inductance should be the most significant part of the inductance because it represents the coupling between the stator and rotor. Our approach to determining the air gap inductance is based on modeling a single tooth. This approach is justified by Fig. 4.4, in which a coil that spans three slots is represented by the magnetic equivalent of three series-connected coils. Note that the net current in the two intermediate slots is zero.

Consistent with the decomposition of the inductance problem into consideration of what happens with a single tooth, we can work with the magnetic equivalent circuit of a single stator tooth as shown in Fig. 4.5. This equivalent magnetic circuit for the kth tooth is comprised of an mmf source, stator tooth reluctance, air gap reluctance, and rotor reluctance. The stator back iron is chosen as the reference point. The quantity F_r represents the mmf of the rotor back iron. The need for the rotor mmf will become apparent in the discussion to follow. The factor S_k is used to denote the stator tooth mmf as follows:

$$S_k = \begin{cases} 1 & \text{for flux directed out of the rotor} \\ 0 & \text{for zero Ampere-turns} \\ -1 & \text{for flux directed into the rotor} \end{cases} . \qquad (4.14)$$

---

[1]See, for example, T. A. Lipo, *Introduction to Ac Machine Design*, Wisconsin Power Electronics Research Center, University of Wisconsin-Madison, Madison, WI, USA, 2004.

4.3. Winding Inductances

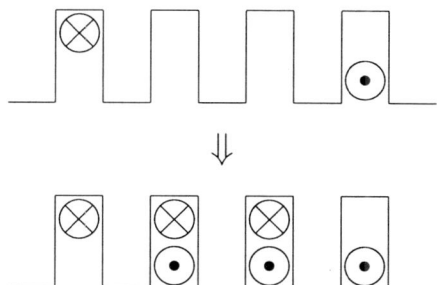

Figure 4.4: A coil that spans three slots and its magnetic equivalent.

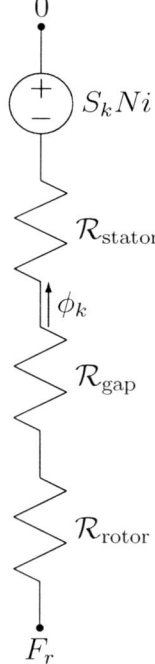

Figure 4.5: The magnetic circuit model for a single stator tooth.

The values of S superimpose for the coils within a winding. This accounts for coils in the same phase that share a tooth. For example, S for each tooth in a distributed winding would increment by one as additional Ampere-turns are added in consecutive slots, and decrease as Ampere-turns are removed.

The flux in each stator tooth is

$$\phi_k = \frac{F_r + S_k N i}{\mathcal{R}_{\text{tooth}}}, \quad (4.15)$$

where

$$\mathcal{R}_{\text{tooth}} = \mathcal{R}_{\text{stator}} + \mathcal{R}_{\text{gap}} + \mathcal{R}_{\text{rotor}}. \quad (4.16)$$

Gauss's law requires that the net flux entering (or leaving) the rotor must be zero, giving

$$\sum_{k=1}^{N_s} \phi_k = \sum_{k=1}^{N_s} \frac{F_r + S_k N i}{\mathcal{R}_{\text{tooth}}} = 0. \quad (4.17)$$

This reduces to

$$F_r = -\frac{Ni}{N_s} \sum_{k=1}^{N_s} S_k. \quad (4.18)$$

For an integral slot winding, F_r will always reduce to zero. For a fractional slot winding, F_r may or may not be zero; F_r could be zero if there are an even number of coils in the winding, depending on the orientation of the coils. Essentially, F_r is compensating for the dc component of the mmf distribution. As we discussed in Chapter 3, only the fractional slot winding could have a dc component to the air gap mmf distribution for a uniform air gap.

Assuming all coils are connected in series, the flux linked by each winding is given by $\lambda = N\phi$ where ϕ is the sum of the tooth fluxes seen by each coil. Using the sign of S_k to keep track of the direction of the flux through tooth k we have

$$\lambda = N \sum_{k=1}^{N_s} \text{sign}(S_k) \phi_k = N \sum_{k=1}^{N_s} \text{sign}(S_k) \left[\frac{F_r + S_k N i}{\mathcal{R}_{\text{tooth}}} \right]. \quad (4.19)$$

Putting in F_r gives

$$\lambda = N \sum_{k=1}^{N_s} \text{sign}(S_k) \left[\frac{S_k N i - \frac{Ni}{N_s} \sum_{m=1}^{N_s} S_m}{\mathcal{R}_{\text{tooth}}} \right]. \quad (4.20)$$

Rearranging gives

$$\lambda = \frac{N^2 i}{\mathcal{R}_{\text{tooth}}} \sum_{k=1}^{N_s} \text{sign}(S_k) \left[S_k - \frac{1}{N_s} \sum_{m=1}^{N_s} S_m \right], \quad (4.21)$$

making the inductance

$$L = \frac{N^2}{\mathcal{R}_{\text{tooth}}} \sum_{k=1}^{N_s} \text{sign}(S_k) \left[S_k - \frac{1}{N_s} \sum_{m=1}^{N_s} S_m \right]. \quad (4.22)$$

4.3. Winding Inductances

We can use this same formulation to determine the mutual air gap inductances. The fluxes get summed in a similar way, except that it is the fluxes produced by the currents of one phase and the directions associated with the coils of another phase. That is, the self inductance for phase a is

$$L_{aa} = \frac{N^2}{\mathcal{R}_{\text{tooth}}} \sum_{k=1}^{N_s} \text{sign}(S_{ak}) \left[S_{ak} - \frac{1}{N_s} \sum_{m=1}^{N_s} S_{am} \right] , \qquad (4.23)$$

while the mutual inductance between phases a and b is

$$L_{ba} = \frac{N^2}{\mathcal{R}_{\text{tooth}}} \sum_{k=1}^{N_s} \text{sign}(S_{bk}) \left[S_{ak} - \frac{1}{N_s} \sum_{m=1}^{N_s} S_{am} \right] . \qquad (4.24)$$

The mutual inductance between phases b and c is

$$L_{cb} = \frac{N^2}{\mathcal{R}_{\text{tooth}}} \sum_{k=1}^{N_s} \text{sign}(S_{ck}) \left[S_{bk} - \frac{1}{N_s} \sum_{m=1}^{N_s} S_{bm} \right] . \qquad (4.25)$$

The development presented here works for both integral slot windings and fractional slot windings. Because of the structure for the integral slot winding, it is possible to work with the inductance per pole rather than for the entire winding. This could be beneficial if the phase windings are formed through some combination of series and parallel connections.

As an example, consider the fractional slot winding developed for a four pole machine using 15 slots. The winding structure for this machine is summarized in Table 3.2. Taking tooth 1 to be between slots 1 and 2, and taking the start of a winding to correspond to trying to drive flux from the rotor to the stator in the corresponding tooth, Table 4.1 gives the S_k for the three phase windings. Note that the S_k for the phases obey the phase offset of 10 slots found when determining the winding layout.

Using the data in Table 4.1, we have

$$F_r = -\frac{Ni}{N_s} \sum_{k=1}^{N_s} S_k = -\frac{3Ni}{15} , \qquad (4.26)$$

$$L_{aa} = \frac{N^2}{\mathcal{R}_{\text{tooth}}} \sum_{k=1}^{N_s} \text{sign}(S_{ak}) \left[S_{ak} - \frac{1}{N_s} \sum_{m=1}^{N_s} S_{am} \right] = \frac{N^2}{\mathcal{R}_{\text{tooth}}} (14.8) , \qquad (4.27)$$

and

$$L_{ba} = \frac{N^2}{\mathcal{R}_{\text{tooth}}} \sum_{k=1}^{N_s} \text{sign}(S_{bk}) \left[S_{ak} - \frac{1}{N_s} \sum_{m=1}^{N_s} S_{am} \right] = \frac{N^2}{\mathcal{R}_{\text{tooth}}} (-6.2) . \qquad (4.28)$$

Table 4.1: The values of S_k for the three phase machine with 4 poles and 15 slots, consistent with the winding structure given in Table 3.2.

Tooth	S_{ak}	S_{bk}	S_{ck}
1	1	-1	1
2	1	-1	-1
3	1	1	-1
4	0	2	-1
5	-1	2	0
6	-1	1	1
7	-1	-1	1
8	1	-1	1
9	2	-1	0
10	2	0	-1
11	1	1	-1
12	-1	1	-1
13	-1	1	1
14	-1	0	2
15	0	-1	2

Ideally we would expect $L_{ba} = -(1/2)L_{aa}$ if our winding were perfectly balanced since $\cos(2\pi/3) = -1/2$. When determining the inductances using the summation, $\text{sign}(S_k = 0) = 0$, since when $S_k = 0$, it indicates that flux through that tooth does not contribute to the flux linked by the coil.

As a second example, consider the inductance per pole pair for a machine with 48 slots, 4 poles, and 3 phases, with each coil spanning 8 slots. The air gap mmf distribution for this configuration is given in Fig. 3.11. Table 4.2 gives the S_k for the associated winding configuration.

Using the data in Table 4.2, we have

$$F_r = -\frac{Ni}{N_s} \sum_{k=1}^{N_s} S_k = 0 \quad , \tag{4.29}$$

$$L_{aa} = \frac{N^2}{\mathcal{R}_{\text{tooth}}} \sum_{k=1}^{N_s} \text{sign}(S_{ak}) \left[S_{ak} - \frac{1}{N_s} \sum_{m=1}^{N_s} S_{am} \right] = \frac{N^2}{\mathcal{R}_{\text{tooth}}}(64) \quad , \tag{4.30}$$

and

$$L_{ba} = \frac{N^2}{\mathcal{R}_{\text{tooth}}} \sum_{k=1}^{N_s} \text{sign}(S_{bk}) \left[S_{ak} - \frac{1}{N_s} \sum_{m=1}^{N_s} S_{am} \right] = \frac{N^2}{\mathcal{R}_{\text{tooth}}}(-32) \quad . \tag{4.31}$$

4.3. Winding Inductances

Table 4.2: The values of S_k for a three phase machine with 4 poles and 48 slots, where each coil spans 8 slots. Only one pole pair is given.

Tooth	S_{ak}	S_{bk}	S_{ck}
1	1	-4	3
2	2	-4	2
3	3	-4	1
4	4	-4	0
5	4	-3	-1
6	4	-2	-2
7	4	-1	-3
8	4	0	-4
9	3	1	-4
10	2	2	-4
11	1	3	-4
12	0	4	-4
13	-1	4	-3
14	-2	4	-2
15	-3	4	-1
16	-4	4	0
17	-4	3	1
18	-4	2	2
19	-4	1	3
20	-4	0	4
21	-3	-1	4
22	-2	-2	4
23	-1	-3	4
24	0	-4	4

In this case we have $L_{ba} = -(1/2)L_{aa}$. Based on the fundamental structure of an integral slot winding relative to a fractional slot winding, it should be appreciated that an integral slot winding should be able to provide at least the same degree of balance, if not higher degree, than a fractional slot winding.

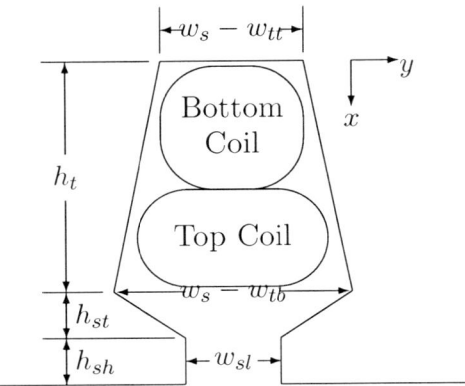

Figure 4.6: A model of a winding slot showing two coil sides for a double-layer winding.

4.3.2 Slot Inductances

The slot inductances account for flux that passes through the winding slots but not across the air gap. Because this flux does not cross the air gap, the slot inductance is considered a component of leakage inductance. Our discussion focuses on a two layer winding, with a coil side occupying either the top (near the air gap) or the bottom of the slot (near the back iron).

Figure 4.6 shows a winding slot occupied by two coils. For the case where the sides of the slot are parallel, such as when $w_{tt} = w_{tb}$, the magnetic field within the slot created by the bottom coil side is shown in Fig. 4.7. We can determine the slot inductance by using the coenergy density integrated over the appropriate volume as discussed in Sec. 2.4.

We will determine part of the self inductance using coenergy equivalence (see the discussion on page 49) and part of the self inductance using an equivalent magnetic circuit. The coenergy density within the slot is

$$w' = \frac{\mu_0 H_y^2(x)}{2} \quad . \tag{4.32}$$

Integrating this over the volume of the slot will give the coenergy stored in the slot field. We can then equate this coenergy with the coenergy stored in an inductor that is carrying the same current as the coil side. An alternative approach is to use the coenergy density where most convenient and permeances where most convenient. The use of the coenergy density is particularly convenient within the active coil where the field intensity is a function of position. The use of permeance is convenient outside of the active coil, and especially in the region of the taper.

4.3. Winding Inductances

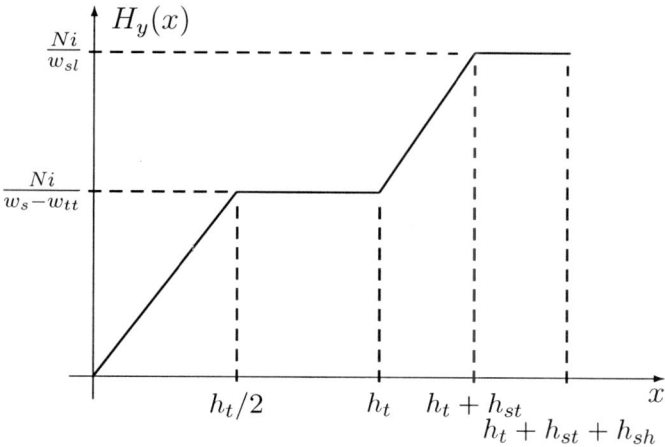

Figure 4.7: The magnetic field in the winding slot due to the bottom coil.

In the coil region, the coenergy is

$$W' = \frac{\mu_0}{2} \int_0^{h_t/2} \left[\frac{Ni}{(w_s - w_{tt})}\right]^2 \left[\frac{2x}{h_t}\right]^2 (w_s - w_{tt}) L_s dx \quad . \tag{4.33}$$

Carrying out the integration and reducing gives

$$W' = \frac{\mu_0 L_s h_t N^2 i^2}{12(w_s - w_{tt})} \quad , \tag{4.34}$$

giving us an inductance (for the coil region only) of

$$L_{\text{bottom coil}} = \frac{\mu_0 L_s h_t N^2}{6(w_s - w_{tt})} \quad . \tag{4.35}$$

Above the bottom coil, the permeance associated with the region for the top coil is simply

$$\mathcal{P}_{\text{top coil}} = \frac{\mu_0 L_s h_t}{2(w_s - w_{tt})} \quad . \tag{4.36}$$

In the region of the taper, the determination of the permeance is complicated by the taper created by the tooth shoe. The incremental permeance is given by

$$d\mathcal{P}_{\text{taper}} = \frac{\mu_0 L_s dx'}{\left[\frac{w_{sl} - (w_s - w_{tt})}{h_{st}} x' + (w_s - w_{tt})\right]} \quad , \tag{4.37}$$

where $x' = x - h_t$. The permeance for the taper region is then

$$\mathcal{P}_{\text{taper}} = \int_0^{h_{st}} \frac{\mu_0 L_s dx'}{\left[\frac{w_{sl} - (w_s - w_{tt})}{h_{st}} x' + (w_s - w_{tt})\right]} \quad . \tag{4.38}$$

This evaluates to

$$\mathcal{P}_{\text{taper}} = \frac{\mu_0 h_{st} L_s}{(w_s - w_{tt} - w_{sl})} \ln\left[\frac{w_s - w_{tt}}{w_{sl}}\right] \quad . \tag{4.39}$$

The permeance associated with the slot opening is

$$\mathcal{P}_{\text{slot opening}} = \frac{\mu_0 L_s h_{sh}}{w_{sl}} \quad . \tag{4.40}$$

Putting these pieces together gives the slot inductance for the bottom coil to be

$$L_{\text{bottom coil}} = \mu_0 L_s N^2 \left[\frac{h_t}{6(w_s - w_{tt})} + \frac{h_t}{2(w_s - w_{tt})} + \frac{h_{st}}{(w_s - w_{tt} - w_{sl})} \ln\left[\frac{w_s - w_{tt}}{w_{sl}}\right] + \frac{h_{sh}}{w_{sl}}\right] , \tag{4.41}$$

where the terms in the square brackets account for the region of the active coil, the inactive region of the top coil, the taper of the tooth shoe, and the slot opening, respectively. Remember that this result is for the special case of parallel slot sides.

We can build off of the result for the bottom coil to quickly determine the slot inductance for the top coil side. Figure 4.8 shows the magnetic field within the slot created by the top coil side. Comparing Figs. 4.7 and 4.8, we see that the field profile within the active coil, the tooth taper, and the slot opening remains the same. The field below the top coil is zero by virtue of the high permeability path provided by the teeth and back iron, suggesting that the slot inductance for the top coil is smaller than that for the bottom coil. That is,

$$L_{\text{top coil}} = \mu_0 L_s N^2 \left[\frac{h_t}{6(w_s - w_{tt})} + \frac{h_{st}}{(w_s - w_{tt} - w_{sl})} \ln\left[\frac{w_s - w_{tt}}{w_{sl}}\right] + \frac{h_{sh}}{w_{sl}}\right] \quad . \tag{4.42}$$

The inductances given by Eqs. 4.41 and 4.42 need proper interpretation. As they stand, these inductances represent additional self inductance for the bottom and top coil sides in the slot, respectively. In addition, there is mutual inductance between the coil sides.

4.3. Winding Inductances

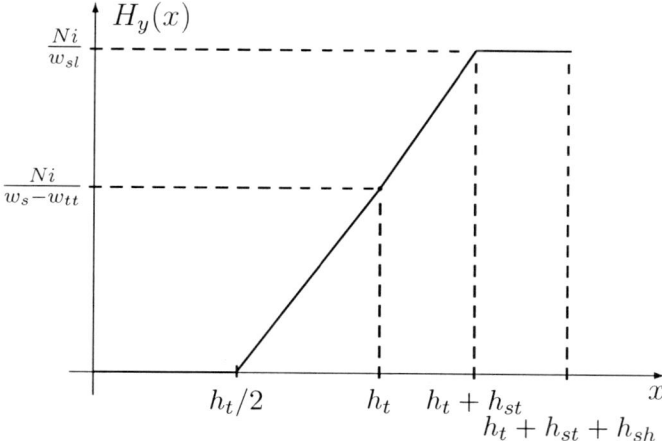

Figure 4.8: The magnetic field in the winding slot due to the top coil.

With regard to mutual inductance, the flux produced by the top coil side will completely link the bottom coil side since the flux will travel through the teeth before crossing the air gap. The flux produced by the bottom coil side, however, will not completely link the top coil side since some of the flux will travel below the bottom of the top coil side. Further, not all of the flux that travels through the top coil will link all turns of the coil. Since the slot opening and the taper are the regions with the smallest reluctance, it is reasonable to approximate the mutual inductance contributed by the bottom coil to be the flux passing through these regions.

Coenergy equivalence can be used to explicitly determine the mutual inductance between the two coil sides. If the two coil sides are from the same phase, the inductance between the coil sides adds to the phase self inductance. If the coil sides belong to different phases, the inductance between the coil sides adds to the mutual inductance between the phases.

4.3.3 Other Inductances

The end turn region of the machine contributes to both phase and mutual inductances. The details of these inductances depend substantially on the way the machine is wound, making a general formulation of end turn inductance difficult. Lap, concentric, and wave windings require different considerations because the winding path in the end regions are different. The design of end turn overlap for machines with concentric windings can vary significantly with regard to inductance determination. End turn inductance is

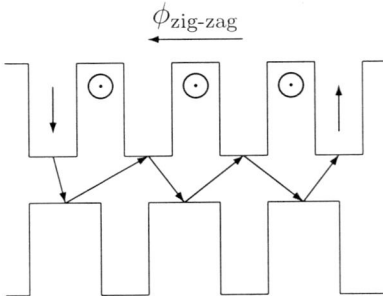

Figure 4.9: The magnetic field behind zig-zag inductance within a phase belt.

best handled on a case-by-case basis.

Zig-zag inductance refers to air gap inductance that is created by magnetic flux that jumps back and forth across the air gap as the flux moves from stator tooth to rotor tooth in the circumferential direction. It is a consequence of distributed windings, where coil sides supporting current flow in the same direction create the circulating flux that has opportunity to bounce back and forth across the air gap between stator and rotor teeth. Figure 4.9 illustrates the issue and suggests that the zig-zag inductance depends on the tooth structure of the stator and rotor. Since the rotor and stator are generally designed to have different numbers of slots, the zig-zag inductance will be periodic with rotor position.

4.4 Back Emfs

Consider the rotary version of Eq. 2.7 for a machine with N_ϕ phases with only one degree of mechanical freedom:

$$v_k = \sum_{j=1}^{N_\phi} \left.\frac{\partial \lambda_k}{\partial i_j}\right|_\theta \frac{di_j}{dt} + \left.\frac{\partial \lambda_k}{\partial \theta}\right|_i \frac{d\theta}{dt} \quad , \tag{4.43}$$

where

$$\lambda_k = \lambda_k\left(i_1, i_2, \ldots, i_k, \ldots, i_{N_\phi}, \theta\right) \quad . \tag{4.44}$$

The first term on the right of Eq. 4.43 represents the induced voltage due to variations in current; in Chapter 2 this was labeled the transformer voltage. The second on the right of Eq. 4.43 represents the induced voltage due to mechanical movement; in Chapter 2 this was labeled the motional voltage. In electric machines, the motional voltage is also commonly referred to as

4.4. Back Emfs

the back emf. The role of the back emf is quite significant in the operation of the machine since it embodies the electromechanical coupling between the stator and the rotor. We take the back emf of the kth phase to be

$$e_k = \left.\frac{\partial \lambda_k}{\partial \theta}\right|_i \frac{d\theta}{dt} \quad . \tag{4.45}$$

Using Eq. 4.43, the electrical power input into the terminals of our lossless machine can be written as

$$\sum_{k=1}^{N_\phi} i_k v_k = \sum_{k=1}^{N_\phi} \left\{ i_k \left[\sum_{j=1}^{N_\phi} \left.\frac{\partial \lambda_k}{\partial i_j}\right|_\theta \frac{di_j}{dt} \right] \right\} + \frac{d\theta}{dt} \sum_{k=1}^{N_\phi} \left[i_k \left.\frac{\partial \lambda_k}{\partial \theta}\right|_i \right] \quad . \tag{4.46}$$

The first term on the right represents the variation in field coenergy as a consequence of time-varying currents. Note that for a linear electric machine the $\partial \lambda_k / \partial i_j$ term is $L_{kj}(\theta)$. The second term on the right corresponds to the mechanical power output through the shaft. It follows that

$$T^e = \sum_{k=1}^{N_\phi} \left[i_k \left.\frac{\partial \lambda_k}{\partial \theta}\right|_i \right] = \frac{1}{\omega} \sum_{k=1}^{N_\phi} i_k e_k \quad . \tag{4.47}$$

The conclusion expressed in Eq. 4.47 may not be obvious. However, it is entirely consistent with our discussion of energy in Sec. 2.4. For a lossless machine with N_ϕ phases, the electrical power either goes into changing the field energy or into shaft power. That is,

$$\sum_{k=1}^{N_\phi} i_k \frac{d\lambda_k}{dt} = \frac{dW_m}{dt} + T^e \frac{d\theta}{dt} \quad . \tag{4.48}$$

Comparing Eq. 4.48 with Eq. 4.46, we conclude that Eq. 4.47 is correct. From this, we see that the power delivered to the rotor for transmission out the shaft is equal to the power delivered to the back emfs. Accordingly, the shape of the back emf waveform and its interaction with the phase current has a lot to say about the quality of the torque produced by the machine. We will address this in the next section.

Consider the back emf of the simple coil of Fig. 3.14 as a source of magnetic flux passes below it, such as by permanent magnets; Fig. 4.10 depicts the situation. We assume that the rotor flux alternates between north and south poles, with each pole of full pitch so it covers 180° electrical measure. Angle θ_e denotes the rotor position relative to the location where the pole boundary on the rotor is centered on the simple coil.

110   Chapter 4. Inductances, Back Emfs, and Space Harmonics

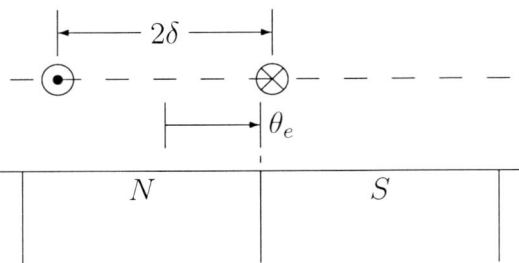

Figure 4.10: The geometry used to determine the back emf induced in a coil.

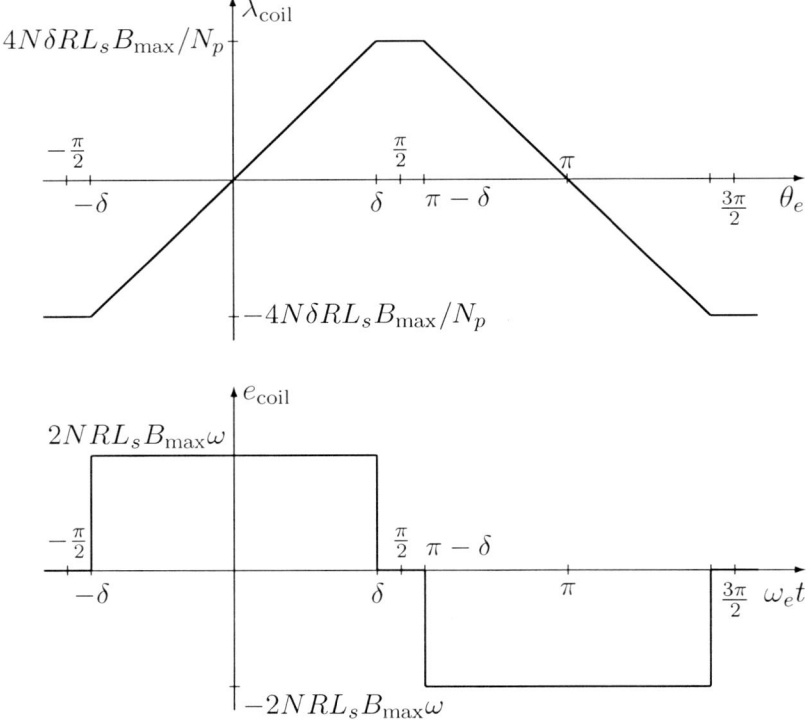

Figure 4.11: The flux linkage and back emf for the coil in Fig. 4.10.

Figure 4.11 shows the flux linkage and induced voltage in the coil of Fig. 4.10. Starting from $\theta_e = 0$, we see that the flux linkage increases linearly as the rotor moves, reaching a peak when the north pole is completely under the coil. The flux linkage remains at its maximum value until the south pole starts moving under the coil at angle $\pi - \delta$.

4.4. Back Emfs

The peak flux linkage is found when the coil sees only one pole of the rotor flux. This peak flux linkage is

$$\lambda_{\text{peak}} = 4N\delta RL_s B_{\max}/N_p \quad, \tag{4.49}$$

where $4\delta RL_s/N_p$ is the physical area of the coil exposed to the rotor flux, N is the number of turns in the coil, and $B_{\max}$ is the flux density produced by the rotor in the plane of the coil. To get the physical coil area we must convert the coil pitch from electrical measure to mechanical measure, thereby introducing a factor of $2/N_p$. We see that the flux linkage inherits attributes of the rotor flux density distribution, filtered by the area of the coil window and the number of turns. That is, an analytic description for the flux linkage is

$$\lambda = \sum_{\substack{n=-\infty \\ n \text{ odd}}}^{\infty} \left\{ -j\frac{2B_{\max}}{n\pi} \exp(jn\theta_e) \right\} \left\{ \frac{4\delta RL_s N}{N_p} \right\} \left\{ \frac{\sin(n\delta)}{n\delta} \right\} \quad. \tag{4.50}$$

The first term in curly brackets represents the rotor flux distribution, in this case a square wave of magnitude $B_{\max}$. The second term in curly brackets is the physical coil area times the number of turns in the coil. The third term in curly brackets is the filter term introduced by the coil in exactly the same manner as slot openings and skew. It is possible to cancel a factor of δ between the second and third terms, but doing so would mask the origin of the terms.

The back emf is the time rate of change of the flux linkage due to the rotor flux, which is piecewise constant for the ideal square wave distribution of $B_{\max}$. The amplitude of the back emf is

$$e = \frac{\partial \lambda}{\partial \theta} \frac{d\theta}{dt} = \frac{\partial \lambda}{\partial \theta_e} \frac{d\theta_e}{dt} \quad, \tag{4.51}$$

which for our single coil becomes

$$e_{\text{peak}} = \frac{4N\delta RL_s}{N_p} B_{\max} \frac{1}{\delta} \omega_e = 2NRL_s B_{\max} \omega \quad, \tag{4.52}$$

where

$$\omega_e = \frac{N_p}{2} \omega \tag{4.53}$$

is the velocity of the time variation in the field. The back emf will contain temporal harmonics that depend on δ and the shape of the field produced by the rotor. It is worth noting that the flux linkage and back emf of the single coil does not contain a dc offset like its mmf does. In addition, it possesses half-wave symmetry for steady rotation. From Eq. 4.50,

$$e = \sum_{\substack{n=-\infty \\ n \text{ odd}}}^{\infty} \left\{ \frac{2B_{\max}}{n\pi} \exp(jn\omega_e t) \right\} \{2n\delta RL_s N\} \left\{ \frac{\sin(n\delta)}{n\delta} \right\} \omega \quad . \quad (4.54)$$

We conclude from this discussion that a single coil sees a time-varying back emf. Within the phase winding, the back emfs of the coils will combine to create the phase back emf. With the phase comprised of coils that are spatially displaced from one another, the back emfs will be temporally displaced from one another. The temporal displacement will cause some temporal harmonics to add while others will cancel. Hopefully the resultant waveform will approach the desired waveform, usually a sinusoid but sometimes a trapezoid.

It is perhaps useful to connect this discussion with the torque production of the simple machine given in Figs. 2.5 and 2.14. In the discussion of Sec. 2.7, the torque for the simple machine was determined to be the product of the stator current and the rotor flux linking the stator winding:

$$T^e = i_s \left(-M i_r \sin \theta \right) \quad . \quad (4.55)$$

This is entirely consistent with our discussion here about the torque being equal to the power delivered to the back emf (divided by the rotor speed). As discussed in Sec. 2.2, the fundamental of the flux linking the stator winding due to rotor current is

$$\lambda_s = M i_r \cos \theta \quad , \quad (4.56)$$

so

$$T^e = i_s \frac{\partial \lambda_s}{\partial \theta} = i_s \left(-M i_r \sin \theta \right) \quad , \quad (4.57)$$

where the mutual inductance M captures the geometry of the stator coil and the radial air gap.

As a more practical example, consider the four pole fractional slot winding based on 15 slots considered in Secs. 3.5 and 4.3. Figure 4.12 shows the back emf generated within the phase a winding due to rotor motion. In Fig. 4.12 the voltage is normalized to speed and a peak voltage of 1 V per coil.

Figure 4.12 makes use of a distribution factor in much the same way we represented the mmf distribution created by coils within an integral slot winding. As shown in Fig. 3.17, the five coils associated with phase a of the machine are symmetrically displaced around 0° electrical, with spacing of 24° electrical. Since the vertical components of the induced voltages cancel,

4.4. Back Emfs

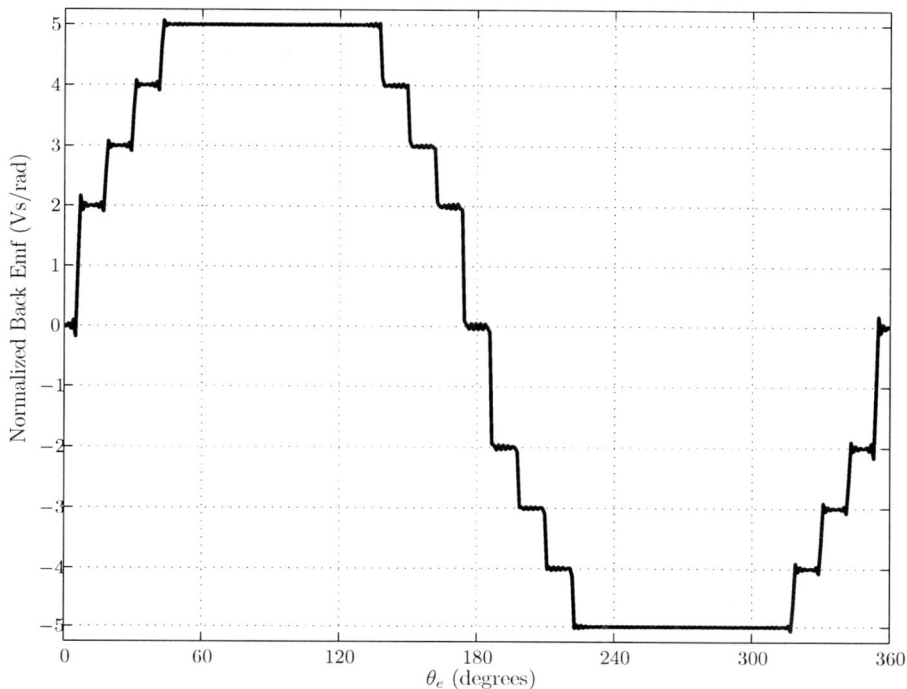

Figure 4.12: The back emf distribution for a machine with a fractional slot winding with $N_s = 15$, $N_p = 4$, and $N_\phi = 3$.

it follows that the appropriate distribution factor for the phase a back emf is

$$k_d = \cos(n\,0°) + 2\cos(n\,24°) + 2\cos(n\,48°) \quad, \tag{4.58}$$

where the factors of 2 are because of the two coils displaced from the origin by the same amount, but in opposite directions. It can be shown using the development of Sec. 3.4 this this can also be represented as

$$k_d = \frac{\sin(5\,n\,12°)}{\sin(n\,12°)} \quad, \tag{4.59}$$

where the factor of 5 is because there are five coils in the set.

For an integral slot winding design, the back emf can be built in a similar manner, but the peak flux linkage for each coil will be expressed differently to reflect the fundamental structure of the coils. To build the back emf for an integral slot winding, we start with the coil structure of Fig. 3.7 as a source of flux moves below it. This is the same situation depicted in Fig. 4.10 except that there are now two coils and the angular span of each coil is $\pi - 2\delta$. The voltages induced in the two coils are in phase by virtue of a spatial displacement of exactly one pole.

114   Chapter 4. Inductances, Back Emfs, and Space Harmonics

From Eq. 4.50, the flux linkage of the basic coil structure is

$$\lambda = \sum_{\substack{n=-\infty \\ n \text{ odd}}}^{\infty} \left\{ -j\frac{2B_{\max}}{n\pi} \exp(jn\theta_e) \right\} \left\{ \frac{4(\pi - 2\delta)RL_sN}{N_p} \right\}$$

$$\left\{ \frac{\sin\left(n\left(\frac{\pi}{2} - \delta\right)\right)}{n\left(\frac{\pi}{2} - \delta\right)} \right\} , \qquad (4.60)$$

where a factor of 2δ in the coil area is replaced by $\pi - 2\delta$ for the basic integral coil, and a factor of 2 has been added to account for the two coils in the basic winding structure; it is assumed that these two windings are in series. The total flux linked by a phase belt having N turns in q coils that are spatially displaced from one another becomes

$$\lambda_{\text{phase belt}} = \sum_{\substack{n=-\infty \\ n \text{ odd}}}^{\infty} \left\{ -j\frac{2B_{\max}}{n\pi} \exp(jn\theta_e) \right\} \left\{ \frac{4(\pi - 2\delta)RL_sN}{N_p} \right\}$$

$$\left\{ \frac{\sin\left(n\left(\frac{\pi}{2} - \delta\right)\right)}{n\left(\frac{\pi}{2} - \delta\right)} \right\} k_d k_{sl} k_\alpha , \qquad (4.61)$$

where k_d, k_{sl}, and k_α are the distribution, slot and skew factors given by Eqs. 3.28, 3.41, and 3.46, respectively. It is important to keep the distribution, slot, and skew effects within the summation because their values are tied to harmonic order.

The corresponding back emf for the phase belt is

$$e_{\text{phase belt}} = \sum_{\substack{n=-\infty \\ n \text{ odd}}}^{\infty} \left\{ \frac{2B_{\max}}{n\pi} \exp(jn\omega_e t) \right\} \left\{ 2n(\pi - 2\delta)RL_sN \right\}$$

$$\left\{ \frac{\sin\left(n\left(\frac{\pi}{2} - \delta\right)\right)}{n\left(\frac{\pi}{2} - \delta\right)} \right\} k_d k_{sl} k_\alpha \omega . \qquad (4.62)$$

This back emf is the voltage induced over two poles, and assumes that all coils within the phase winding over these two poles are connected in series. Depending on the connections of the phase belts, the voltage measured at the terminals may be larger than this if multiple phase belts are connected in series.

The development here was based on a rectangular flux source as shown in Fig. 4.10. The development was careful to explicitly show the structure of this flux source in the flux linkage and back emf. If the flux source is not rectangular, the expressions for flux linkage and back emf can be updated using appropriate Fourier coefficients for the flux source. In the ideal case of

4.5 Space Harmonics

a sinusoidal flux source, all of the harmonic terms would disappear leaving only the fundamental. That is, the spatial harmonics of the flux linkage and back emf are driven by the structure of the flux source that is linking the windings.

4.5 Space Harmonics

Our discussions of windings in Chapter 3 and back emfs in the previous section show that there are harmonic fields in the air gap of the machine, even if the phase currents are perfectly sinusoidal. In practice it is common for ac machines to be fed with currents that are synthesized electronically, suggesting time harmonics in the current waveforms. The interaction between the space harmonics of the back emf and the temporal harmonics of the current waveforms result in time-varying torque, or torque ripple. Torque ripple is undesirable because it can cause speed fluctuations, structural vibrations, and generate acoustic noise.

Based on the discussion in the last section, consider the back emf for phase a to be

$$e_a = \sum_{\substack{n=-\infty \\ n \text{ odd}}}^{\infty} E_{an} \exp[\jmath n \omega_e t] \quad . \tag{4.63}$$

The back emf waveforms for phases b and c will be similar but displaced in time:

$$e_b = \sum_{\substack{n=-\infty \\ n \text{ odd}}}^{\infty} E_{bn} \exp\left[\jmath n \left(\omega_e t - \frac{2\pi}{3}\right)\right] \quad ; \tag{4.64}$$

$$e_c = \sum_{\substack{n=-\infty \\ n \text{ odd}}}^{\infty} E_{cn} \exp\left[\jmath n \left(\omega_e t + \frac{2\pi}{3}\right)\right] \quad . \tag{4.65}$$

If the machine is symmetrically wound, $E_{an} = E_{bn} = E_{cn}$. We will assume this is the case.

Similarly, consider the current for phase a to be

$$i_a = \sum_{\substack{n=-\infty \\ n \text{ odd}}}^{\infty} I_{an} \exp[\jmath n \omega_s t] \quad . \tag{4.66}$$

The currents for phases b and c will be similar but displaced in time:

$$i_b = \sum_{\substack{n=-\infty \\ n \text{ odd}}}^{\infty} I_{bn} \exp\left[jn\left(\omega_s t - \frac{2\pi}{3}\right)\right] \quad ; \quad (4.67)$$

$$i_c = \sum_{\substack{n=-\infty \\ n \text{ odd}}}^{\infty} I_{cn} \exp\left[jn\left(\omega_s t + \frac{2\pi}{3}\right)\right] \quad . \quad (4.68)$$

In order for average power conversion, we require that $\omega_e = \omega_s$ thereby synchronizing the rotor field with the stator field as we discussed in Sec. 2.7. Accordingly, the power delivered to the phase a back emf is

$$p_a(t) = \left\{ \sum_{\substack{n=-\infty \\ n \text{ odd}}}^{\infty} E_{an} \exp\left[jn\omega_e t\right] \right\} \left\{ \sum_{\substack{m=-\infty \\ m \text{ odd}}}^{\infty} I_{am} \exp\left[jm\omega_s t\right] \right\} \quad . \quad (4.69)$$

Recognizing that multiplication in the time domain gives convolution in the frequency domain, this reduces to

$$p_a(t) = \sum_{\substack{n=-\infty \\ n \text{ even}}}^{\infty} \sum_{\substack{m=-\infty \\ m \text{ odd}}}^{\infty} E_{am} I_{a(n-m)} \exp\left[jn\omega_e t\right] \quad . \quad (4.70)$$

There are some subtleties in Eq. 4.70 that are worth noting. The inner summation is over only odd harmonic orders because it inherits this from the properties of the back emf. Also, the properties of the current stipulate that $n - m$ must be odd for the harmonic coefficient to be nonzero. Given that m is odd, the only way for $n - m$ to be odd is for n to be even.

Our convolution process has taken two waveforms with only odd harmonic orders and converted it into a waveform with only even harmonic orders, though it is based on inheriting the odd-order coefficients of the underlying waveforms. A careful comparison between the orders of the coefficients and the resulting frequency will show that the resulting frequency is equal to the sum or difference of the harmonic orders in the coefficients. For example, the frequencies corresponding to the coefficient $E_5 I_1$ will be $4\omega_e t$ and $6\omega_e t$.

Similarly, for the other phases we have

$$p_b(t) = \sum_{\substack{n=-\infty \\ n \text{ even}}}^{\infty} \sum_{\substack{m=-\infty \\ m \text{ odd}}}^{\infty} E_{am} I_{a(n-m)} \exp\left[jn\left(\omega_e t - \frac{2\pi}{3}\right)\right] \quad ; \quad (4.71)$$

4.5. Space Harmonics

$$p_c(t) = \sum_{\substack{n=-\infty \\ n \text{ even}}}^{\infty} \sum_{\substack{m=-\infty \\ m \text{ odd}}}^{\infty} E_{am} I_{a(n-m)} \exp\left[jn\left(\omega_e t + \frac{2\pi}{3}\right)\right] \quad . \tag{4.72}$$

Summing up the power over the three phases gives

$$p_{\text{total}}(t) = \sum_{\substack{n=-\infty \\ n \text{ even}}}^{\infty} \sum_{\substack{m=-\infty \\ m \text{ odd}}}^{\infty} E_{am} I_{a(n-m)} \left\{ \exp\left[jn\left(\omega_e t - \frac{2\pi}{3}\right)\right] + \right.$$

$$\left. \exp\left[jn(\omega_e t)\right] + \exp\left[jn\left(\omega_e t + \frac{2\pi}{3}\right)\right] \right\} \quad . \tag{4.73}$$

The phase terms contained in the exponentials tell us how the contributions from the three phases combine. These terms can be viewed as the summation of three sines or cosines of the form

$$\sin[n\omega_e t] + \sin\left[n\left(\omega_e t - \frac{2\pi}{3}\right)\right] + \sin\left[n\left(\omega_e t + \frac{2\pi}{3}\right)\right] \quad , \tag{4.74}$$

or

$$\cos[n\omega_e t] + \cos\left[n\left(\omega_e t - \frac{2\pi}{3}\right)\right] + \cos\left[n\left(\omega_e t + \frac{2\pi}{3}\right)\right] \quad . \tag{4.75}$$

Recognizing that these expressions sum to a constant except for when n is a multiple of 6, we conclude that any combination of harmonic orders that combine to give a multiple of 6 contributes to a time-varying component of the torque. Table 4.3 considers the combinations of harmonic orders to produce time-varying torque components. Table 4.3 does not include triplen current harmonics because these currents cannot flow in a three phase machine with the windings connected in an ungrounded Y, thereby preventing triplen harmonics in back emf from contributing to time-varying torque. For a machine with the phase windings connected in Δ, it is extremely important to structure the windings so that triplen harmonics in back emf are minimized to prevent triplen harmonic currents from circulating around the Δ.

If we were to look deeper into the time-varying torque components, we would determine that some of the terms are created by traveling mmf waves moving in the same direction of rotor travel. Other components are created by traveling mmf waves moving in the direction opposite to rotor travel. Regardless of the direction of travel for the underlying magnetic fields, both types of terms are objectionable.

Table 4.3: The combinations of harmonic orders in current and back emf that give rise to time-varying torque.

		E_n						
		1	3	5	7	9	11	13
	1	–	–	6	6	–	12	12
	5	6	–	–	12	–	6	18
I_m	7	6	–	12	–	–	18	6
	11	12	–	6	18	–	–	24
	13	12	–	18	6	–	24	–

4.6 Summary

This chapter has examined additional issues associated with the winding structures presented in Chapter 3. In particular, air gap and slot inductances have been treated in detail. The importance of the back emf has been discussed, and we have looked at how to determine the back emf for a coil within a phase winding.

Finally, we acknowledged that both phase currents and back emfs contain harmonics. The interaction among these harmonics is responsible for creating time-varying torques that can result in speed fluctuations, mechanical vibrations, and acoustic noise. The ability to control the time-varying torques is dependent on the ability to shape the back emf waveform and regulate the phase currents.

In the chapters to follow, we are going to take more of a terminal-based approach to examining the operating principles of electric machines in which the emphasis is going to be on the fundamental components of inductances, back emfs, and phase currents. With the foundation presented in Chapter 3 and this chapter it should be possible to read on the discussions to follow a deeper appreciation of some of the consequences of space harmonic components in the inductance and back emf, particularly in conjunction with temporal harmonics of phase currents.

Chapter 5

Elements of Electric Machine Design

5.1 Introduction

While this book does not intend to treat the subject of design in detail, the material covered in Chapters 1 through 4 is sufficiently fundamental to the description of electric machines that a brief discussion of machine design can serve as a summary of the material that has been presented to this point. This chapter pulls together the material in previous chapters, projecting it onto the machine design problem. While not as complete as a text on machine design, it should provide an appreciation of the challenges that the machine designer must confront[1].

Machine design can be viewed as a constrained optimization. The optimization objectives are generally expressed in part through the specification which captures the electromechanical requirements for the machine. Naturally there is always a desire to minimize cost subject to meeting the other performance requirements. The performance requirements will likely include:

- Meeting average torque requirements at one or more speeds.
- The maximum amount of torque ripple the machine can create at one

---

[1]For a more complete discussion of the design problem, see, for example, T. A. Lipo, *Introduction to Ac Machine Design*, Wisconsin Power Electronics Research Center, University of Wisconsin-Madison, Madison, WI, USA, 2004, D. C. Hanselman, *Brushless Permanent Magnet Motor Design*, The Writer's Collective, 2003, and T. J. E. Miller and J. R. Hendershot, *Design of Brushless Permanent-Magnet Motors*, Oxford University Press, 1994.

or more operating points.

- Meeting efficiency requirements at one or more operating points.
- The duration of time the machine must operate at critical operating points.
- The maximum weight of the machine.
- The maximum volume the machine can occupy.
- The maximum current the machine can draw from its supply.
- The voltage available to excite the machine.
- The time rate of change of voltage applied to the machine. This is a particularly important parameter in machines that are excited through inverters (see Chapter 7) that impose very large rates of dv/dt that can be stressful on the insulation system.
- The allowable acoustic noise that the machine can produce.
- The allowable structure-borne noise that the machine can produce.
- The thermal environment that the machine must operate within. This is likely to include not only minimum and maximum operating temperatures, but also the storage temperature and the type of cooling available to the machine.
- The details of how the machine will be mechanically mounted in relation to the electrical and mechanical terminals.
- Other environmental factors, such as moisture, altitude, confinement, etc.

The design of an electric machine is inherently multi-disciplinary. The magnetic field descriptions and how they translate into the terminal behavior of the machine have been the focus of the preceding chapters. While central to machine design, the magnetic fields are only a portion of the story. The generation of the magnetic fields is going to involve currents, and the flow of current is going to result in the creation of heat. The amount of heat generated is going to be tied to the space made available to the windings, the frequency of the currents, and the strength of the magnetic fields. The heat being generated will have to be managed with the available cooling methods. As with many things electrical, thermal considerations often impose severe limitations on how compact a device can be made.

The windings of the machine must be made compatible with the electric supply. This will impose constraints on the magnitude of the back emf, thereby limiting the number of turns in each phase winding. Since torque is tied to Ampere-turns, limiting the number of turns has consequences on the current drawn from the supply. This becomes particularly challenging at low voltages because the currents will become quite large.

The machine must be compatible with its physical environment. At its lowest level, this requires the machine to mate compatibly through its shaft. This may require a standard housing or a custom configuration. In addition, the machine must be designed considering cooling and other environmental factors.

Overarching the electrical, magnetic, mechanical, and thermal considerations are the issues of reliable and inexpensive manufacture. To succeed, the machine must ultimately be built from materials using processes that are suited to the application.

The discussion in this chapter is oriented toward radial field machines. These machines still dominate the landscape, though there continue to be applications where axial field machines create useful design options. The concepts in this chapter apply equally to radial and axial field machines.

5.2 Electromagnetic Design

For most electric machine designs, the electromagnetic design is the logical place to start. Rough sizing of the machine is often accomplished through the use of a sizing equation that relates the volume of the rotor to how heavily the machine is loaded. There are two elements to loading: electric and magnetic. The electric loading deals with how hard the current carrying conductors are stressed. The magnetic loading deals with how hard the magnetic materials are stressed.

Recall that in Sec. 2.3 we developed a field description for the force responsible for creating torque on the machine rotor. Equation 2.57 captured the tractive force on the rotor as

$$\tau_\theta = \mu_0 H_\theta H_r \quad , \tag{5.1}$$

where μ_0 is appropriate because it is assumed we are determining the traction in the air gap. The dimension of the traction is force per unit area, consistent with our discussion in Sec. 2.3.

We can view Eq. 5.1 as being the product of the tangential magnetic field intensity and the radial flux density if we associate μ_0 with the H_r

term. When viewed in this way, we begin to see the motivation behind the electric and magnetic loadings. The electric loading is associated with the creation of the H_θ term and the magnetic loading is associated with the B_r term.

5.2.1 Electric Loading

The electric loading is usually given the symbol A, and defined as the total rms current per unit length of circumferential periphery of the machine. As such, the electric loading is tied very tightly to the production of winding mmf. In fact, the winding mmf is responsible for generating H_θ. This may seem to contradict some of our earlier analysis, but it is actually quite consistent.

In Sec. 2.2 we developed an elementary field description for the simple electric machine of Fig. 2.5. In our analysis of this machine we considered the magnetic fields within the air gap to be only radially directed. We extended our analysis of this machine in Sec. 2.7 where we determined the torque produced by the simple machine before determining the required relationship among stator frequency, rotor frequency, and rotational speed to produce average torque. The torque expression,

$$T^e = -M i_s i_r \sin\theta \quad , \tag{5.2}$$

captured the interaction among the stator current, rotor current, and the angular displacement between the stator and rotor. The mutual inductance M captured the relevant geometry of the machine and the number of turns in each winding.

Within Eq. 5.2, we can decompose the expression into the product of stator Ampere-turns and the radial air gap field produced by the rotor. The $\sin\theta$ term continues to account for the spatial displacement between the stator and rotor excitation. When viewed in this manner, Eq. 5.2 is consistent with Eq. 5.1.

Another approach is worth considering as well since it is the most fundamental. Figure 5.1 shows the separation between two regions, allowing for the possibility of surface currents $\vec{K}_s$ at the interface between the two regions. The magnetic field intensity in the region above the interface is $\vec{H}^a$ and the magnetic field intensity in the region below the interface is $\vec{H}^b$. The magnetic field intensity in each region can be decomposed into a component that is tangential to the interface and a component that is normal to the interface. The subscripts n and t are used to denote the normal and tangential components, respectively.

5.2. Electromagnetic Design

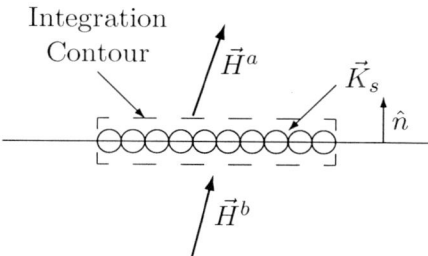

Figure 5.1: The interface between two regions, where there may be surface currents at the interface.

Application of the integral version of Ampere's law around the contour shown in Fig. 5.1 in the limit as the contour segments normal to the surface shrink to enclose only the interface requires

$$\vec{H}_t^a - \vec{H}_t^b = \vec{K}_s \quad , \tag{5.3}$$

or

$$\hat{n} \times \left(\vec{H}^a - \vec{H}^b\right) = \vec{K}_s \quad , \tag{5.4}$$

where $\hat{n}$ is a vector normal to the interface, directed from region b toward region a. Equation 5.4 is the boundary condition imposed by Ampere's law on the description of the magnetic field intensity within regions a and b.

In the analysis of electric machinery, it is common to model the stator windings as being responsible for establishing a surface current at the inside periphery of the stator bore. A similar view can be taken of rotor currents at the periphery of the rotor. Consistent with Eq. 5.4, we can take the magnetic field intensity on the air gap side of the stator winding to be equal to the surface current within the winding. We put all of the field into the air gap because typically the winding is backed by highly permeable material that would have negligible magnetic field intensity within it.

In a three phase machine where each phase has N turns per pole, there is a total of $3N$ turns carrying rms current I_{rms}. Since each slot has two sides there is a total slot rms current of $6NI_{\text{rms}}$. It follows that the electric loading is

$$A = \frac{6N_p N I_{\text{rms}}}{\pi D} \quad , \tag{5.5}$$

where D is the diameter of the machine, typically taken at the inside of the stator bore.

To increase the electric loading, more current must be put into the machine. This can be accomplished by:

1. Increasing the slot depth. This will come at the price of increasing the slot leakage inductance, increasing the mmf dropped in the stator teeth, and pushing out the overall diameter of the machine.

2. Widening the slots. This will tend to decrease the slot leakage inductance, but it will serve to further limit the magnetic loading as discussed in the next subsection.

3. Increasing the current density. This would increase the dissipation within the windings. Hence, it must be countered with improvements in cooling, a higher temperature classification for the insulation system, or some combination of the two.

Increasing the electric loading also increases the forces on the conductors, which magnifies the importance of holding the windings in place to prevent vibration during operation and the resultant wear on the insulation system. The time variation in the winding currents create forces that pulsate at twice the excitation frequency. The repeated contraction and relaxation can be a source of vibration and small winding motions if the windings are not adequately bonded together.

The tangential field in the air gap is directly related to the electric loading. Consider the mmf created by a three phase winding, one phase of which is shown in Fig. 3.8. First, we note that the current distribution and the mmf distribution are orthogonal to one another in space. The current distribution depends on $\cos\theta_e$ and the mmf depends on $\sin\theta_e$. The tangential magnetic field intensity H_θ has the magnitude of the mmf but with the spatial phase shift.

A single phase winding produces a fundamental mmf of

$$\mathcal{F} = \frac{4Ni}{\pi} k_{w1} \sin\theta_e \quad , \tag{5.6}$$

where k_{w1} is the fundamental winding factor that includes the effects of pitch, distribution, finite slot opening, and skew. In a three phase machine, the air gap mmf is created by the superposition of the three phases, resulting in an mmf amplitude that is 3/2 that of a single phase mmf. Expressing the peak current in terms of the rms current, we conclude that the tangential air gap field is

$$H_\theta = \frac{3}{2} \frac{4\sqrt{2} N I_{\text{rms}} k_{w1}}{\pi} \cos\theta_e = \frac{6\sqrt{2} N I_{\text{rms}} k_{w1}}{\pi} \cos\theta_e \quad . \tag{5.7}$$

Using the definition of the electric loading from Eq. 5.5 gives

$$H_\theta = \frac{\sqrt{2} D A k_{w1}}{N_p} \cos\theta_e \quad . \tag{5.8}$$

We will use this expression for the tangential air gap field below as we develop a generic output equation.

5.2.2 Magnetic Loading

The magnetic loading is defined as the peak radial magnetic flux density in the air gap of the electric machine, and is given the symbol $B_{g\mathrm{max}}$. This definition allows us to neglect any ripple in the air gap flux density caused by slotting. The achievable magnetic loading is a function of the magnetic materials used in the machine.

As discussed in Sec. 1.4 the magnetic materials used within electric machines exhibit saturation. Typically for most soft magnetic materials this occurs at a flux density of $1.5-1.8\,\mathrm{T}$. Operation above this flux density requires a substantial increase in the current with only limited increase in torque production.

In a conventional electric machine with windings that are inserted into slots in the inside periphery of the stator and the outside periphery of the rotor, the space taken up by the teeth is comparable to the space made available for the winding slots. This suggests that the teeth become the limiting factor in increasing the magnetic loading. Because of tooth saturation, the magnetic loading is generally limited to $0.9\,\mathrm{T}$ or less. Some permanent magnet machines will increase the magnetic loading at the expense of the electric loading to have the magnets do as much of the work as possible. This can help to increase machine efficiency, but core losses in the teeth will need to be watched closely.

The limitation imposed by tooth saturation can be removed by eliminating the winding slots, placing the winding directly in the air gap at the inside bore of the stator. However, putting the windings directly in the air gap effectively increases the air gap length, requiring additional magnet material or increased Ampere-turns on the rotor for the same magnetic loading. In addition, placing the windings directly in the air gap also exposes the windings directly to the time-varying magnetic fields of the air gap, necessitating the use of Litz wire to avoid excessive eddy current losses within the conductors.

There may be situations where going to a slotless design is advantageous, but it depends largely on the details of the specification. A clear advantage of slotless designs is the elimination of torque created by the interaction between stator and rotor teeth (generally known as cogging torque), but there are potentially more complications than there are benefits in many applications. The biggest complication is forming the windings and holding them in place, followed closely by effectively cooling the windings.

5.2.3 The Output Equation

The output equation is often the starting point for the electromagnetic design because it ties together the electric loading, the magnetic loading and the volume of the rotor into a single equation so that one is able to develop an estimate of machine size. Assume sinusoidal variation in the tangential magnetic field intensity according to Eq. 5.8, and sinusoidal variation in the air gap radial flux density of the form

$$B_r = B_{g\max} \cos(\theta_e + \delta) \quad , \tag{5.9}$$

where δ allows for displacement between the radial magnetic flux density and the tangential magnetic field intensity. Using Eqs. 5.8 and 5.9 the tangential component of the traction becomes

$$\tau_\theta = \int_0^\pi \left\{ \frac{\sqrt{2} D A k_{w1}}{N_p} \cos \theta_e \right\} \{B_{g\max} \cos(\theta_e + \delta)\} \, d\theta_e \quad , \tag{5.10}$$

where the integration is carried out over one magnetic pole. Multiplying the traction by the area $(\pi D L_s)$ and the number of poles gives the total torque, which is

$$T^e = \frac{\pi^2 L_s D^2 k_{w1} A B_{g\max}}{\sqrt{2}} \cos \delta \quad . \tag{5.11}$$

This is the torque applied to the rotor. We see the torque depends directly on the rotor volume, the magnetic loading, and the electric loading. There is also a constant in Eq. 5.11 that includes the composite winding factor.

It may be counter-intuitive that the torque depends on the rotor volume, but this dependence on the rotor volume comes from two places. The rotor area times the magnetic shear stress gives the tangential force acting on the rotor. This magnetic shear stress acts on the rotor surface through the moment arm of the rotor radius. The product of the rotor surface area and the moment arm combine to give the rotor volume. The dependence of torque on rotor volume is frequently used as a way to scale a design to another rating. Equation 5.11 shows that the torque drives the physical size of the machine.

The factor of $\cos \delta$ in Eq. 5.11 suggests that maximizing torque production requires synchronizing the position of the air gap radial flux density with the tangential magnetic field intensity. Since the air gap mmf is orthogonal to the magnetic field intensity, this is consistent with our earlier observations that torque production is maximized when the stator and rotor

5.2. Electromagnetic Design

fields are orthogonal. (See, for example, Eq. 2.110 and the associated discussion.) The ability to synchronize the air gap fields is limited by leakage inductances and other parasitics.

It is worth noting that the torque does not depend directly on the number of poles. However, the number of poles is an important design parameter because through the number of poles we can make the machine compatible with both its electric supply and its required mechanical rotation. Further, the required thickness of the stator and rotor back iron is inversely proportional to the number of poles.

5.2.4 Detailed Magnetic Design

Equation 5.11 is a useful starting point in determining preliminary dimensions for a candidate design, but there are many elements left in the magnetic design. Some of these elements are discussed here. Most of these elements interact with the mechanical and thermal design elements discussed in subsequent sections. Self-consistently satisfying the specification, while producing a design that can be manufactured, inevitably requires a substantial amount of iteration.

Winding Design: Winding design is going to involve determining the most appropriate style of winding consistent with the specification. As discussed in Chapter 3, possibilities include single- and double-layer windings that are based on either integral slot or fractional slot coil structures. The objective is to determine the distribution of conductors that balances ease of manufacture with supporting the performance specification.

Stator Lamination Design: For a conventional machine that has the stator on the outside and the rotor on the inside, the stator lamination starts at the air gap and proceeds out to the interface between the stator and the housing. Most of the attention will be focused on the details of the teeth and the shoes put on the teeth. The tooth width requires balancing magnetic saturation on the one hand with making as much room for windings as possible. The core losses can be reduced by allowing the teeth to get wider as they approach the root allowing the peak flux density to decrease. This also helps with stiffening the teeth, as discussed in the next section. However, doing so reduces the space available for windings.

Rotor Magnetic Design: The design of the rotor can vary significantly with the type of machine. An induction machine involves the design of slots into which the rotor bars are placed. Shaping the rotor slots has a profound influence on the torque speed characteristics of the machine because of the frequency-dependent rotor resistance, as discussed in Chapter 9.

A permanent magnet machine will focus on balancing the performance requirements with the manufacturing cost. The selection of the magnet material will be driven by balancing environmental, cost, and electromagnetic performance.

Finite Element Analysis: It is now rare to build an electric machine without numerical field analysis, typically through finite element analysis[2]. With the increased emphasis on power density, it is virtually impossible to maximize power density by using only lumped parameter and analytic models. Even the flux tube analysis discussed in Sec. 4.2 benefits from finite element analysis, through guidance in how best to break up different regions of the machine, particularly as the magnetic materials approach saturation.

While the time-intensive nature of finite element analysis has eased considerably over the years, it is still not a tool that supports the rapid development of models useful for design. It is generally reserved for fine-tuning a design, evaluating the machine under heavy load or fault conditions, and for generating high fidelity terminal models to support the development of control algorithms.

Terminal Models: The models discussed for inductance and back emf in Chapter 4 are the basis for terminal models that allow for the evaluation of a candidate design within the intended application. Integrating the terminal models into dynamic simulations allows for assessing whether or not the machine will satisfy the performance specifications. As the design process continues, it may become important for the terminal models to integrate magnetic nonlinearities that will ultimately limit the electromechanical output of the machine. Consideration of saturation is aided considerably by finite element analysis.

5.3 Mechanical Design

The mechanical design focuses on the structure of the machine. An inadequate structure, or inadequate attention to the assembly of the machine, will

---

[2] See, for example, S. J. Salon, *Finite Element Analysis of Electrical Machines*, Kluwer Academic Publishers, 1995 for an introduction to finite element analysis applied to electric machines.

5.3. Mechanical Design

undermine the attention to detail in the magnetic design. Inadequate structure or inherent asymmetries in the construction will result in a machine that is noisy and vibrates, possibly to the point where the machine cannot be used. A danger in having electrical engineers design electric machinery is the tendency to focus on the magnetic performance of the machine, with structural issues receiving less attention. Many a noisy machine has been the result of back iron that is sufficient for magnetic purposes but not structural purposes!

A mechanical design that is robust and easy to manufacture addresses the following issues:

Rotor/stator Concentricity: A principal objective of the machine housing is to maintain concentricity between the stator and the rotor on both a static and dynamic basis. A rotor that is offset within the stator bore will have unbalanced tangential and radial forces acting on it which will tend to exacerbate the offset. Often the responsibility for maintaining concentricity falls to the alignment among the stator housing, the end bells, and the bearing seats.

The end bells usually locate on the stator housing. The seats on the housing are machined using the stator bore as a reference to enforce concentricity. The bearing seats are within the end bells. This type of arrangement provides acceptable alignment for machines with reasonably large radial air gaps.

In machines with small radial air gaps, it may be necessary to maintain rotor/stator concentricity by banking the bearing seats into the ends of the stator bore. This requires that the stator stack be longer than the rotor stack to allow axial room for the bearing seats to engage the stator stack. This technique has been used in variable-reluctance machines where good machine performance depends heavily on a small radial air gap.

Rotor Balance: The rotor needs to be dynamically balanced so that its principal axis of rotation is coincident with the axis formed by the shaft and bearings. In induction machines the rotor is often balanced by removing some material from the rotor end rings. Other machines may use putty to add small amounts of weight in the right locations to bring the rotor into balance.

Although slightly different than balance, the rotor should be sufficiently stiff that longitudinal flexure is kept to a minimum. Otherwise the rotor is likely to encounter dynamic problems as it tries to accelerate through certain speeds that naturally excite these flexing modes.

Deformation of Magnetic Cores: Section 2.3 showed that there are axial, tangential, and radial forces acting on the rotor. These forces not only cause the rotor to spin, but they can also cause the rotor to deform if there is insufficient mechanical structure to the rotor. Rotor deformation is not usually as significant a problem as stator deformation. Because the stator sees the reaction forces applied to the rotor, the stator sees the same forces as the rotor.

The tangential reaction forces applied to the stator can cause the stator teeth to deflect in the tangential direction. This can become more significant as the winding slots are made deeper to increase the electric loading. Adding some small amount of taper to the teeth can help make them substantially stiffer without giving up too much winding space. As noted in the previous section, some taper can also improve machine efficiency, reducing core losses in the teeth by allowing the flux density to decrease along the tooth.

The radial forces will try to deform the stator, causing it to try and take a shape that is not round because their amplitude varies around the periphery of the machine. As the rotor spins, the location of the deformation moves and the deflections end up traveling around the periphery. These deflections transfer mechanical motion into acoustic energy and structure-borne vibrations. To reduce the acoustic noise of the machine, additional thickness can be added to the back iron so that it is more difficult for the radial forces to distort the stator. Radial forces can be an order of magnitude larger than the tangential forces that make the rotor spin, so they demand respect.

Increasing the number of poles in the machine is another way to reduce its acoustic noise. As the number of poles increases, so do the number of places where the stator is being deformed. Since the arc length of the section being deformed has been shortened, the machine appears to be stiffer for the same amount of stator back iron. The desire to increase the poles for mechanical reasons must be balanced with the electromechanical implications of doing so.

The natural tendency for the machine to deform should be evaluated from the perspective of identifying the natural modes and frequencies associated with the structure, to determine if any of these natural modes are excited by operation within the intended duty of the machine. If so, steps need to be taken to either avoid those natural frequencies, or to move them. There are many ways of changing the natural modes, such as by adding some additional material to the stator lamination. Doing this by creating a varying thickness in back iron can interrupt the natural modes of the structure. For this reason many

5.3. Mechanical Design

electric machines have stator laminations that are not a simple cylinder. Most analytic methods for determining the natural modes and frequencies of an electric machine are approximate at best. A finite element structural analysis may be useful, but is complicated by the stator core being a laminated structure.

Winding Vibrations: As time-varying currents pass through the phase windings, the windings experience time-varying forces. Current flow through a bundle of conductors will create Lorentz forces on those conductors. If the conductors are carrying the same current, the forces will tend to pull the conductors together. As the current is reduced, the bundle is able to relax.

It follows that the phase windings experience forces that cause the conductors within the windings to pull together, then relax. If the windings are not encapsulated and bonded together the windings will be able to move due to these time-varying forces. Winding movement is undesirable because it creates vibrations that can lead to acoustic noise. In addition, winding movement may abrade and degrade the insulation system over time.

Thrust Loads: The intended motion of the rotor is rotational, but some allowance must be made for dealing with thrust loads that could cause the rotor to bind. It is common in electric machines to allow a small amount of axial play (measured in fractions of a millimeter) in the rotor. This is accommodated by putting a wave spring on one end of the machine, often between a bearing and its seat. Proper mechanical alignment is needed to center the rotor magnetically within the stator, thereby preventing the machine from creating internal thrust loads.

Bearings: The type of bearing must match the intended duty and environment of the electric machine. The wear surfaces reside within the bearing. There are many types of bearings that provide different types of protection from liquids that might enter the machine, thrust loads, and shock. There are also bearings that are specially designed for electric machines that are to be fed by adjustable speed drives to prevent the flow of bearing currents. Bearing currents can degrade the rolling elements of the bearings by electric discharge machining, thereby causing premature failure. Shock loads and ball bearings do not mix well since the small contact area causes the shock forces on the bearings to deform (Brinell) the races. Where shock is a significant issue, roller bearings could be a logical choice. Roller bearings replace the point contact of a ball bearing with a line contact along the surface of a roller.

Table 5.1: The temperature rating associated with common insulation classes.

Insulation Class	Temperature °C
A	105
B	130
F	155
H	180

End Bells: The end bells are responsible for supporting the rotor in operation, and must be sufficiently rigid to prevent movement of the rotor away from its intended principal axis. As discussed above, the end bells are often an integral part of maintaining concentricity between the stator and the rotor. In most electric machines the end bells originate as castings that are subsequently machined to positively register on the stator housing. The bearing seats are machined to maintain concentricity with the stator bore once the end bells are put in place. Ribs on the end bells can add stiffness without excessive weight.

5.4 Thermal Design

The thermal design of the machine ultimately limits its torque rating. The temperature of the machine will rise until the heat being rejected balances the heat being generated. High performance machines are more power dense because they combine efficient heat removal with reduced heat generation. The temperature of the machine dictates the insulation class of the machine. Table 5.1 gives the temperature ratings for standard insulation classes. These ratings assume a nominal insulation life of 20,000 operating hours.

There are three mechanisms associated with the flow of thermal energy: conduction, convection, and radiation. Thermal design of most electric machines relies only on conduction and convection heat transfer; an obvious exception would be machines designed for use in space in which radiation heat transfer is the limiting factor in heat rejection. Under normal conditions, heat transfer starts as a conduction problem originating from the heat generated in the windings and core. Removal of the heat from the machine involves conducting the heat to a solid/fluid interface, at which point convection heat transfer takes over.

Conduction heat transfer takes place within solids, and is facilitated by

5.4. Thermal Design

the tight thermal coupling of the molecules which form the solid. Conduction is the mechanism by which the thermal energy of each molecule is shared with its adjacent neighbors. In the absence of a source of heat, the natural tendency is for the thermal energy to evenly distribute throughout the body of the solid.

Conduction heat transfer also takes place in fluids, but it is generally far more complicated because fluids tend to move when subjected to a thermal gradient. When the fluid begins moving, the problem makes the transition from a conduction problem to a convection problem. The description of heat transfer in the presence of a moving fluid is referred to as convection heat transfer. Some heat transfer authorities consider the convection mode to be an extension of the conduction mode. We will consider them to be different, owing to the fluid motion in convection.

It can be shown from conservation of energy that the flow of thermal energy within solids satisfies

$$-\nabla \cdot \vec{q} + \dot{q} = \rho \frac{\partial u}{\partial t} \quad , \tag{5.12}$$

where $\vec{q}$ is the heat flux, $\dot{q}$ is the rate of internal heat generation per unit volume, ρ is the mass density of the material, and u is the intensive internal energy (energy per unit volume) of the material. This is the most primitive form of the conduction equation. Its vector expression makes it geometry invariant. That is, Eq. 5.12 holds for any right-hand coordinate system.

We can now introduce the temperature through Fourier's law of heat conduction:

$$\vec{q} = -k\nabla T \quad , \tag{5.13}$$

where k is the thermal conductivity and T is temperature. The negative sign accounts for the flow of heat in the direction of the negative gradient. That is, heat flows from higher temperatures to lower temperatures. In addition, we know from thermodynamics and the chain rule that

$$\frac{\partial u}{\partial t} = \frac{du}{dT}\frac{\partial T}{\partial t} = c\frac{\partial T}{\partial t} \tag{5.14}$$

for an incompressible substance with heat capacity c. Putting Eqs. 5.13 and 5.14 into Eq. 5.12 gives

$$\nabla \cdot (k\nabla T) + \dot{q} = c\rho \frac{\partial T}{\partial t} \quad . \tag{5.15}$$

If k is constant, then we can pull k outside of the vector operations and write

$$\nabla^2 T + \frac{\dot{q}}{k} = \frac{1}{\alpha}\frac{\partial T}{\partial t} \quad , \tag{5.16}$$

Table 5.2: A summary of the electrical and thermal analogy.

Electrical				Thermal
Quantity	Symbol	Symbol	Quantity	
Current	i	q	Heat	
Potential Difference	$v_1 - v_2$	$T_i - T_o$	Temperature Difference	
Resistance	R_e	R	Resistance	
Conductivity	σ	k	Conductivity	
Current Density	J	$\hat{q}$	Heat Flux	

where $\alpha = k/(\rho c)$ is called the thermal diffusivity. One interpretation of α is that it represents the competition between the conduction process and the thermal inertia of the conducting material.

Equation 5.16 is often called the conduction equation. In Eq. 5.16, temperature is the independent variable. This equation is in the general form of a diffusion equation. It describes how temperature evolves spatially and temporally. That is, $T = T(x, y, z, t)$ in a Cartesian coordinate system. Because T varies as a function of position and time, we can refer to T as a scalar field. Similarly, we can refer to $\vec{\hat{q}}$ as a vector field.

In many situations it is possible to view conduction heat transfer through an analogy with electric circuits, particularly when conduction is dominated by heat flow in one dimension. This analogy is summarized in Table 5.2. As current flows spontaneously through a resistor subjected to a potential difference, heat flows spontaneously in the face of a temperature difference.

Convective heat transfer involves a moving fluid. This implies a coupling between the fluid mechanics which describe the motion of the fluid and the heat transfer which is taking place within the fluid. This coupling makes the heat transfer process extremely difficult to describe analytically within the boundary layers.

A boundary layer refers to a region along a rigid surface. It is within this boundary layer that we find things happening, both thermally and in a fluid mechanical sense. That is, in the boundary layer the fluid velocity profile is varying as we move away from the rigid surface. The fluid temperature is also varying as we move away from the surface. The boundary layer thickness delimits the boundary layer. Further, the boundary layer thickness grows as the fluid and the rigid surface remain in contact. There are two boundary layers which we must concern ourselves with: the momentum boundary layer and the thermal boundary layer.

5.4. Thermal Design

Fluid properties such as viscosity and density are functions of temperature. When a stationary fluid is subjected to heat flow, a gradient in temperature gives rise to gradients in viscosity and density. These gradients give rise to buoyancy forces which cause the fluid to begin moving. Fluid motion in the absence of active flow promotion is known as free or natural convection. Thus, there is a strong connection between the heat being delivered to the fluid and its resulting motion.

In a fluid which is forced to move, there is also strong coupling between the temperature profile and the fluid velocity profile, except that now the fluid is cause and the temperature is effect. Whether forced or free, however, there is strong coupling between the evolution of the momentum and thermal boundary layers. It follows then that the description of the momentum boundary layer requires a description of the thermal boundary layer, and vice versa.

Most convection heat transfer problems involve the use of empirical correlations to describe the heat transfer coefficient. In order to make the correlations apply to as many different problems as possible, most correlations are tied to dimensionless groups, or numbers, which describe the major fluid mechanical behavior associated with the problem. The dimensionless groups usually describe the competition between two processes. For example, the Prantl number describes the relationship between the thickness of the thermal and momentum boundary layers. The Reynolds number represents the competition between inertial and viscous forces.

Convection heat transfer is described by Newton's law of cooling:

$$q = hA(T_w - T_\infty) \quad , \qquad (5.17)$$

or

$$\hat{q} = h(T_w - T_\infty) \quad , \qquad (5.18)$$

where A is the surface area of the wall normal to the direction of heat transfer, T_w is the wall temperature of the solid surface, T_∞ is the free-stream fluid temperature and h is the heat transfer coefficient. Our previous discussion has motivated the dependency of h on fluid mechanics, fluid properties, and geometry, suggesting that h is difficult to determine for realistic situations. The structure of Eq. 5.17 suggests that we can view convection heat transfer as an extension of conduction heat transfer since

$$R_{\text{convection}} = \frac{T_w - T_\infty}{q} = \frac{1}{hA} \quad . \qquad (5.19)$$

The combination of conduction and convection resistances is sometimes referred to as an overall heat transfer coefficient.

Radiation heat transfer is characterized by energy transport between two bodies via electromagnetic waves. These waves travel at the speed of light and can thermally connect two bodies which are separated by a vacuum. The most important example of radiation is that of the sun which delivers energy to the earth across vast evacuated space.

The starting point in the study of radiation heat transfer is the Stefan-Boltzmann law for blackbody radiation:

$$E_b = \sigma T^4 \quad . \tag{5.20}$$

In Eq. 5.20, E_b is the emissive blackbody flux and σ is the Stefan-Boltzmann constant. In reality there are very few blackbodies, but it is possible to develop models that allow us to relate realistic "gray" surfaces to a blackbody at the same temperature. Gray objects can be incorporated into models by treating them as an ideal blackbody acting behind resistances that summarize the surface attributes. This approach is very much parallel to the representation of realistic voltage sources through the use of a Thévenin equivalent circuit.

Unlike conduction and convection, radiation heat transfer depends on how well two or more objects see one another. Geometry plays just as important a role in radiation heat transfer as the characteristics of the surfaces involved. Models must account for all of the radiant energy leaving each surface, and we will complete our radiation model by quantifying the resistance to direct heat transfer based on the ability of the objects involved to see one another.

For conventional electric machines, the principal heat transfer mechanisms are conduction and convection. If there is radiation heat transfer, it is considered an added bonus in limiting the temperature rise. Heat is generated in the electric machine in three regions:

1. In windings due to resistive and eddy current losses.

2. In magnetic materials due to hysteresis and eddy current losses.

3. In bearings due to frictional losses.

Heat generated within the stator is much easier to deal with than heat generated within the rotor. Removing heat from a rotating body is tricky business.

The simplest way of dealing with stator losses is to use free convection for cooling. With this approach, the losses within the windings and core will be conducted to the surface of the machine to be carried away by natural convection. This approach imposes the greatest limitation on the electric loading. In a well-designed machine, the hottest spot will be the end turns of the stator windings since the end turns are commonly surrounded by air. Since air is such a poor conductor of heat (particularly still air), heat generated in the end turns must be conducted into the stator stack before being transferred to the stator core.

As far as the stator is concerned, forced air cooling is substantially more aggressive than free convection, and liquid cooling is still more aggressive. Liquid cooling can be very effective particularly if the end turns are in good thermal contact with the fluid. Liquid cooling is generally implemented in smaller machines (up to hundreds of kilowatts) by using a cooling jacket around the periphery of the stator. Very large machines may use hollow phase winding conductors through which cooling fluid is passed. In terms of electric loading, forced convection supports an increase of about an order of magnitude over free convection; liquid cooling supports an increase of about one more order of magnitude. In terms of rms current density in the winding slot, a machine with free convection cooling may use 2-3 MA/m^2, while a liquid cooled machine may use 15 MA/m^2 or more.

Rotor losses deserve significant attention, even if it appears the rotor losses are small. For example, a machine that uses permanent magnets on the rotor will not have winding losses, but there will be hysteresis and eddy current losses in the magnets, particularly if the machine is driven by an inverter to form an adjustable speed drive (see Chapter 7). Because of the inverter, rotor losses may be nearly independent of speed. At low speeds the rotor will lose less heat due to convection, so the self heating of the magnets could get to a level where some loss of magnetization could occur.

In electric machines that operate over a relatively narrow speed range, it is possible to integrate a fan onto the shaft to provide a flow of cooling air. On electric machines driven by an adjustable speed drive, it may be necessary to have the drive control the fan speed so that cooling effort and machine speed are independent of one another. Some permanent magnet machines are cooled by fan-driven blowers on one or two ends of the machine. One blower pulls air into the machine, forcing it into the air gap. The second blower pulls the air through the air gap and exhausts it to the surroundings. This type of cooling is very effective, but can be very noisy and not practical for all applications.

5.5 Summary

This chapter has identified a number of issues associated with the design of electric machines. The discussion of magnetic design was intended to reinforce the material covered in Chapters 1 through 4. As such, this chapter has pulled together the material in previous chapters, projecting it onto the machine design problem. The intent was to provide an appreciation of the challenges that the machine designer must confront.

The discussion of thermal design raises some issues associated with the intended duty of the electric machine. In particular, machines intended to be driven by inverters as adjustable speed drives must consider this early in the design process to ensure that adequate attention is being given to thermal management over the intended duty of the machine. As we move forward, our attention is going to turn from the details of what is in the machine, and the details of the magnetic fields producing torque to a more terminal-centric view of the machine. We are going to be particularly interested in models of electric machines that describe the electromechanical dynamics in a form suitable to understand their dynamic control.

Chapter 6

Transformations

6.1 Introduction

In the chapters to follow, we are interested in generating dynamic models for various ac electric machines so that we can discuss their control. As we will see, the natural model for most machines is inconvenient for purposes of control. To manipulate the natural model, we will need, for example, to move quantities from the rotor reference frame to the stator reference frame. In another case, we will want to move both stator and rotor quantities to a common reference frame. In general, we use transformations to move quantities from one reference frame to another.

There are two basic transformations of interest to us. Other transformations we use will be combinations of the two basic transformations. One transformation is used to take three coupled phase windings into three decoupled phase windings. This is very valuable in three-phase machines in which the phases are symmetrically displaced by 120° electrical measure. While the symmetrical displacement is valuable in terms of symmetry and field generation, it results in a coupling among the three phases that complicates the structure of the resulting models.

A second transformation allows us to move quantities in one reference frame to a second reference frame that is displaced from the first by an angle ψ. That is, this is a rotary transformation between two reference frames that share a common origin.

An important aspect of the transformations we develop is power invariance. We want power to be naturally expressed in each coordinate system without having to include multiplication factors. Power invariance also makes the inverse transformation the transpose of the basic transformation.

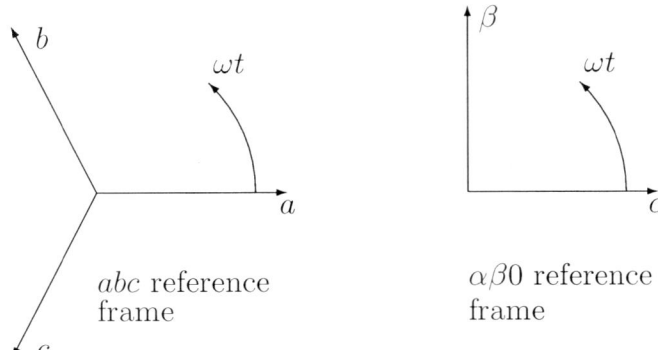

Figure 6.1: Space vectors for the *abc* reference frame and the $\alpha\beta 0$ reference frame.

Manipulation is enhanced when

$$T^{-1} = T^T \quad , \tag{6.1}$$

where T is one of our transformations.

6.2 Three Phases to Two Phases ($abc \leftrightarrow \alpha\beta 0$)

The space vectors for a three-phase machine and an equivalent two-phase machine are shown in Fig. 6.1. These space vectors represent the orientation of the winding fluxes which, when considered in conjunction with time-varying currents of proper phase, combine to create rotating magnetic fields in the direction shown in Fig. 6.1. A balanced set of three-phase currents in a balanced three-phase winding sets up a magnetomotive force (mmf) of constant amplitude and angular velocity. In the three phase windings these currents will produce an mmf of amplitude $3N\hat{I}/2$ where N is the effective number of turns and $\hat{I}$ is the amplitude of the phase currents. As we have seen, a balanced two-phase system can also set up a rotating mmf. In the two phase winding the mmf will have an amplitude of $N\hat{I}$.

Now let the phase a and phase α axes be coincident. Let N_3 be the number of turns of the three-phase winding and let N_2 be the number of turns for the two-phase winding. Resolving the three mmfs of the *abc* frame along the α and β axes, and equating the three-phase and two-phase quantities gives

$$N_2 i_\alpha = N_3 i_a + N_3 i_b \cos\left(\frac{2\pi}{3}\right) + N_3 i_c \cos\left(\frac{4\pi}{3}\right) \tag{6.2}$$

6.2. Three Phases to Two Phases ($abc \leftrightarrow \alpha\beta 0$)

$$N_2 i_\beta = \qquad N_3 i_b \sin\left(\frac{2\pi}{3}\right) + N_3 i_c \sin\left(\frac{4\pi}{3}\right) \quad . \tag{6.3}$$

For completeness, we need a third variable in addition to i_α and i_β which is independent of them:

$$N_2 i_0 = k N_3 i_a + k N_3 i_b + k N_3 i_c \quad . \tag{6.4}$$

The independence of phase 0 suggests it is orthogonal to both the α and β axes. These relationships can be summarized in vector form as

$$\begin{bmatrix} i_\alpha \\ i_\beta \\ i_0 \end{bmatrix} = \frac{N_3}{N_2} \begin{bmatrix} 1 & -1/2 & -1/2 \\ 0 & \sqrt{3}/2 & -\sqrt{3}/2 \\ k & k & k \end{bmatrix} \begin{bmatrix} i_a \\ i_b \\ i_c \end{bmatrix} = T \begin{bmatrix} i_a \\ i_b \\ i_c \end{bmatrix} \quad . \tag{6.5}$$

If we invert T we get

$$T^{-1} = \frac{2}{3} \frac{N_2}{N_3} \begin{bmatrix} 1 & 0 & 1/2k \\ -1/2 & \sqrt{3}/2 & 1/2k \\ -1/2 & -\sqrt{3}/2 & 1/2k \end{bmatrix} \quad . \tag{6.6}$$

It follows that

$$\left(T^{-1}\right)^T = T^{-T} = \frac{2}{3} \frac{N_2}{N_3} \begin{bmatrix} 1 & -1/2 & -1/2 \\ 0 & \sqrt{3}/2 & -\sqrt{3}/2 \\ 1/2k & 1/2k & 1/2k \end{bmatrix} \quad . \tag{6.7}$$

We want $T = T^{-T}$ so that the transformation for all variables is the same, and also so that we have invariance of power. This is satisfied if $N_3/N_2 = \sqrt{2/3}$ and $k = 1/\sqrt{2}$, giving

$$T = \sqrt{\frac{2}{3}} \begin{bmatrix} 1 & -1/2 & -1/2 \\ 0 & \sqrt{3}/2 & -\sqrt{3}/2 \\ 1/\sqrt{2} & 1/\sqrt{2} & 1/\sqrt{2} \end{bmatrix} \quad . \tag{6.8}$$

There will be times when the dynamics on the 0 axis are of no interest to us. Under these conditions, our transformation will be reduced by dropping a row ($abc \to \alpha\beta$) or a column ($\alpha\beta \to abc$).

To accommodate the transformation between the abc and the $\alpha\beta$ reference frames, we modify our transformation T by leaving out the 0 component. This leaves us with T_{23} converting abc quantities to $\alpha\beta$ quantities and T_{32} performing the inverse transformation:

$$T_{23} = \sqrt{\frac{2}{3}} \begin{bmatrix} 1 & -1/2 & -1/2 \\ 0 & \sqrt{3}/2 & -\sqrt{3}/2 \end{bmatrix} \quad ; \tag{6.9}$$

$$T_{32} = \sqrt{\frac{2}{3}} \begin{bmatrix} 1 & 0 \\ -1/2 & \sqrt{3}/2 \\ -1/2 & -\sqrt{3}/2 \end{bmatrix} . \quad (6.10)$$

Despite leaving out the 0 phase component, we still have $T_{32} = T_{23}^T$. Further, $T_{23}T_{32} = I_2$.

The simplification of working with orthogonal phases is helpful, particularly when dealing with induced voltages. We will see that flux in the α axis gives rise to voltage in the β axis. Flux in the β axis gives rise to voltage in the negative α axis. These relationships would be difficult to see in a coupled three phase system.

6.3 Rotation ($\alpha\beta 0 \leftrightarrow dq0$)

Figure 6.2 shows an arbitrary vector $\vec{a}$ decomposed into two reference frames where one frame is displaced from the other by an angle ψ. As such, Fig. 6.2 forms the basis of a transformation between reference frames that can rotate relative to one another. This is particularly useful in modeling electric machines, because it allows us to look at the dynamics from the perspective of an observer riding along with the rotating magnetic field within the air gap, or to move quantities from the rotor to the stator. In Fig. 6.2, each reference frame is denoted by a direct axis and a quadrature axis; the direct and quadrature axes are orthogonal within each reference frame. The vector $\vec{a}$ is taken to only have two components consistent with abc or $\alpha\beta$ reference frames discussed in the last section. Inclusion of the 0 phase in the transformation is discussed below.

We can show[1] that

$$\begin{bmatrix} a_{d_1} \\ a_{q_1} \end{bmatrix} = \begin{bmatrix} \cos(\psi) & -\sin(\psi) \\ \sin(\psi) & \cos(\psi) \end{bmatrix} \begin{bmatrix} a_{d_2} \\ a_{q_2} \end{bmatrix} . \quad (6.11)$$

---

[1] Begin by taking β to be the angle between $\vec{a}$ and the d_2 axis. Next, recall the trigonometric identity
$$\cos(\beta + \psi) = \cos\beta\cos\psi - \sin\beta\sin\psi .$$
Multiplying both sides by $|\vec{a}|$ we see that
$$a_{d_1} = \cos\psi\, a_{d_2} - \sin\psi\, a_{q_2} .$$
The second line of the transformation is found starting with the identity
$$\sin(\beta + \psi) = \cos\beta\sin\psi + \sin\beta\cos\psi .$$

6.3. Rotation ($\alpha\beta 0 \leftrightarrow dq0$)

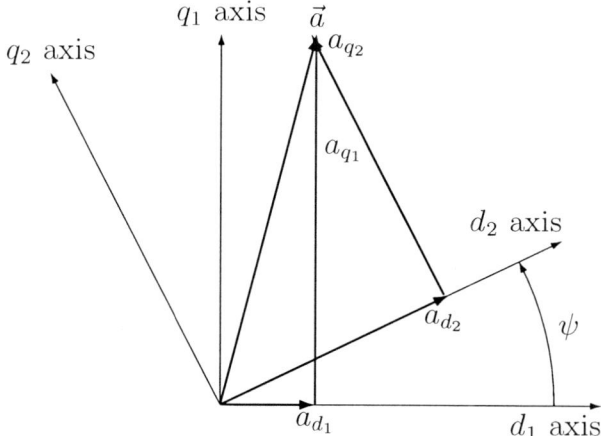

Figure 6.2: The vector $\vec{a}$ decomposed into two reference frames, with angular displacement ψ between them.

It follows that

$$\begin{bmatrix} a_{d_2} \\ a_{q_2} \end{bmatrix} = \begin{bmatrix} \cos(\psi) & \sin(\psi) \\ -\sin(\psi) & \cos(\psi) \end{bmatrix} \begin{bmatrix} a_{d_1} \\ a_{q_1} \end{bmatrix} . \qquad (6.12)$$

It is significant to note that

$$\begin{bmatrix} \cos(\psi) & \sin(\psi) \\ -\sin(\psi) & \cos(\psi) \end{bmatrix} = e^{J\psi} , \qquad (6.13)$$

where

$$J = \begin{bmatrix} 0 & 1 \\ -1 & 0 \end{bmatrix} . \qquad (6.14)$$

It is also worth observing that $J^2 = -I$ where I is the identity matrix, similar to $\jmath^2 = -1$, suggesting $e^{J\psi}$ is analogous to $e^{\jmath\psi} = \cos\psi + \jmath\sin\psi$. Since $e^{J\psi} = (e^{-J\psi})^T = (e^{J\psi})^{-T}$, $e^{J\psi}$ is a unitary transformation. Hence, $e^{J\psi}$ represents a power invariant transformation from one reference frame to another reference frame through rotation of angle ψ.[2]

With this rotary transformation, we can project the rotor and stator dynamics onto a common reference frame. The common reference frame is

---

[2]The matrix $e^{J\psi}$ is referred to as the matrix exponential and is an important concept in linear system theory. See, for example, R. Brockett, *Finite Dimensional Linear Systems*, Wiley, 1970 or T. Kailath, *Linear Systems*, Prentice-Hall, 1980, for more details. Our interest in the matrix exponential is partially in its state transition properties (as in transitioning quantities from one reference frame to another reference frame), but also in streamlining our notation.

normally chosen to rotate in synchronism with the magnetic field in the air gap and is described by a direct and a quadrature axis, referred to hereafter as the dq reference frame. Other uses of this transformation also exist, in which case the axes will be appropriately labeled.

In situations where we have three phases that need to be rotated, as in the $\alpha\beta 0$ reference frame, we can augment our original transformation so that we have

$$R_\psi = \begin{bmatrix} \cos\psi & \sin\psi & 0 \\ -\sin\psi & \cos\psi & 0 \\ 0 & 0 & 1 \end{bmatrix}, \quad (6.15)$$

reflecting the fact that the 0 phase is not altered in going from one reference frame to the other.

6.4 Useful Combinations

Through combination of T and R_ψ we can add to our collection of transformations. As a specific example, consider the case where we have abc quantities expressed in the rotor reference frame that we would like to express as abc quantities in the stator reference frame. That is, we would like to express x_{abc}^r in terms of x_{abc}^s where the rotor and stator are displaced from one another by electrical angle θ_e. In this notation, the subscript indicates the type of coordinate system and the superscript indicates the reference frame. Right now x is some generic quantity; it could represent a vector of voltages, flux linkages, or currents.

We can move x_{abc}^s to the rotor reference frame in the following steps:

1. Convert the abc quantities into $\alpha\beta 0$ quantities using transformation T.

2. Rotate the stator $\alpha\beta 0$ quantities through angle θ_e using R_{θ_e} to put them into the rotor frame.

3. Convert the $\alpha\beta 0$ quantities back to abc quantities by using T^T.

Accordingly, we have

$$x_{abc}^r = T^T R_{\theta_e} T x_{abc}^s \quad . \quad (6.16)$$

Carrying out the indicated multiplication gives

$T^T R_{\theta_e} T =$
$$\frac{2}{3}\begin{bmatrix} \cos\theta_e + \frac{1}{2} & \cos\left(\theta_e - \frac{2\pi}{3}\right) + \frac{1}{2} & \cos\left(\theta_e + \frac{2\pi}{3}\right) + \frac{1}{2} \\ \cos\left(\theta_e + \frac{2\pi}{3}\right) + \frac{1}{2} & \cos\theta_e + \frac{1}{2} & \cos\left(\theta_e - \frac{2\pi}{3}\right) + \frac{1}{2} \\ \cos\left(\theta_e - \frac{2\pi}{3}\right) + \frac{1}{2} & \cos\left(\theta_e + \frac{2\pi}{3}\right) + \frac{1}{2} & \cos\theta_e + \frac{1}{2} \end{bmatrix} (6.17)$$

We will use this transformation when we examine the induction machine in Chapter 9.

Our process of developing the transformation may seem somewhat backward. We are interested in moving rotor quantities to the stator, and we formed the transformation by apparently moving stator quantities to the rotor. However, in this way we ended up with the rotor quantities in terms of the stator quantities, ready for substitution and further manipulation.

6.5 Summary

This chapter has presented two fundamental transformations that are extremely useful in manipulating models of electric machines for purposes of control. The first transformation allows three-phase machines to be viewed as equivalent two-phase machines with the phases decoupled. The second transformation allows rotation of quantities between reference frames that share the same origin but are displaced from one another by an angle. There are many ways to combine these transformations into variations on a theme. One example showed how quantities in the rotor reference frame can be expressed equivalently in the stator reference frame.

The transformations presented in the Chapter have been in use for a long time. Variants of the transformations presented here are sometimes referred to as Park's transformation[3], or the Blondel-Park transformation since Park's work was built on that of Blondel[4]. The $abc \rightarrow \alpha\beta$ transformation is also sometimes known as the Clarke transformation.

---

[3] See R.H. Park, "Two reaction theory of synchronous machines," *AIEE Trans.*, Vol. 48, pp. 716-730, 1929.

[4] See A. Blondel, *Synchronous Motors and Converters, Part III*, New York: McGraw-Hill, 1913.

Chapter 7

Inverters

7.1 Introduction

In the chapters to come, our interest is going to be in operating ac machines at adjustable speeds. To accomplish this, we need to synthesize an ac source over which we have control of frequency and voltage amplitude. The common way to do this is through electronic power converters. These converters are either fed directly with dc, or make dc by rectifying the fixed-frequency, fixed-amplitude ac source.

Inverters are used to convert dc into ac. This is accomplished through alternating application of the source to the load, achieved through proper use of controllable semiconductor switches. This chapter reviews the basic principles of inverter circuits and their control. While our principal interests in inverters are for the control of electric machines, inverter circuits are applied to uninterruptable power supplies (commonly referred to as UPSs), distributed generation systems, and active power filters.

Both voltage- and current-source inverters are used in practice. The trend, however, is to use voltage-source inverters for the vast majority of applications. Current-source inverters are still used at extremely high power levels, though voltage-source inverters are gradually filling even these applications. Because of the dominance of voltage-source inverters, this chapter focuses exclusively on this type of inverter.

There are many issues involved in the design of an inverter. The more prominent issues involve the interactions among the source, the power circuit, the load, and the control. Other subtle issues involve the control of parasitics and the protection of the controllable switches through the use of snubber and clamp circuits, and the juxtaposition of controller speed with

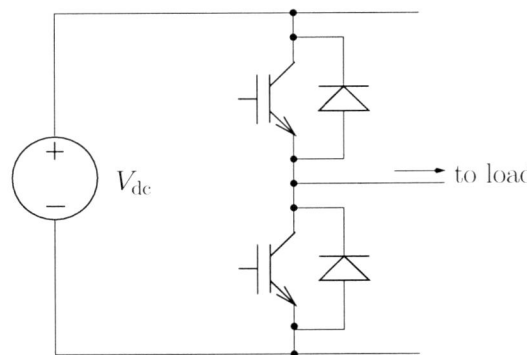

Figure 7.1: An inverter phase leg.

the desire for increased switching frequency while maintaining high efficiency.

The technical literature contains abundant information on inverters, as do specialized books on power electronics[1]. Many of the ancillary issues associated with the design of power converters are not discussed here, but the basic principles presented should be sufficient to appreciate the use of inverters as ac sources with adjustable frequency, amplitude, and wave shape.

7.2 An Inverter Phase Leg

An inverter phase leg is shown in Fig. 7.1. It is comprised of two fully controllable switches and two diodes in antiparallel to the controllable switches. This phase-leg is placed in parallel with a voltage source. The center of the phase-leg is taken to the load. The basic circuit shown in Fig. 7.1 is usually augmented with a snubber circuit or clamp to shape the switching locus of the controllable switches. Insulated gate bipolar transistors (IGBTs) are shown as the controllable switches in Fig. 7.1. While the IGBT finds significant application in inverters, any fully controllable device, such as a field-effect transistor (FET) or a gate turn-off thyristor (GTO), may be used in its place.

---

[1] See, for example, N. Mohan, T. M. Undeland and W. P. Robbins, *Power Electronics: Converters, Applications and Design*, 2nd ed., New York: John Wiley & Sons, 1995; J. G. Kassakian, M. F. Schlecht and G. C. Verghese, *Principles of Power Electronics*, Reading, MA: Addison Wesley, 1991; P. T. Krein, *Elements of Power Electronics*, New York: Oxford University Press, 1998; or B. K. Bose, ed., *Power Electronics and Variable Frequency Drives*, New York: IEEE Press, 1997.

7.2.1 Basic Principles

In the phase-leg of Fig. 7.1, there are two restrictions on the use of the controllable switches. First, at most one controllable switch may be conducting at any time. The dc supply is shorted if both switches are conducting. In practice one switch is turned off before the other is turned on. This blanking time, also known as dead time, compensates for the tendency of power devices to turn on faster than they turn off. Second, at least one controllable switch (or an associated diode) must be on at all times if the load current is to be continuous, required for inductive loads.

If the upper switch is conducting, the load is connected to the positive side of V_{dc}. If the lower switch is conducting, the load is connected to the negative side of V_{dc}. It follows that the voltage applied to the load will, on average, fall somewhere between 0 and V_{dc}. When the phase-leg of Fig. 7.1 is used with one or more additional phase legs, the load voltage can be made to alternate. The details of how the voltage alternates is the responsibility of the controller.

The peak voltage seen by each switch is the total dc voltage across the phase-leg. This is determined by recognizing that one of the two switches is always conducting. The peak current which must be supported by each switch is the peak current of the load. Under balanced control of the two switches, each switch must support the same peak current; each diode must support the same peak current as the switches.

In an effort to improve the efficiency and spectral performance of inverters, the use of inverters with a resonant dc link has been reported in the technical literature[2]. Through periodic resonance of the dc bus voltage to zero, or the dc bus current to zero, the inverter switches can change states in synchronism with these zero crossings in order to reduce the switching losses in the power devices. Figure 7.2 shows a schematic for the basic resonant dc link converter. The resonance of L_r and C_r forces the bus voltage (the voltage applied to the controllable switches) to swing between zero and $2V_{\text{dc}}$. The inverter switches change states when the voltage across C_r is zero. It is often necessary to hold the bus voltage at zero for a brief time to insure that sufficient energy has been put into L_r to force resonance back to zero voltage across C_r. The bus can be clamped at zero voltage by turning on both switches in an inverter phase leg. One drawback of the resonant dc link is the increased voltage or current imposed on the power devices. Auxiliary

---

[2]See D. M. Divan, "The resonant dc link converter–a new concept in static power conversion," *IEEE Trans. on Industry Applications*, Vol. 25, pp. 317-325, 1989, and Y. Murai and T. A. Lipo, "High frequency series resonant dc link power conversion," *IEEE/IAS Annual Meeting Conference Record*, pp. 772-779, 1998.

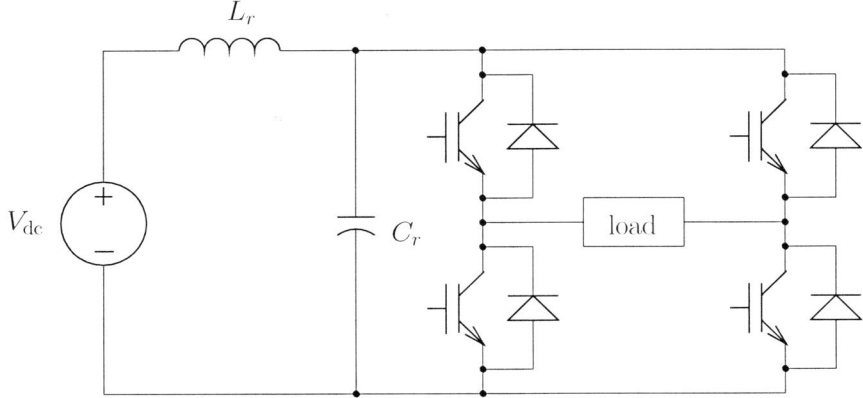

Figure 7.2: A basic resonant dc link inverter system.

clamp circuits have been implemented to minimize this drawback[3]. The control of a resonant link inverter is complicated by the simultaneous need to manage energy flow to the load while managing energy in the resonant link.

7.2.2 Snubber Circuits and Clamps

Snubber circuits are used to control the voltage across and the current through a controllable switch as that device is turning on or off. A complete snubber will typically limit the rate of rise in current as the device is turning on and limit the rate of rise in voltage as the device is turning off. Additional circuit components are used to accomplish this shaping of the switching locus. Figures 7.3 through 7.5 show three snubber circuits that are used with inverter legs, one of which has auxiliary switches[4]. Snubber circuits become increasingly important as the power rating of the inverter increases, where the additional cost of small auxiliary devices is justified by improved efficiency and spectral performance.

Clamps differ from snubbers in that a clamp is used only to limit a switch

---

[3] See D. M. Divan and G. Skibinski, "Zero-switching-loss inverters for high-power applications," *IEEE Trans. on Industry Applications*, Vol. 25, pp. 634-643, 1989; J. He and N. Mohan, "Parallel resonant dc link circuit–a novel zero switching loss topology with minimum voltage stresses," *IEEE Trans. on Power Electronics*, Vol. 6, pp. 687-694, 1991; or J. M. Simonelli and D. A. Torrey, "An alternative bus clamp for resonant dc link converters," *IEEE Trans. on Power Electronics*, Vol. 9, pp. 56-63, 1994.

[4] See W. McMurray, "Efficient snubbers for voltage-source GTO inverters," *IEEE Trans. on Power Electronics*, Vol. PE-2, pp. 264-272, 1987; W. McMurray, "Resonant snubbers with auxiliary switches," *IEEE/IAS Annual Meeting Conference Record*, pp. 829-834, 1989; and T. M. Undeland, "Switching stress reduction in power transistor converters," *IEEE/IAS Annual Meeting Conference Record*, pp. 383-392, 1976.

7.2. An Inverter Phase Leg

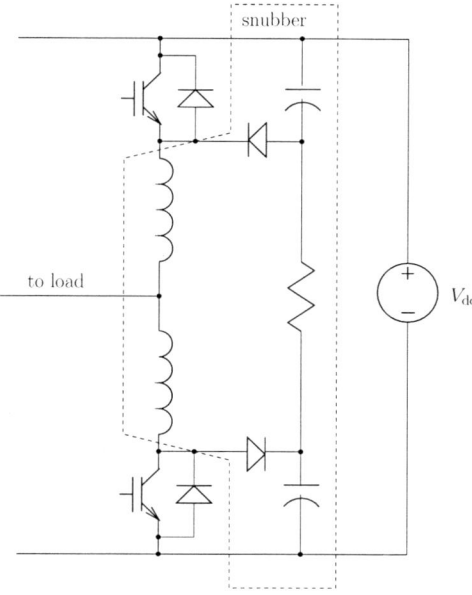

Figure 7.3: A McMurray snubber implemented on an inverter phase leg.

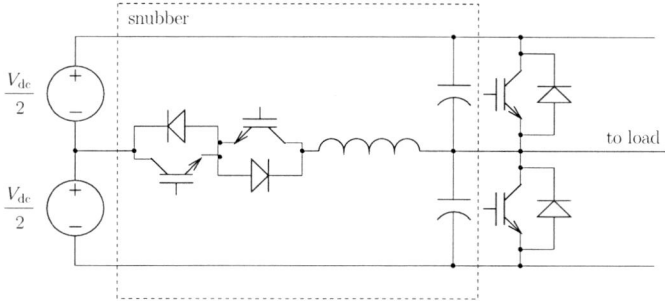

Figure 7.4: A resonant snubber with auxiliary switches implemented on a phase leg.

variable, usually the voltage, to some maximum value. The clamp does not dictate how quickly this maximum value is attained. Figures 7.6 through 7.8 show three clamp circuits that are commonly used with inverter legs. The clamp circuits become more complex with increasing power level.

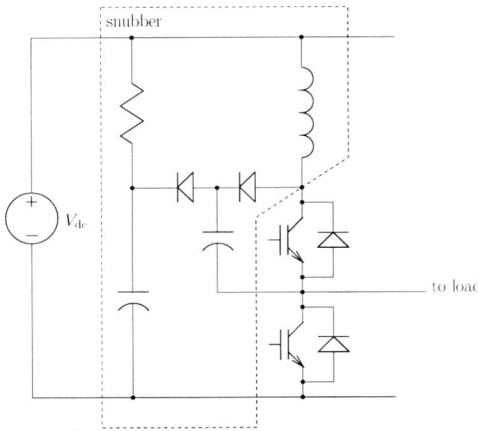

Figure 7.5: An Undeland snubber implemented on an inverter phase leg.

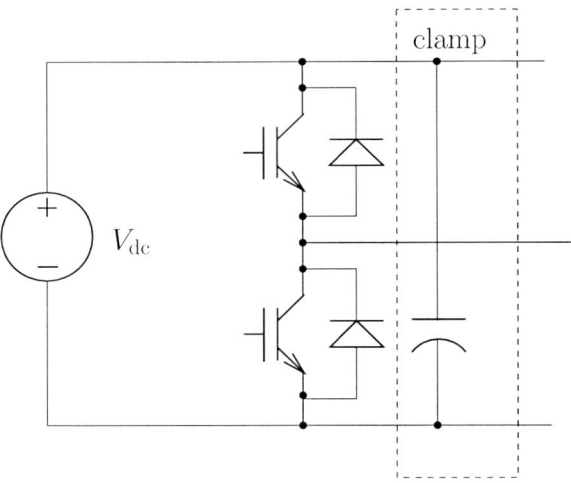

Figure 7.6: A clamp circuit for a low power ($\sim 50\,\text{A}$) inverter phase leg.

7.2.3 Interfacing to Controllable Switches

The interface to the controllable switches within the inverter phase leg of Fig. 7.1 requires careful attention. Power semiconductor devices are generally controlled by manipulating the control terminal, usually relative to one of the power terminals. For example, Fig. 7.1 shows insulated gate bipolar transistors (IGBTs) as the controllable switches. These devices are turned on and off through the voltage level applied to the gate relative to the emitter. Because the emitter voltage of the upper IGBT moves around as the devices switch, the control of the upper IGBT must accommodate this move-

7.2. An Inverter Phase Leg

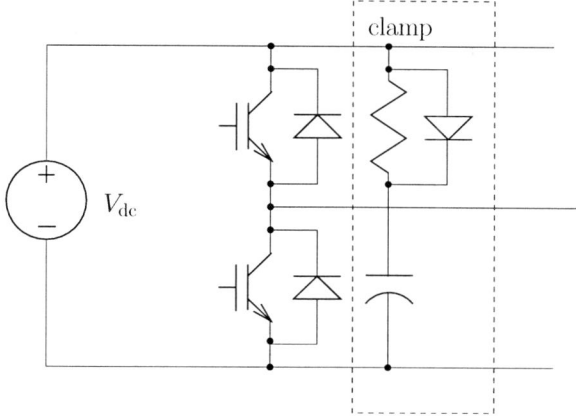

Figure 7.7: A clamp circuit for a medium power ($\sim 200\,\text{A}$) inverter phase leg.

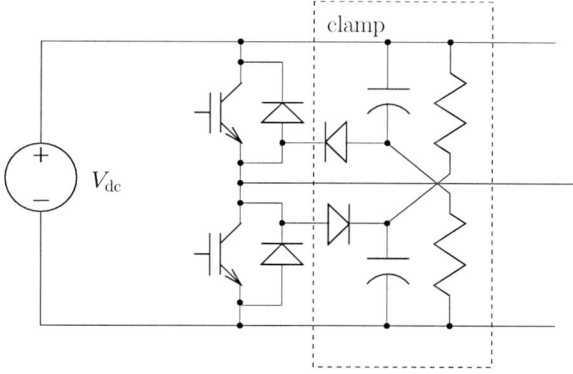

Figure 7.8: A clamp circuit for a high power ($\sim 400\,\text{A}$) inverter phase leg.

ment. The emitter voltage of the upper IGBT moves from the positive side of the dc bus when the IGBT is conducting to the negative side of the dc bus with the lower IGBT is conducting. The circuit responsible for turning the power semiconductor on and off as commanded by a controller is usually known as a gate drive circuit. Some gate drive circuits incorporate protection mechanisms for over current or over temperature.

There are a number of approaches that are used to control the semiconductor devices within the inverter. The choice among these approaches is often dictated by the power levels involved, the switching frequency supported by the switches, and the preference of the designer, among others. It is possible to purchase high-voltage integrated circuits (HVICs) that perform the level shifting necessary to take a logic signal referenced to the negative side of the dc bus and control the upper controllable switch in the phase leg. This approach is generally limited to applications where the controllable switches do not require a negative bias to hold them in the blocking state. High frequency applications may use transformer-coupled gate drives that are insensitive to the common-mode voltage between the primary and secondary. This approach runs into problems at lower frequencies because the size of the transformer core begins to get large. High power applications may use optocouplers to optically couple the control information to the gate drive, where the gate drive is supported by an isolated power supply. Often the power supply is the dominant factor in the overall cost of the gate drive.

Figure 7.9(a) shows how a HVIC is interfaced to a phase leg. The capacitors are used as local power supplies for the upper and lower gate drives. In this implementation, the upper capacitor is charged through the diode while the lower switch is conducting. Figure 7.9(b) shows a transformer-coupled gate drive for a high frequency application[5]. It is important to design a mechanism for resetting the core in a transformer coupled gate drive. In Fig. 7.9(b), the core is reset by driving the transformer primary with a bipolar voltage that provides sufficient Volt-seconds to drive the transformer into saturation. This need for core reset may place unacceptable limitations on the duty ratio of the switches for some applications. Figure 7.9(c) shows the use of an optocoupler to provide isolation of the control signal going to the gate drive. The isolated power supply required to support the use of the optocoupler is not shown.

7.3 Single-Phase Inverters

There are two ways to form a single-phase inverter. The first method is shown in Fig. 7.10, where the phase leg of Fig. 7.1 is used to control the voltage applied to one side of the load. The other side of the load is connected to the common node of two voltage sources. The half-bridge inverter of Fig. 7.10 applies positive voltage to the load when the upper switch is conducting and negative voltage to the load when the lower switch is conducting. It is not possible for the half-bridge inverter to apply zero voltage to the load.

A single-phase inverter is also formed by placing the load between two inverter phase legs, as shown in Fig. 7.11. This circuit is often referred to

---

[5]See, for example, International Rectifier, "Transformer-isolated gate driver provides very large duty cycle ratios," Application Note AN-950. Available through URL http://www.irf.com/.

7.3. Single-Phase Inverters

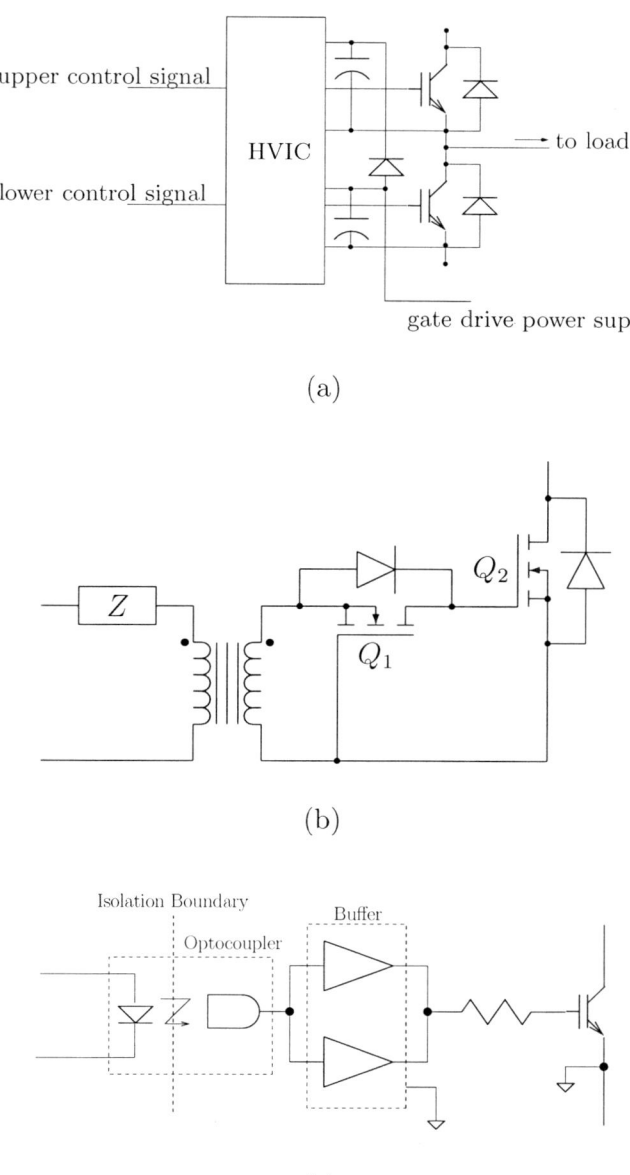

Figure 7.9: Three common approaches to interfacing to controllable switches: (a) the use of a HVIC; (b) the use of transformer coupling; and (c) the use of an optocoupler.

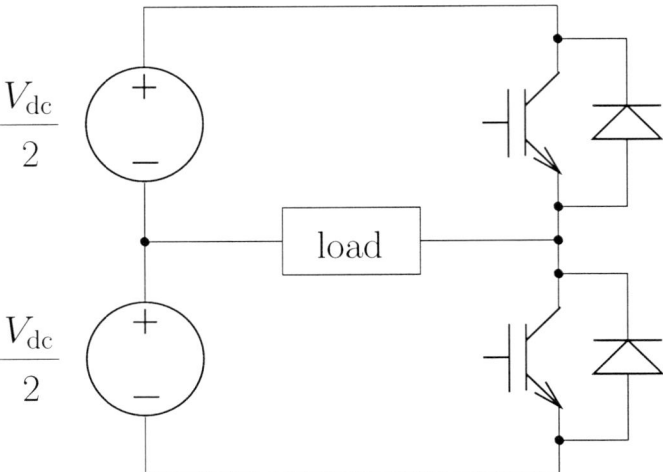

Figure 7.10: A single-phase inverter using one phase-leg and two dc voltage sources.

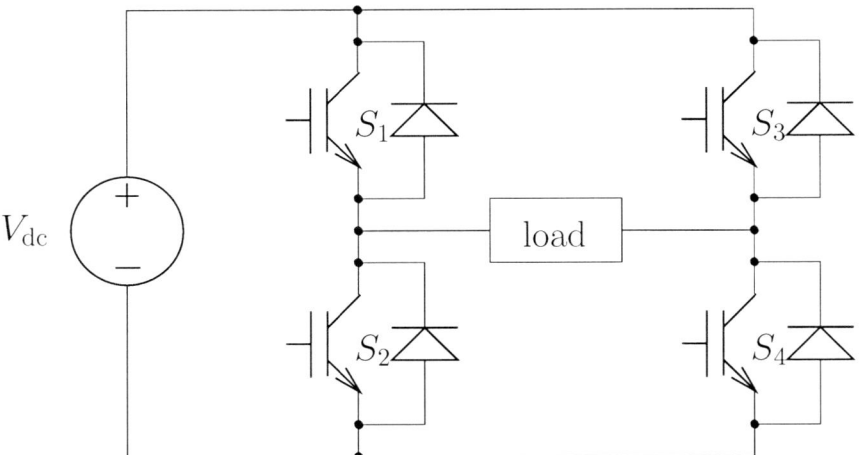

Figure 7.11: A single-phase inverter using two phase-legs and one dc voltage source.

as a full- or H-bridge inverter. Through appropriate control of the inverter switches, positive, negative and zero voltage can be applied to the load. The zero voltage state is achieved by having both upper switches, or both lower switches, conducting at the same time. The zero-voltage state requires one switch and one diode to be supporting the load current.

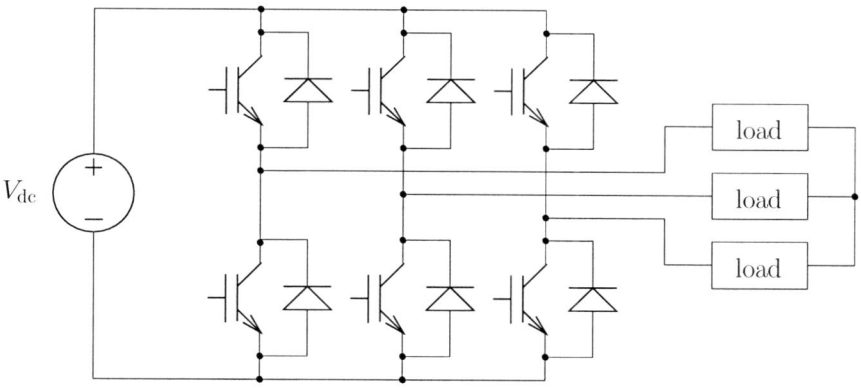

Figure 7.12: The three-phase inverter topology.

7.4 Three-Phase Inverters

A three-phase inverter is used to support both Δ- and Y-connected three-phase loads. The three-phase inverter topology can be derived by using three single-phase full-bridge inverters, with each inverter supporting one phase of the load. Upon careful examination of the resulting connection of the twelve controllable switches with the load, it is seen that there are six redundant switches because inverter phase-legs are connected in parallel. Elimination of the six redundant switches yields the topology shown in Fig. 7.12.

There are six switches in the three-phase inverter topology of Fig. 7.12. Considering all combinations of the switch states, there are seven possible voltages that may be applied to the load; the cases of all three upper switches being closed and all three lower switches being closed are functionally indistinguishable. The controller is responsible for applying the appropriate voltage to the load according to the method used for synthesizing the output voltage.

7.5 Multilevel Inverters

The three-phase inverter of Fig. 7.12 applies one of two voltages to each output load terminal. The output voltage is V_{dc} when the upper switch is conducting, and it is zero when the lower switch is conducting. Accordingly, this inverter could be called a two-level inverter. Multilevel inverters are based on extending this concept, and are becoming the inverter of choice for higher-voltage and higher-power applications. The discussion here focuses on a three-level inverter; extensions to five or more levels can be found in the technical literature[6].

Among their advantages, multilevel inverters allow the synthesis of voltage waveforms that have a lower harmonic content than a two-level inverter for the same switching frequency. This is because each output terminal is switched among at least three voltages, not just two. In addition, the input dc bus voltage can be higher because multiple devices are connected in series to support the full bus voltage.

Figure 7.13 shows one phase leg of a three-level inverter. In the three-level inverter the dc bus is partitioned into two equal levels of $V_{dc}/2$. The four controllable switches with anti-parallel diodes, S_1 through S_4, are connected in series to form a phase leg. In addition, two steering diodes, D_5 and D_6, are used to support current flow to and from the mid-point of the dc bus. Additional phase legs would be connected in parallel across the full dc bus.

The operation of the four switches is used to connect the load output terminal to either V_{dc}, $V_{dc}/2$ or zero. Switches S_1 and S_2 are used to connect the load to V_{dc}, switches S_2 and S_3 are used to connect the load to $V_{dc}/2$, and switches S_3 and S_4 are used to connect the load to zero. Two switches (or their anti-parallel diodes) are always in operation, and these switches are always adjacent.

While switches S_1 and S_2 are conducting, diode D_6 ensures that the voltage across switch S_4 does not exceed $V_{dc}/2$. Similarly, when switches S_3 and S_4 are conducting, diode D_5 ensures that the voltage across switch S_1 does not exceed $V_{dc}/2$. While switches S_2 and S_3 are conducting, the voltage across both S_1 and S_4 is clamped at $V_{dc}/2$; diode D_5 and switch S_2 support positive load current while diode D_6 and switch S_3 support negative load current.

The use of steering diodes D_5 and D_6 clamps the voltage across the power semiconductors to $V_{dc}/2$. For this reason, this topology is known as a diode-clamped multilevel inverter. It is also possible to use capacitors for clamping device voltages, but this is not the preferred method in machine drives.

A three-phase, three-level inverter provides substantially increased flexibility for voltage synthesis over the conventional three-phase inverter of Fig. 7.12. The three-phase inverter of Fig. 7.12 offers eight switch combinations that support seven different voltage combinations among the three output terminals. A three-phase, three-level inverter offers twenty-seven switch

---

[6]See, for example, J.-S. Lai and F. Z. Peng, "Multilevel converters–a new breed of power converters," *IEEE Trans. on Industry Applications*, Vol. 32, pp. 509-517, 1996, and F. Z. Peng, "A generalized multilevel inverter topology with self voltage balancing," *IEEE Trans. on Industry Applications*, Vol. 37, pp. 611-618, 2001.

7.5. Multilevel Inverters

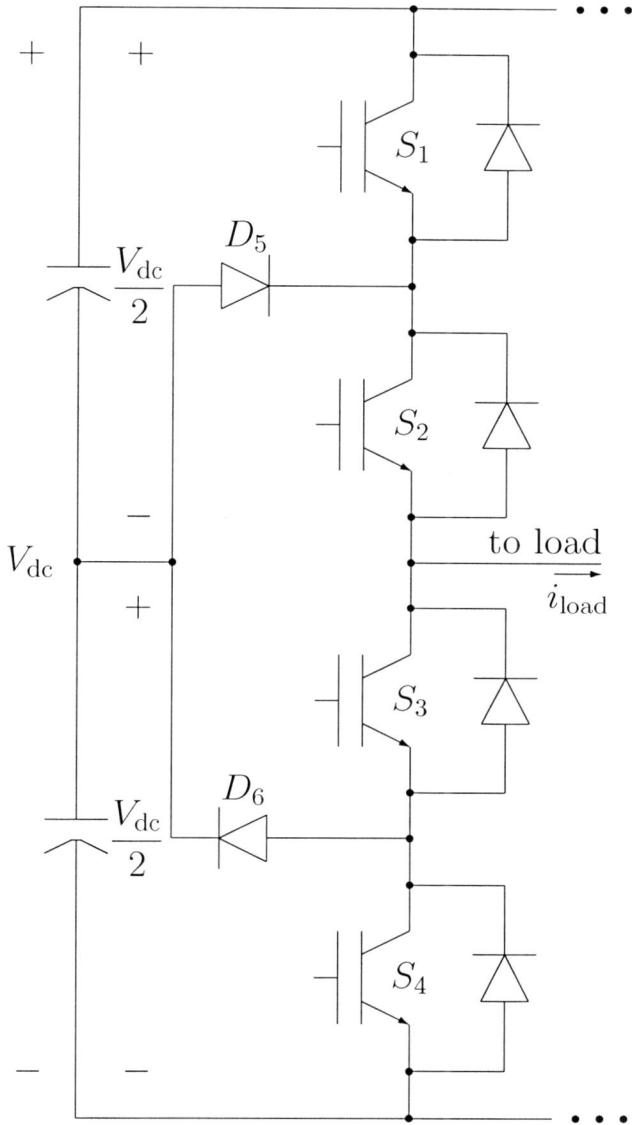

Figure 7.13: A phase leg for a three-level inverter.

combinations, supporting nineteen different voltage combinations among the three output terminals; there are seven voltage vectors that can be created in multiple ways. The redundancy of certain voltages provides additional degrees of freedom in designing the voltage synthesis algorithm. These additional degrees of freedom could, for example, be used to minimize the common mode voltage between the three outputs.

Issues within the design and control of the multilevel inverter include the dynamic balancing of the voltages within each level of the dc bus and the switching patterns needed to best synthesize the desired output voltage. One would expect that symmetric operation of the phase leg should be sufficient for maintaining balanced voltages across each level of the dc bus. The variation in capacitor values, however, will cause the mid-point of the dc bus to move to a voltage other than $V_{dc}/2$ for symmetric load currents. This shift in voltage will have repercussions on the synthesis of the output voltage.

7.6 Voltage Waveform Synthesis Techniques

There are three principal ways to synthesize the output voltage waveform in an inverter: harmonic elimination, harmonic cancellation, and pulse-width modulation. The synthesis technique that is applied is generally driven by consideration of the required output quality, the inverter power rating (which is closely tied to the speed of the controllable switches), the computational power of the available controller, and the acceptable cost of the inverter. This subsection reviews some of the common techniques used to synthesize the inverter output voltage.

7.6.1 Harmonic Elimination

Harmonic elimination implies that the output waveform shape is controlled to be free of specific harmonics through the selection of switch transitions[7]. That is, the switches are controlled so that one or more harmonics are never generated. This is often accomplished by notching the output waveform. Examples of harmonic elimination are shown in Figs. 7.14 and 7.15, which respectively show the elimination of only the third harmonic and simultaneous elimination of the third and fifth harmonics from the output of a single-phase inverter.

As suggested by Figs. 7.14 and 7.15, additional switch transitions must be inserted in the output waveform for each harmonic that is to be eliminated. As the number of notches gets very large, the output voltage waveform begins to resemble something that could be produced by pulse-width modulation.

---

[7]See H. S. Patel and R. G. Hoft, "Generalized techniques of harmonic elimination and voltage control in thyristor inverters: Part I–Harmonic elimination techniques," *IEEE Trans. on Industry Applications*, Vol. IA-9, pp. 310-317, 1973, and H. S. Patel and R. G. Hoft, "Generalized techniques of harmonic elimination and voltage control in thyristor inverters: Part II–Voltage control techniques," *IEEE Trans. on Industry Applications*, Vol. IA-10, pp. 666-673, 1974.

7.6. Voltage Waveform Synthesis Techniques

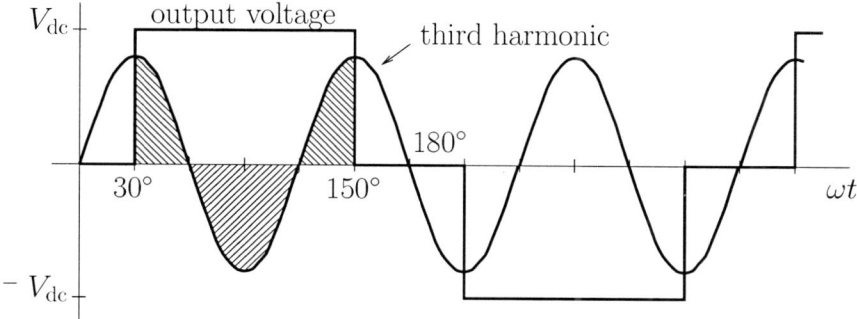

Figure 7.14: Elimination of the third harmonic.

Harmonic elimination is sometimes referred to as programmed PWM because the switching angles of the output voltage are programmed according to the intended harmonic content[8].

7.6.2 Harmonic Cancellation

Harmonic cancellation uses the superposition of two or more waveforms to cancel undesired harmonics. Figure 7.16 shows how two waveforms which contain the third harmonic may be phase-shifted and superimposed in order to create a waveform that is free of the third harmonic. The circuit of Fig. 7.17 can be used to synthesize the waveform of Fig. 7.16.

By combining harmonic cancellation and harmonic elimination, it is possible to create relatively high quality voltage waveforms. This quality requires a more complicated circuit. This additional complexity may be warranted depending on the power level and the specified quality.

7.6.3 Pulse-Width Modulation

Pulse-width modulation (PWM) is a method of voltage synthesis through which high frequency voltage pulses are applied to the inverter load[9]. The width of the pulses are made to vary at the desired frequency of the output voltage. Successful application of PWM generally involves a wide frequency separation between the carrier frequency (the desired output frequency) and

---

[8]See P. N. Enjeti, P. D. Ziogas, and J. L. Lindsay, "Programmed PWM techniques to eliminate harmonics: A critical evaluation," *IEEE Trans. on Industry Applications*, Vol. 26, pp. 302-316, 1985.

[9]See J. Holtz, "Pulsewidth modulation for electronic power conversion," *IEEE Proceedings*, Vol. 82, pp. 1194-1214, 1994.

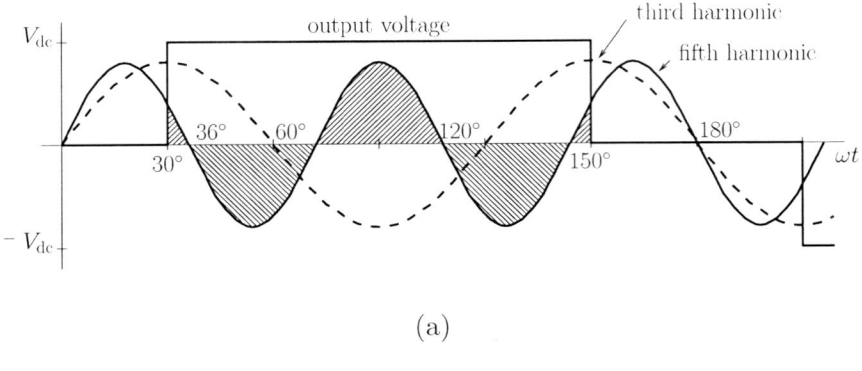

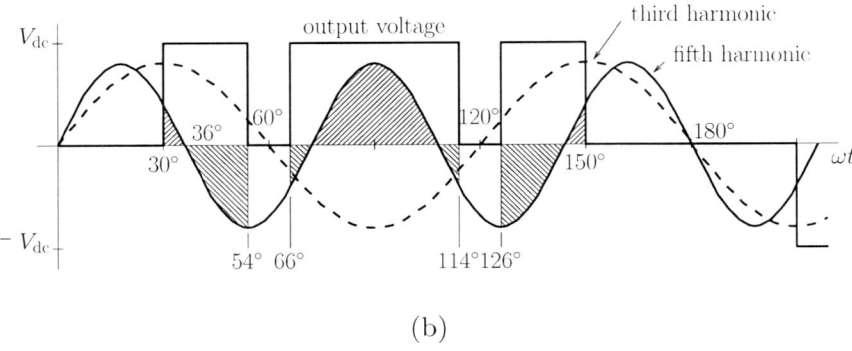

Figure 7.15: Simultaneous elimination of the third and fifth harmonics. Part (a) shows the fifth harmonic superimposed on the waveform of Fig. 7.14; part (b) shows the switch transitions introduced to eliminate the fifth harmonic without reintroducing the third harmonic.

the modulation frequency (the switching frequency). This frequency separation moves the distortion in the output voltage to high frequencies, thereby simplifying the required filtering. In many commercial machine drives the machine itself is used as the output filter. This subsection reviews two of the more common techniques used for synthesizing voltage waveforms using modulation.

Sinusoidal PWM

Sinusoidal PWM implies that the pulse widths of the output voltage are distributed sinusoidally. The pulse widths are generally determined by comparing a sinusoidal reference waveform with a triangular waveform. The

7.6. Voltage Waveform Synthesis Techniques

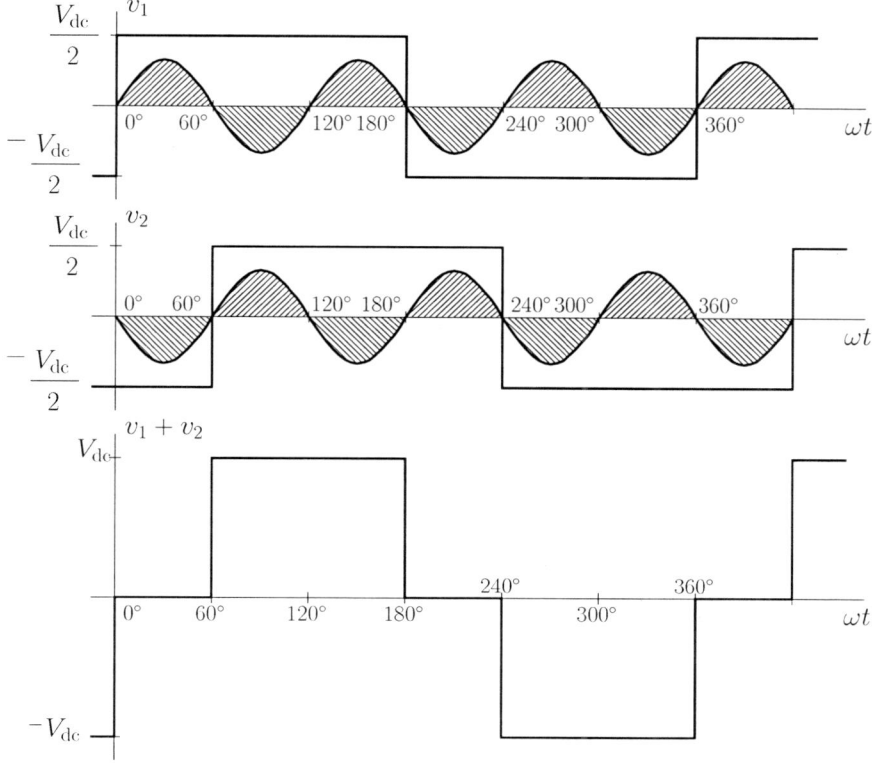

Figure 7.16: The superposition of two waveforms to cancel the third harmonic.

sinusoidal waveform sets the modulation (output) frequency and the triangular waveform sets the switching frequency. Sinusoidal PWM is routinely applied to single- and three-phase inverters.

In a single-phase inverter, the implementation of sinusoidal PWM depends on whether or not both phase-legs are operated at high frequency. Referring to Fig. 7.11, we see that it is not necessary for both phase-legs to operate at high frequency. We could, for example, operate switches S_1 and S_2 at high frequency to control the shape of the voltage, while switches S_3 and S_4 are operated at the frequency of the reference sinusoid to dictate the polarity of the output voltage. One advantage of this approach is that the inverter is more efficient because only two of the switches are operated at high frequency. Figure 7.18 shows how the sinusoidal pulse widths are created by a comparator and a unipolar triangular carrier. An alternative approach is to use switches S_2 and S_4 to control the polarity of the output voltage. While switch S_4 is conducting, switches S_1 and S_2 are used to control the

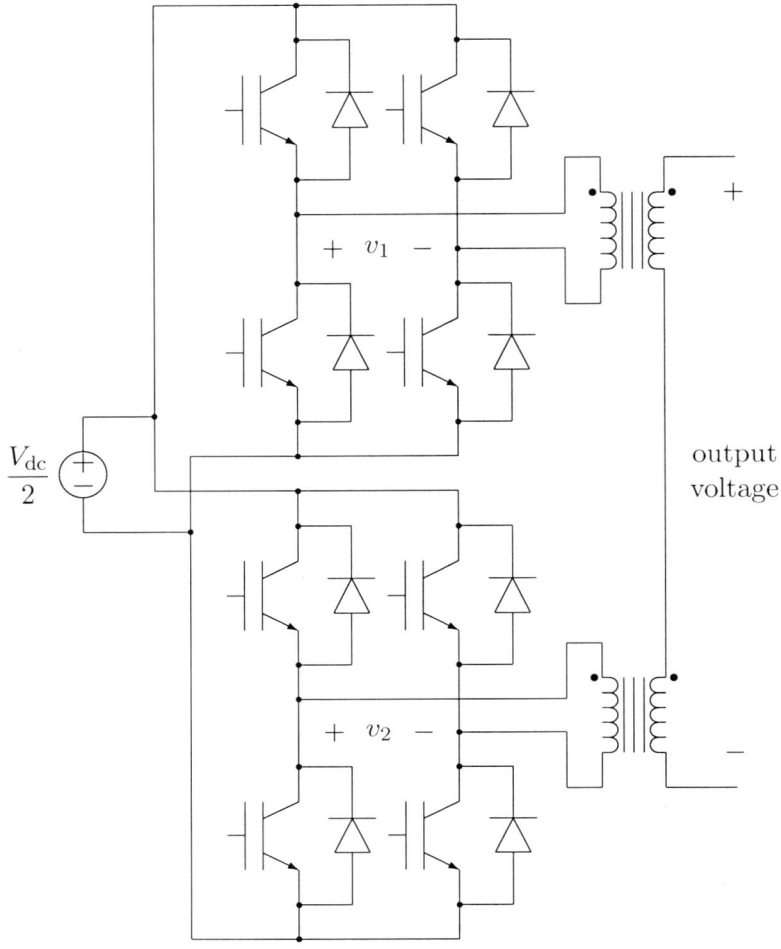

Figure 7.17: A circuit capable of synthesizing the waveforms of Fig. 7.16.

shape of the voltage. Similarly, switches S_3 and S_4 are used to control the shape of the voltage while switch S_2 is conducting. This approach tends to equalize the stress on the two phase legs, and can simplify the control logic necessary to implement the PWM.

A second way of implementing sinusoidal PWM in a single-phase inverter is to operate both phase-legs at high frequency. This method of control gives the same output voltage waveform as the high frequency/low frequency approach. The basic difference in control structures between the two is that the first method uses a unipolar triangular carrier, while the second way uses a bipolar triangular carrier. Figure 7.19 shows how the bipolar triangular

7.6. Voltage Waveform Synthesis Techniques

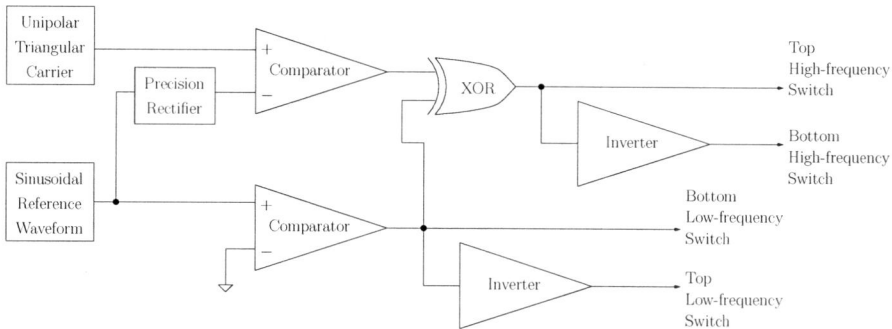

Figure 7.18: The generation of sinusoidally distributed pulse widths using a unipolar triangular carrier.

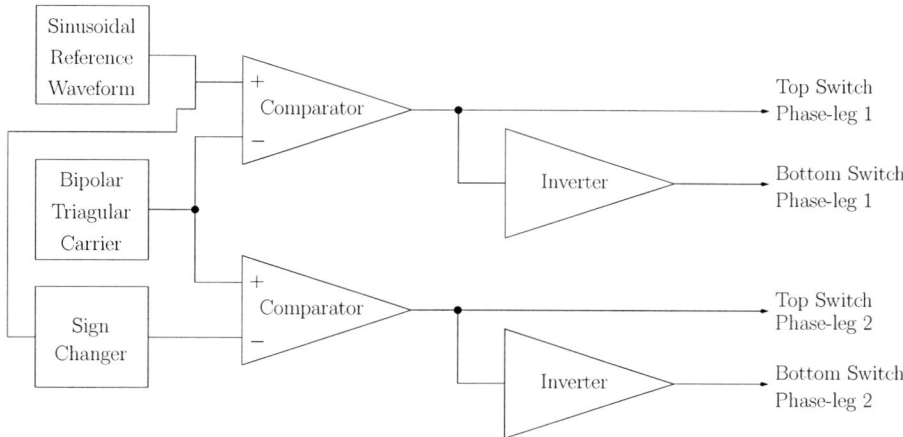

Figure 7.19: The generation of sinusoidally distributed pulse widths using a bipolar triangular carrier.

carrier is used to create the sinusoidally distributed pulse widths.

In a three-phase inverter, typically all phase legs share the same modulation signal. This results in the switching actions for all three phases being centered on the peak of the modulation signal. When viewed in this way, the similarities between sinusoidal PWM and space-vector modulation become visible. Many digital signal processors (DSPs) are available for machine control and include a rich set of peripherals to support real-time control. These DSPs typically have a programmable timer unit that is designed to support a variety of modulation algorithms.

Space-Vector PWM

Space-vector modulation has become the standard method for controlling the output voltage of three-phase inverters. The technique bears great similarity to the field-oriented control techniques that are applied to ac electric machines[10]. Chapters 10 and 12 discuss field oriented control of induction and permanent magnet machines, respectively.

As discussed in Chapter 6, a balanced set of three-phase quantities can be projected onto orthogonal α and β axes through the transformation

$$\begin{bmatrix} x_\alpha \\ x_\beta \end{bmatrix} = \sqrt{\frac{2}{3}} \cdot \begin{bmatrix} 1 & -1/2 & -1/2 \\ 0 & \sqrt{3}/2 & -\sqrt{3}/2 \end{bmatrix} \begin{bmatrix} x_a \\ x_b \\ x_c \end{bmatrix} \quad , \tag{7.1}$$

where x is a voltage or current. A similar transformation exists for taking direct and quadrature components back to phase quantities, though space-vector modulation does not require use of the inverse transformation.

Applying Eq. 7.1 to the three-phase inverter of Fig. 7.12, we see that each distinct switch state of the inverter corresponds with a different space vector. The zero vector can be produced by the electrically equivalent topologies of all upper switches conducting and all lower switches conducting. It is important to note that the six non-zero space vectors created by the inverter states are of the same magnitude ($\sqrt{2/3}V_{\text{dc}}$) and are symmetrically displaced. Figure 7.20 shows the connection between the three-phase inverter of Fig. 7.12 and the generation of the eight space vectors.

Any desired output voltage, up to the magnitude of $V_{\text{dc}}/\sqrt{2}$, may be synthesized by taking the three adjacent space vectors in proper proportion. Figure 7.21 shows how the desired voltage vector, $\vec{V}^*$, is synthesized from the space vectors $\vec{V}_1$, $\vec{V}_2$ and $\vec{V}_0$. The desired voltage makes an angle γ relative to voltage vector $\vec{V}_1$. Over one sampling interval, the duty ratios of vectors $\vec{V}_1$, $\vec{V}_2$ and $\vec{V}_0$ are, respectively,

$$d_1 = \frac{2/\sqrt{3}\,|V^*|\sin(60° - \gamma)}{\sqrt{2/3}V_{\text{dc}}} \quad , \tag{7.2}$$

$$d_2 = \frac{2/\sqrt{3}\,|V^*|\sin\gamma}{\sqrt{2/3}V_{\text{dc}}} \quad , \tag{7.3}$$

---

[10] See H. W. Van der Broeck, H. C. Skudelny, and G. Stanke, "Analysis and realization of a pulse width modulator based on space vector theory," *IEEE Trans. on Industry Applications*, Vol. IA-24, pp. 142-150, 1988; J. Holtz, P. Lammert, and W. Lotzkat, "High-speed drive system with ultrasonic MOSFET PWM inverter and single-chip microprocessor control," *IEEE/IAS Annual Meeting Conference Record*, pp. 12-17, 1986; and A. M. Trzynadlowski and S. Legowski, "Minimum-loss vector PWM strategy for three-phase inverters," *IEEE Trans. on Power Electronics*, Vol. 9, pp. 26-34, 1994.

7.6. Voltage Waveform Synthesis Techniques

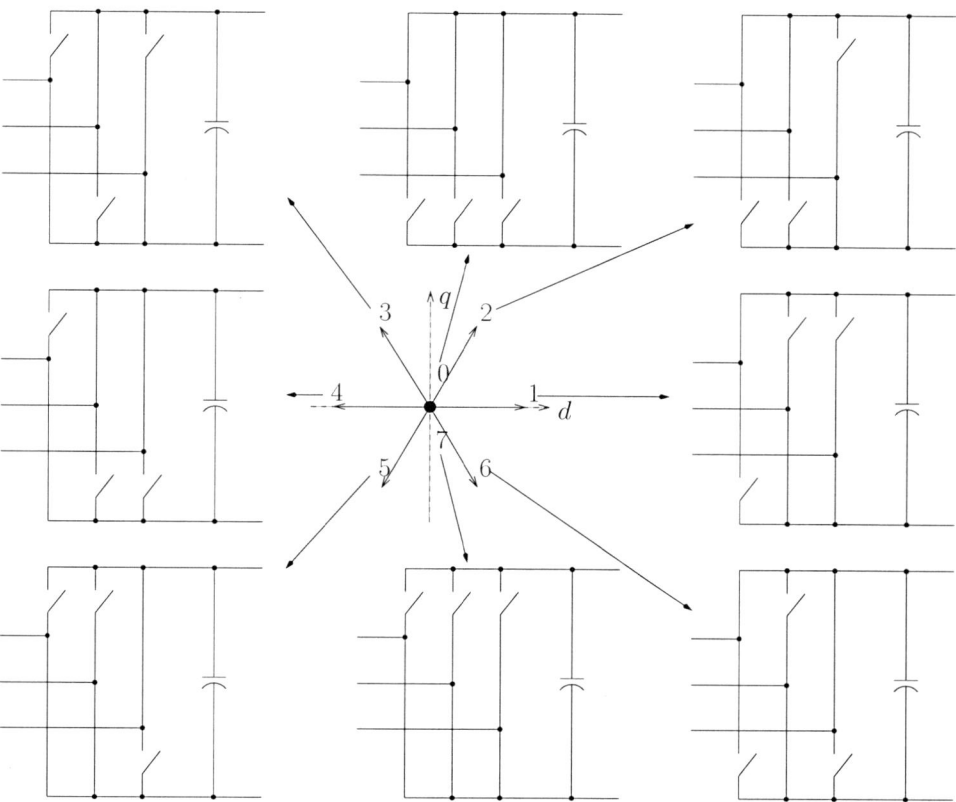

Figure 7.20: The eight space vectors which can be produced by the three-phase inverter of Fig. 7.12.

$$d_0 = 1 - d_1 - d_2 \quad . \tag{7.4}$$

The order used in implementing the space vectors is driven by the desire to minimize the number of switching operations. Careful examination of Figs. 7.20 and 7.21 reveals that within any of the six segments delimited by space vectors, the move from one vector to the next requires changing the state of only one switch. In practice, the switching in adjacent sampling intervals would apply the sequence $\cdots |V_1|V_2|V_0|V_0|V_2|V_1|\cdots$. Different approaches use different criteria for selecting the best implementation of V_0.

Extension of the space-vector concept is possible with multilevel inverters[11]. Multilevel inverters, however, offer a substantially greater number of

---

[11] See D. G. Holmes and B. P. McGrath, "Opportunities for harmonic cancellation with carrier-based PWM for two-level and multilevel cascaded inverters," *IEEE Trans. on Industry Applications*, Vol. 37, pp. 574-582, 2001, and L. M. Tolbert, F. Z. Peng, and T. G.

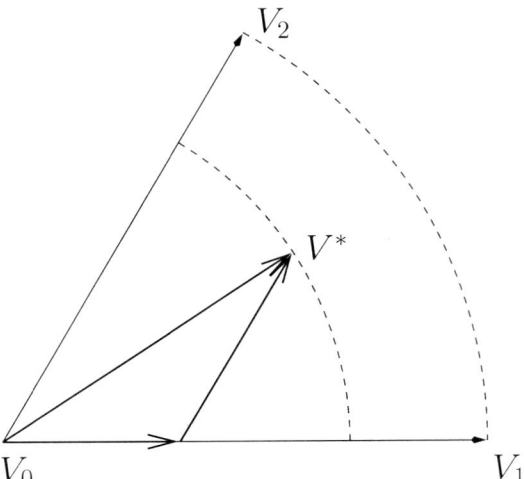

Figure 7.21: The synthesis of voltage V^* using space vector modulation.

space vectors from which to choose. With three-level inverters, for example, there are now nineteen different space vectors. Additional levels would increase the number of space vectors still further. The number of redundant space vectors also increases in multilevel inverters, thereby offering additional degrees of freedom within the voltage synthesis algorithm. Figure 7.22 shows the space vectors that can be created by a three-phase three-level inverter based on the phase leg of Fig. 7.13. The numbers adjacent to each space vector represent the switch configuration of phases a, b and c, respectively. Space vectors with more than one set of numbers can be achieved with any of the switch combinations indicated. Referring to Fig. 7.13, a 0 indicates that switches S_3 and S_4 are connecting the output terminal to the negative side of the dc bus. Similarly, a 1 indicates switches S_2 and S_3 are connecting the output terminal to the midpoint of the dc bus. Finally, a 2 indicates that switches S_1 and S_2 are connecting the load terminal to the positive side of the dc bus.

7.7 Current Waveform Synthesis Techniques

While voltage source inverters always output a voltage, there are many applications where the details of the voltage waveform are driven by the creation of a current with a specific shape. In this context, the controller is de-

---

Habetler, "Multilevel PWM methods at low modulation indices," *IEEE Trans. on Power Electronics*, Vol. 15, pp. 719-725, 2000.

7.7. Current Waveform Synthesis Techniques

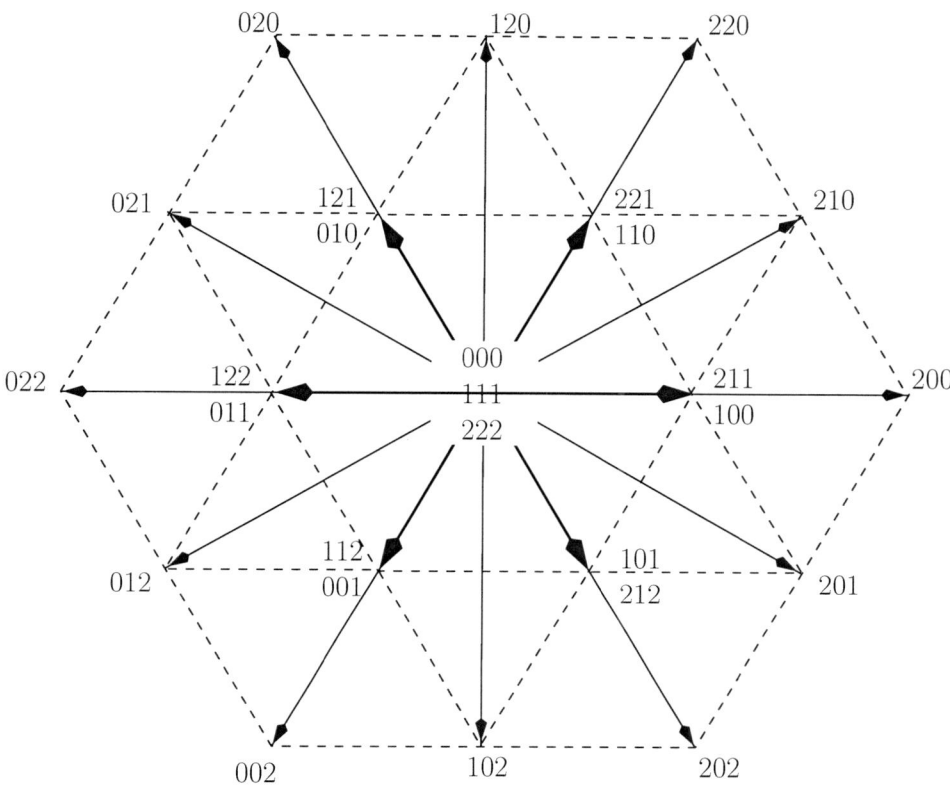

Figure 7.22: The achievable space vectors associated with a three-phase three-level inverter.

termining the state of each inverter switch based on how well the inverter output currents are tracking commanded output currents. While the PWM techniques of the previous subsection can be applied to transform a voltage-source inverter into a controlled current source through feedback control[12], there are some additional techniques that are useful in this type of operation. The control techniques described in this subsection are amenable to synthesizing current waveforms with inverters.

---

[12] See, for example, D. M. Brod and D. W. Novotny, "Current control of VSI-PWM inverters," *IEEE Trans. on Industry Applications*, Vol. IA-21, pp. 562-570, 1985, and T. G. Habetler, "A space vector-based rectifier regulator for ac/dc/ac converters," *IEEE Trans. on Power Electronics*, Vol. 8, pp. 30-36, 1993.

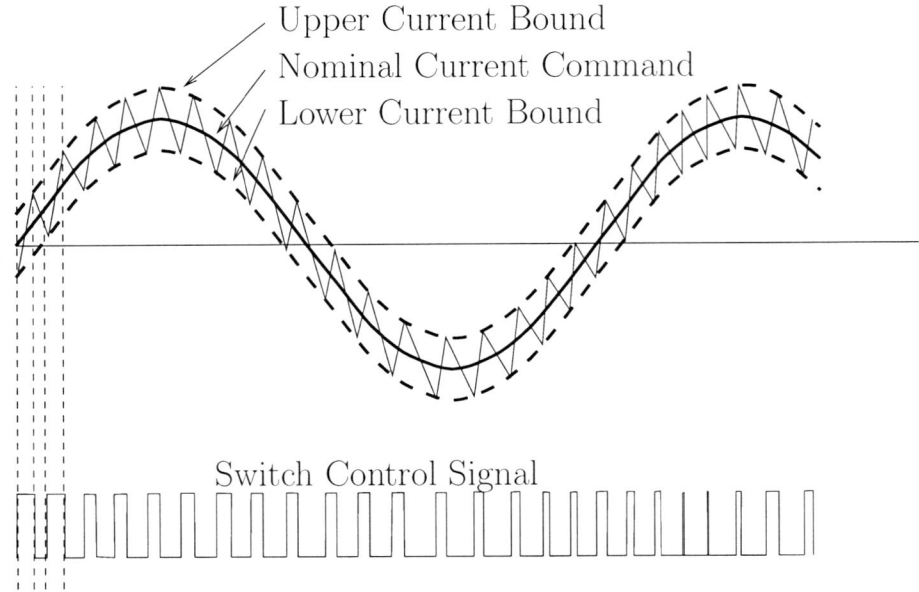

Figure 7.23: The principles of hysteretic current control.

7.7.1 Hysteresis and Sliding-Mode Control

Hysteresis and sliding-mode control are very similar in nature. In both of these control approaches, a reference current waveform is established and the switching of the inverter is tied to location of the actual current relative to the reference waveform[13]. Under hysteretic control, a hysteresis band is introduced around the reference waveform in order to limit the switching frequency. Figure 7.23 illustrates the principles of hysteretic control. Sliding-mode control can be implemented in a manner that is indistinguishable from hysteretic control, or it can be implemented as shown in Fig. 7.24 where there is a known upper limit on the switching frequency.

A common problem with hysteretic control is that the switching frequency is not fixed and may vary widely over one cycle of the output. This can complicate the design of filters and may raise reliability concerns relative to the safe operation of the switches. Fixing the switching frequency is possible[14], at the expense of a hysteresis band that changes width throughout

---

[13] See D. M. Brod and D. W. Novotny, "Current control of VSI-PWM inverters," *IEEE Trans. on Industry Applications*, Vol. IA-21, pp. 562-570, 1985; J. J. Slotine and W. Li, *Applied Nonlinear Control*, Englewood Cliffs, NJ: Prentice-Hall, 1991; or D. A. Torrey and A. M. Al-Zamel, "Single-phase active power filters for multiple nonlinear loads," *IEEE Trans. on Power Electronics*, Vol. 10, pp. 263-272, 1995.

[14] See M. Kazerani, P. D. Ziogas, and G. Joos, "A novel active current waveshaping

7.8. Summary

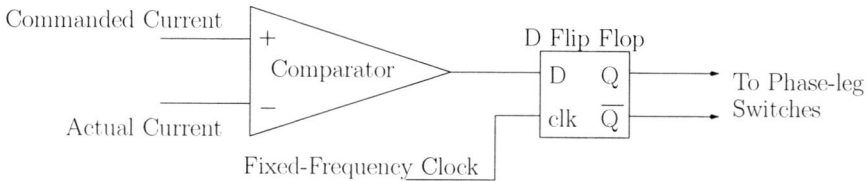

Figure 7.24: One method of implementing sliding-mode control.

each cycle of the output.

7.7.2 Predictive Current Regulation

Predictive current regulation is similar to hysteresis and sliding-mode control in the establishment of a reference current and acceptable error bounds. The predictive controller uses a model of the system in conjunction with measurements of the system state in order to predict how long the next switching state is to be maintained, so that the actual current remains within the established error bounds. In contrast to hysteresis and sliding-mode control, the predictive controller is always looking ahead one sampling interval into the future.

7.8 Summary

The voltage source inverters discussed in this chapter are applied in a very wide range of power levels, from fractional horsepower up to thousands of horsepower, and the upper limit on power continues to increase. These applications range from very precise motion control to adjustable speed operation of pumps and compressors for saving energy. In a machine drive, the inverter is used to provide an adjustable frequency ac voltage to the machine, thereby enabling the machine to operate over a wide range of speeds without derating the torque production of the machine. As we will discuss in Chapter 10, to prevent the machine from being pushed into magnetic saturation, the amplitude of the synthesized voltage is usually tied to the output frequency. In the simplest adjustable speed drives, the ratio of peak output voltage to output frequency is maintained nominally constant, with a boost at low frequencies to compensate for the resistance of the machine windings. More sophisticated adjustable speed drives implement sensorless flux vector control.

---

technique for solid-state input power factor conditioners," *IEEE Trans. on Industrial Electronics*, Vol. 38, pp. 72-78, 1991.

This chapter has provided an overview of voltage source inverters and the basic principles of their control. Chapter 8 is going to expand on the operation of inverters as we look at more of the details of regulating machine phase currents, commonly used to achieve a high level of machine performance.

Chapter 8

Current Regulators

8.1 Introduction

The ability to regulate phase currents is at the heart of precisely controlling the dynamics of electric machines. After all, electric machines are converters of current into torque. It follows that an electric machine can only produce torque to the precision with which the currents within the windings are controlled.

This chapter examines current regulation. Fundamentally, most electric machine drives are based on the voltage source inverters discussed in Chapter 7. Regulation of phase currents is accomplished through closing a feedback loop around a voltage source inverter. Feedback control is used to force the voltage source inverter to behave as a controlled current source.

8.2 The Structure of Current Regulators

A basic current regulator is shown in Fig. 8.1. On the right is a three-phase inverter that interfaces a dc source with a three-phase ac load. (This could be done with a load of any number of phases also, but generally we are dealing with three-phase machine drives.) The inverter is supplied with control signals, one for each controllable switch. Consistent with Fig. 7.12, there are six controllable switches in the inverter, hence six control signals. This could be expanded to support multi-level inverters.

The modulator takes input voltage commands, one for each phase, and converts them into the appropriate commands for the inverter switches. The voltage commands are created by compensators that operate on the current

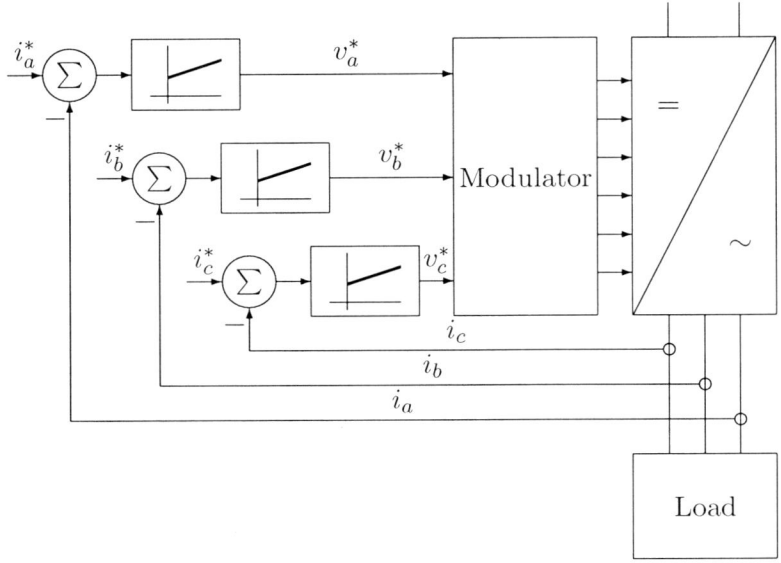

Figure 8.1: A block diagram of a basic current regulator. Commands for each phase are processed individually.

errors, which are the differences between the actual phase currents and the commanded phase currents. In an electric machine drive the commanded currents are frequently the output of a controller that is responsible for regulating speed or torque.

Figure 8.1 shows separate consideration of the three phase currents. It is also possible to work with equivalent two-phase currents using the $abc \rightarrow \alpha\beta$ transformation given by Eq. 6.8. The appropriate approach depends in part on the structure of the modulator. If sinusoidal PWM is used, each phase would naturally be considered separately. If space vector modulation is used, it would be natural to consider equivalent two-phase currents and voltages. Figure 8.2 shows the current regulator structure when working with equivalent two-phase currents and voltages. In this case it is necessary to use T_{23} from Eq. 6.8 to convert the currents measured in the abc reference frame and project them into the $\alpha\beta$ reference frame.

Consider the system shown in Fig. 8.3 that is comprised of a three-phase inverter used to interface a dc voltage source to three ac voltage sources. Inductances are used to support the voltage difference between the inverter output and the ac sources. The inductances could be separate inductors designed to provide adequate filtering of the phase currents, or they could be integral with the voltage sources. If the voltage sources represent, for

8.2. The Structure of Current Regulators

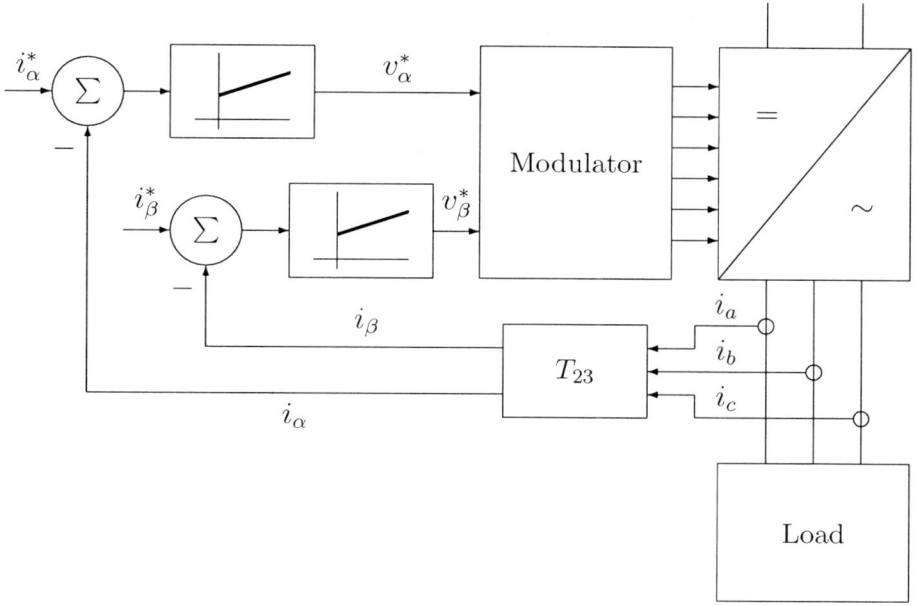

Figure 8.2: A block diagram of a current regulator in which an equivalent two-phase system is used.

example, the motional voltage associated with the phase windings of an electric machine, the inductances would naturally be the phase inductances. The resistances are not desired, but parasitic resistance will exist in the inductances.

Consider the application of the current regulator of Fig. 8.2 to the inverter of Fig. 8.3. We desire to control the line currents such that they are in phase with the load voltages v_{an}, v_{bn}, and v_{cn}. In this way, we can get maximum torque for every Ampere of current fed to the machine, consistent with the discussion in Sec. 4.4. Shaping the phase currents is tantamount to controlling the voltage across each phase inductance since:

$$\frac{d}{dt}\begin{bmatrix} i_a \\ i_b \\ i_c \end{bmatrix} = \frac{1}{L}\left\{\begin{bmatrix} v'_a \\ v'_b \\ v'_c \end{bmatrix} - \begin{bmatrix} v_{an} \\ v_{bn} \\ v_{cn} \end{bmatrix}\right\} - \frac{R}{L}\begin{bmatrix} i_a \\ i_b \\ i_c \end{bmatrix} . \qquad (8.1)$$

Premultiplying each term by T_{23} gives

$$\frac{d}{dt}\begin{bmatrix} i_\alpha \\ i_\beta \end{bmatrix} = \frac{1}{L}\left\{\begin{bmatrix} v'_\alpha \\ v'_\beta \end{bmatrix} - \begin{bmatrix} v_{\alpha n} \\ v_{\beta n} \end{bmatrix}\right\} - \frac{R}{L}\begin{bmatrix} i_\alpha \\ i_\beta \end{bmatrix} . \qquad (8.2)$$

It is the job of the current regulator to generate v^*_α and v^*_β, the voltage commands that will cause the desired phase currents. It is the job of the

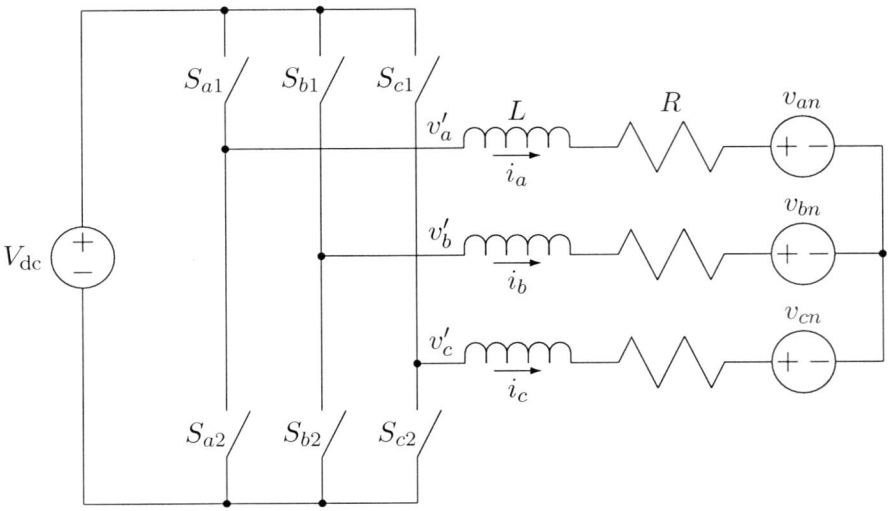

Figure 8.3: A three phase inverter interfaced to three ac sources through phase inductors.

modulator to correctly implement those commanded voltages. The following sections will consider first the current regulator dynamics and then the modulator.

8.3 The Current Regulator Dynamics

Consistent with Fig. 8.2, the input to each compensator is the error between the commanded current and the actual current:

$$e_\alpha = i_\alpha^* - i_\alpha \quad ; \tag{8.3}$$

$$e_\beta = i_\beta^* - i_\beta \quad . \tag{8.4}$$

If we assume each compensator is a simple proportional-integral structure, then the commanded voltages are

$$v_\alpha^* = K_i \int e_\alpha \, dt + K_p e_\alpha \quad ; \tag{8.5}$$

$$v_\beta^* = K_i \int e_\beta \, dt + K_p e_\beta \quad , \tag{8.6}$$

8.3. The Current Regulator Dynamics

where K_i and K_p are the integral and proportional gains, respectively. We see that Eqs. 8.5 and 8.6 introduce two more state equations to the system description. The additional state equations are

$$\frac{dx_\alpha}{dt} = e_\alpha \quad ; \tag{8.7}$$

$$\frac{dx_\beta}{dt} = e_\beta \quad , \tag{8.8}$$

so Eqs. 8.5 and 8.6 become

$$v_\alpha^* = K_i\, x_\alpha + K_p\, e_\alpha \quad , \tag{8.9}$$

and

$$v_\beta^* = K_i\, x_\beta + K_p\, e_\beta \quad , \tag{8.10}$$

respectively.

The state equations that describe the current regulator are:

$$\frac{d}{dt}\begin{bmatrix} i_\alpha \\ i_\beta \\ x_\alpha \\ x_\beta \end{bmatrix} = \begin{bmatrix} \frac{1}{L}\left(v'_\alpha - v_{\alpha n}\right) - \frac{R}{L} i_\alpha \\ \frac{1}{L}\left(v'_\beta - v_{\beta n}\right) - \frac{R}{L} i_\beta \\ \left(i_\alpha^* - i_\alpha\right) \\ \left(i_\beta^* - i_\beta\right) \end{bmatrix} \quad , \tag{8.11}$$

where we use Eqs. 8.9 and 8.10 and the modulator to determine v'_α and v'_β. These state equations can be used to simulate the operation of the current regulator.

Beyond the state equations, we can also get insight into the operation of the current regulator by developing an averaged circuit model under the assumption that the modulator operates as intended[1]. For ideal modulator operation, $\overline{v}'_\alpha = \overline{v}_\alpha^*$ and $\overline{v}'_\beta = \overline{v}_\beta^*$. The averaged circuit model is based on replacing the instantaneous circuit of Fig. 8.3 with a circuit in which instantaneous voltages and currents are replaced by their local average. The local average of a quantity is defined as

$$\overline{x}(t) = \frac{1}{T}\int_{t-T}^{t} x(\tau)d\tau \quad , \tag{8.12}$$

where T is the period of one switching cycle. The running average is sometimes known as a moving average, since it is a function of time. Quantities

---

[1] See, for example, Chapter 11 of Kassakian, Schlecht, and Verghese, *Principles of Power Electronics*, Addison-Wesley, 1991, for a much more detailed description of developing averaged circuit models for power converters.

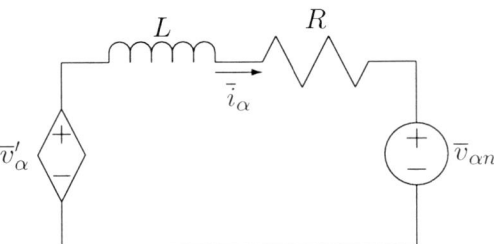

Figure 8.4: An averaged circuit model for the α axis of the inverter-driven load shown in Fig. 8.3.

that are averaged include voltages and currents. The averaging process discards dynamic information at the time scale T, but allows us to construct a dynamic model for the inverter that is useful for control design, particularly when the intended period of the ac load currents is long compared to T. This is typically the case. The averaging process is justified in circuits where the load responds to the moving average value of a quantity rather than the instantaneous value of a quantity. For the inverter of Fig. 8.3, the inductive nature of the load means that the load currents are well described by the inverter output voltages averaged over each switching cycle, without resorting to detailed modeling of the instantaneous inverter output voltages.

Adopting an averaged circuit model for the inverter results in the circuit shown in Fig. 8.4. The dependent voltage source represents the output of the inverter, depending accordingly on the actions of the modulator and the dc bus voltage. The next section deals with the details of the modulator. There is a similar circuit for the β axis.

If we assume a proportional-integral control law for generation of $\bar{v}'_\alpha$, we have

$$\bar{v}'_\alpha = K_i \int \left(\bar{i}^*_\alpha - \bar{i}_\alpha \right) dt + K_p \left(\bar{i}^*_\alpha - \bar{i}_\alpha \right) \quad . \tag{8.13}$$

This representation of $\bar{v}'_\alpha$ can be included in our averaged circuit model as shown in Fig. 8.5. We see that the commanded current acting through the integrator and proportional terms behave as voltage sources that act in opposition to the source voltage $\bar{v}_{an}$.

The integral and proportional terms acting on current $\bar{i}_\alpha$ can be represented in a more physical way than the dependent sources shown in Fig. 8.5. The dependent source involving the integral of the current behaves as a capacitor, in that the voltage is proportional to the integral of the current through it. The equivalent capacitance is $1/K_i$. The dependent source in-

8.3. The Current Regulator Dynamics

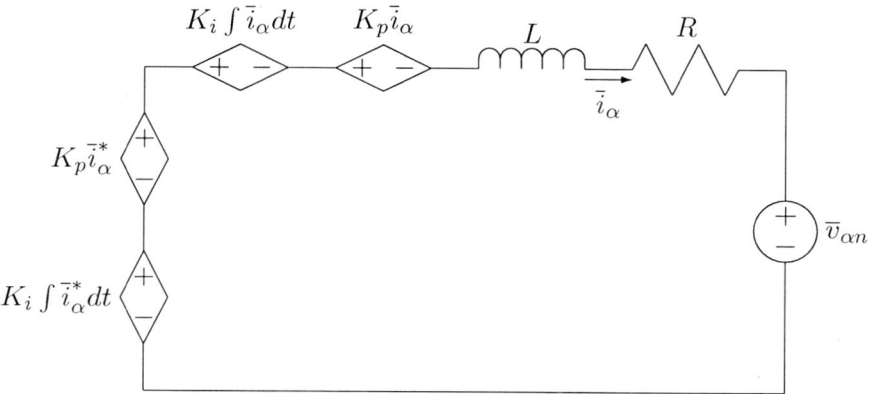

Figure 8.5: The averaged circuit model for the α axis of the inverter-driven load showing the implementation of a proportional-integral control law.

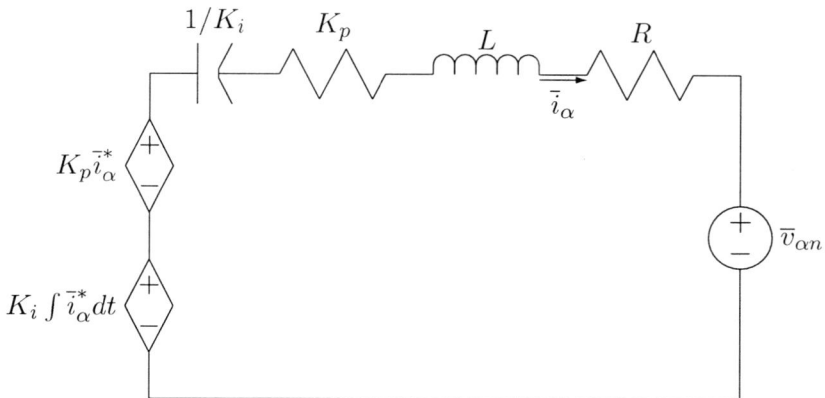

Figure 8.6: The circuit model of Fig. 8.5, replacing two dependent sources with virtual components created by the control law. There is a comparable circuit for the β axis.

volving the current behaves as a resistor since the voltage depends on the current through it. The equivalent resistance is K_p. Substituting these virtual linear elements into the averaged circuit model gives the circuit of Fig. 8.6.

We see that the integral control is responsible for two things. First, the integral of the current command creates a voltage source that is phase shifted from the current $\bar{i}_\alpha$. This phase shift is an effective way of driving current between two ac sources. Second, the integral of the average current behaves as a capacitor. Since the average current through a capacitor in periodic steady state must be zero, the presence of this capacitor shows that integral control results in zero average steady state error.

The linearity of the averaged circuit model of Fig. 8.6 means that the dynamics are governed by the differential equation

$$\frac{d^2\bar{i}_\alpha}{dt^2} + \frac{K_p + R}{L}\frac{d\bar{i}_\alpha}{dt} + \frac{K_i}{L}\bar{i}_\alpha = \frac{K_i}{L}\bar{i}_\alpha^* + \frac{K_p}{L}\frac{d\bar{i}_\alpha^*}{dt} - \frac{1}{L}\frac{d\bar{v}_{an}}{dt} \quad . \tag{8.14}$$

The zeroth order term again shows that the integral term forces zero average error in the inverter output current. Without this term, there could be non-zero steady state error. The characteristic polynomial of the dynamics is

$$s^2 + \frac{K_p + R}{L}s + \frac{K_i}{L} = 0 \quad . \tag{8.15}$$

The roots of the characteristic polynomial are

$$s_{1,2} = -\frac{K_p + R}{2L} \pm \jmath\sqrt{\frac{K_i}{L} - \left(\frac{K_p + R}{2L}\right)^2} \quad , \tag{8.16}$$

assuming an underdamped response. By virtue of the proportional and integral gains in the characteristic polynomial terms, the transient response of the current regulator can be tuned to provide the desired dynamics. There are similar dynamics on the β axis. Employing different gains on the α and β axes allows separate tailoring of the dynamics. However, it is not clear that this would be particularly advantageous, especially since we would usually want symmetric behavior on the α and β axes.

An alternative but equivalent perspective can be developed for the current regulator dynamics. A block diagram describing the averaged circuit dynamics is shown in Fig. 8.7 for the proportional-integral control law of Eq. 8.13. This block diagram shows the generation of the current error, the implementation of the control law, and the conversion of the inverter output voltage into phase current. The role of the modulator is suppressed in this model, since it is assumed that the modulator operates on a much shorter time scale.

The forward loop gain of the dynamics is

$$H(s)G(s) = \frac{K_i + K_p s}{s}\frac{1}{sL + R} = \frac{K_i + K_p s}{s(sL + R)} \quad . \tag{8.17}$$

At crossover the forward loop gain needs to be rolling off at 20 db/decade

8.3. The Current Regulator Dynamics

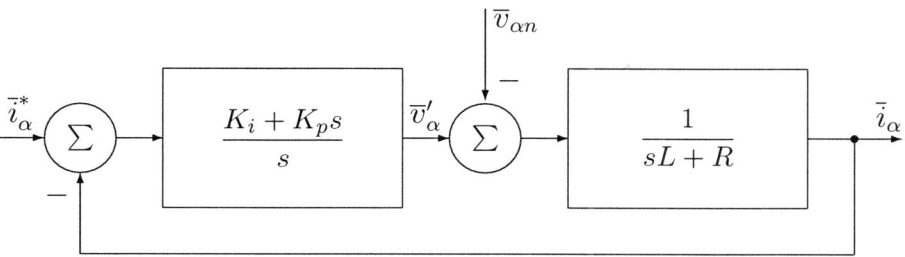

Figure 8.7: A block diagram for the averaged current regulator using proportional-integral control.

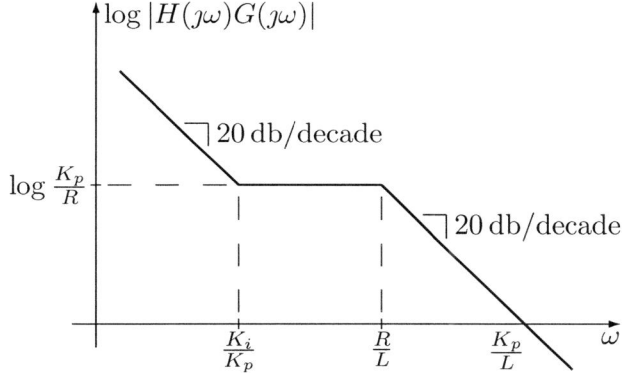

Figure 8.8: The magnitude Bode plot for the current regulator perturbation dynamics when using proportional-integral control.

to provide adequate phase margin[2]. This means that we need the crossover frequency to be above the zero in the numerator and above both poles in the denominator of Eq. 8.17. A qualitative Bode magnitude plot of the forward loop gain is shown in Fig. 8.8 for the case of proportional-integral control. The Bode plot assumes the pole at $-R/L$ occurs at a higher frequency than the zero at $-K_i/K_p$. An alternative would be to place the zero above the pole to accelerate the system toward steady state. It will be appreciated that a more sophisticated compensator will provide additional degrees of freedom in shaping the current regulator dynamics.

By definition, crossover takes place when $|H(\jmath\omega_c)G(\jmath\omega_c)| = 1$, where ω_c

---

[2] See, for example, Chapter 14 of Kassakian, Schlecht, and Verghese, *Principles of Power Electronics*, Addison-Wesley, 1991, for a more complete discussion of why the loop gain must roll off at 20 db/decade.

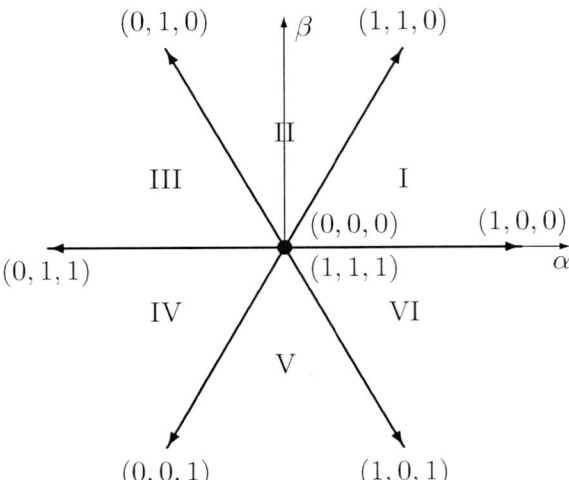

Figure 8.9: The space vectors associated with the inverter of Fig. 8.3. The numbers in parentheses indicate the switching functions used to implement the vector.

is the crossover frequency. If crossover occurs sufficiently far above the pole at $-R/L$, we have

$$|H(\jmath\omega_c)G(\jmath\omega_c)| \approx \frac{K_i}{R}\frac{\frac{K_p}{K_i}\omega_c}{\frac{L}{R}\omega_c^2} = 1 \quad \Rightarrow \quad \omega_c \approx \frac{K_p}{L} \quad . \tag{8.18}$$

8.4 The Modulator

This section considers the implementation of the modulator that is used within the current regulator. Our development assumes space-vector modulation. Modulators based on techniques such as sinusoidal pulse-width modulation can be developed similarly. Figure 8.9 shows the space vectors that can be created by the two-level inverter of Fig. 8.3. (The inverter of Fig. 8.3 is a two level inverter because each phase output can be connected to either the positive side of the dc bus, or the negative side of the dc bus. That is, there are only two possible positions in each phase leg. As discussed in Sec. 7.5, multilevel inverters offer many more possibilities through the use of more switches and subdivision of the dc bus.)

The vector space shown in Fig. 8.9 is divided into six segments of identical size. The vertices of each segment correspond to space vectors. It is possible

8.4. The Modulator

Table 8.1: The switching functions for the eight space vectors created by all valid combinations of switch states for a two-level inverter. The switching functions give the switch states for phases a, b, and c, respectively, where a 1 indicates the upper switch in the phase leg is conducting; a 0 indicates the lower switch in the phase leg is conducting.

Space Vector	Switching Function
$\vec{V_0}$	(0,0,0)
$\vec{V_1}$	(1,0,0)
$\vec{V_2}$	(1,1,0)
$\vec{V_3}$	(0,1,0)
$\vec{V_4}$	(0,1,1)
$\vec{V_5}$	(0,0,1)
$\vec{V_6}$	(1,0,1)
$\vec{V_7}$	(1,1,1)

to move from any vector to any other adjacent vector by changing the state of the switches on only one phase leg. We will exploit this in implementing space vector modulation to minimize the number of switching events for maximum efficiency. Table 8.1 gives the switching functions for the eight space vectors. Vectors $\vec{V_0}$ and $\vec{V_7}$ are electrically equivalent.

Figure 8.10 shows one segment of the vector space, defined by space vectors $\vec{V_a}$, $\vec{V_b}$, and $\vec{V_z}$. While the sector shown in Fig. 8.10 looks like segment I in Fig. 8.9, it is intended to be a generic segment. The connection between the specific segments of Fig. 8.9 and the generic segment of Fig. 8.10 is rotation through the appropriate angle; this will be established shortly. The desired voltage is $\vec{V}^*$, which makes an angle γ' relative to vector $\vec{V_a}$. The essence of space vector modulation is to implement a properly time-weighted combination of vectors $\vec{V_a}$, $\vec{V_b}$, and $\vec{V_z}$ over each switching cycle T so as to produce the same Volt-seconds as vector $\vec{V}^*$ implemented over the same interval T.

To begin we need the magnitude of the voltage vectors. Consider the vector $\vec{V_1}$ defined by the switching functions (1,0,0) in Fig. 8.9. In this state it can be shown that $v'_a = 2V_{dc}/3$ and $v'_b = v'_c = -V_{dc}/3$ by requiring that the inverter output voltages form a balanced set of voltages that sum to zero. Applying T_{23} from Eq. 6.8 gives $v'_\alpha = \sqrt{2/3}V_{dc}$ and $v'_\beta = 0$. It can be similarly shown that all non-zero space vectors have a magnitude of

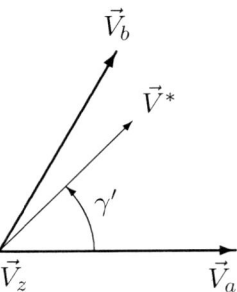

Figure 8.10: One segment of the vector space given in Fig. 8.9. The vector $\vec{V}^*$ is the desired voltage to be synthesized over switching period T.

$\sqrt{2/3}V_{\text{dc}}$ but with different angles.

We associate duty ratio d_a with $\vec{V}_a$, d_b with $\vec{V}_b$, and d_z with $\vec{V}_z$. We require

$$d_a + d_b + d_z = 1 \quad, \tag{8.19}$$

so that the entire switching cycle is covered. We see that vector $\vec{V}_b$ is responsible for contributing all of the vertical axis component of $\vec{V}^*$. Vector $\vec{V}_a$ is then used to add additional horizontal voltage as required. Vector $\vec{V}_z$ is used to complete the switching period. Considering the vertical and horizontal components individually we have

$$\text{Vertical}: \quad \left|\vec{V}^*\right|\sin\gamma' = \sqrt{\frac{2}{3}}V_{\text{dc}} d_b \sin 60° \quad ; \tag{8.20}$$

$$\text{Horizontal}: \quad \left|\vec{V}^*\right|\cos\gamma' = \sqrt{\frac{2}{3}}V_{\text{dc}}\left(d_a + d_b \cos 60°\right) \quad . \tag{8.21}$$

Self-consistently solving for d_a, d_b, and d_z gives the results stated in Eqs. 7.2, 7.3, and 7.4, respectively:

$$d_a = \frac{2/\sqrt{3}\left|\vec{V}^*\right|\sin(60° - \gamma')}{\sqrt{2/3}V_{\text{dc}}} = \sqrt{2}\frac{\left|\vec{V}^*\right|}{V_{\text{dc}}}\sin(60° - \gamma') \quad, \tag{8.22}$$

$$d_b = \frac{2/\sqrt{3}\left|\vec{V}^*\right|\sin\gamma'}{\sqrt{2/3}V_{\text{dc}}} = \sqrt{2}\frac{\left|\vec{V}^*\right|}{V_{\text{dc}}}\sin\gamma' \quad, \tag{8.23}$$

$$d_z = 1 - d_a - d_b \quad . \tag{8.24}$$

Note that the duty ratios d_a and d_b trade off as the desired voltage $\vec{V}^*$ rotates from $\vec{V}_a$ to $\vec{V}_b$. The maximum voltage that can be produced in this manner

8.4. The Modulator

corresponds to $\left|\vec{V}^*\right| = V_{\mathrm{dc}}/\sqrt{2}$. This is the radius of a circle inscribed within the hexagon formed by the non-zero voltage vectors.

Synthesizing voltages with higher fundamental magnitude is possible, but at the expense of introducing low-order harmonic content. Operation in this region is known as overmodulation. Ultimately, maximum fundamental voltage output is obtained when the implemented vector travels sequentially from one non-zero space vector to the next. Each cycle of the output voltages consists of six steps, each step moving from one non-zero space vector to the next. This mode is known as six-step operation. The next section considers overmodulation and six-step operation.

The logical choice of zero vector alternates with the non-zero space vectors. This is seen in Fig. 8.9 where the logical choice of zero vector depends on the non-zero vector active prior to moving to the zero vector. This suggests that if the implementation of the space vectors takes place in a circular fashion,

$$\cdots |\vec{V}_a|\vec{V}_b|\vec{V}_z|\vec{V}_a|\vec{V}_b|\vec{V}_z|\vec{V}_a|\vec{V}_b|\vec{V}_z| \cdots \quad ,$$

then there will be one vector transition that requires changing the state of switches on more than one phase leg. However, if the implementation of the vectors alternates as in

$$\cdots |\vec{V}_a|\vec{V}_b|\vec{V}_z|\vec{V}_z|\vec{V}_b|\vec{V}_a|\vec{V}_a|\vec{V}_b|\vec{V}_z| \cdots \quad ,$$

then we will minimize the number of switch transitions. This approach is known as minimum-loss space vector modulation[3]. It follows that minimum-loss space vector modulation selects the zero vector consistent with $\vec{V}_b$ in each sector. Table 8.2 summarizes the selection of vectors $\vec{V}_a$, $\vec{V}_b$, and $\vec{V}_z$ for each sector in the vector space.

Digital implementation of space vector modulation would proceed like this for each switching period T:

1. Determine the sector in which $\vec{V}^*$ sits. This can be accomplished by determining

$$\vec{V}^* = \left|\vec{V}^*\right| \angle \gamma = \sqrt{v_\alpha^{*2} + v_\beta^{*2}} \angle \tan^{-1} \frac{v_\beta^*}{v_\alpha^*} \quad , \tag{8.25}$$

from which

$$\text{Sector} = 1 + \text{fix}(\gamma, 60°) \quad , \tag{8.26}$$

where $\text{fix}(x, y)$ computes x/y and rounds toward zero.

---

[3]See A. M. Trzynadlowski and S. Legowski, "Minimum-loss vector PWM strategy for three-phase inverters," *IEEE Trans. on Power Electronics*, Vol. 9, pp. 26-34, 1994.

Table 8.2: Identification of vectors $\vec{V}_a$, $\vec{V}_b$, and $\vec{V}_z$ for each sector in the vector space of Fig. 8.9. The switching functions for the zero vectors are $\vec{V}_0 = (0,0,0)$ and $\vec{V}_7 = (1,1,1)$.

Sector	$\vec{V}_a$	$\vec{V}_b$	$\vec{V}_z$
I	$\vec{V}_1$	$\vec{V}_2$	$\vec{V}_7$
II	$\vec{V}_2$	$\vec{V}_3$	$\vec{V}_0$
III	$\vec{V}_3$	$\vec{V}_4$	$\vec{V}_7$
IV	$\vec{V}_4$	$\vec{V}_5$	$\vec{V}_0$
V	$\vec{V}_5$	$\vec{V}_6$	$\vec{V}_7$
VI	$\vec{V}_6$	$\vec{V}_1$	$\vec{V}_0$

2. Select $\vec{V}_a$, $\vec{V}_b$, and $\vec{V}_z$ based on the sector using Table 8.2.

3. Determine the angle γ' within the sector using

$$\gamma' = \mathrm{mod}\,(\gamma, 60°) \quad , \tag{8.27}$$

where $\mathrm{mod}(x,y)$ is the modulus function and returns $x - \mathrm{fix}(x,y) \cdot y$. This is the angle that γ extends into the active sector.

4. Compute d_a, d_b, and d_0 according to Eqs. 8.22, 8.23, and 8.24, respectively.

5. If the previous switching period is finishing with $\vec{V}_z$, the order of implementation is $\vec{V}_z$, $\vec{V}_b$, and $\vec{V}_a$. Otherwise, the vectors are implemented in order of $\vec{V}_a$, $\vec{V}_b$, and $\vec{V}_z$. This ensures that a minimum number of switching transitions are used to create the desired $\vec{V}^*$.

6. Implement the space vectors in proper sequence for the amount of time prescribed by the corresponding duty ratio.

Figure 8.11 shows the duty ratios required to implement space vector modulation where the commanded voltage magnitude is within the linear modulation region. Figure 8.12 shows the commanded voltage trajectory for the duty ratios of Fig. 8.11.

8.5 Overmodulation

Overmodulation occurs when $\left|\vec{V}^*\right| > V_{\mathrm{dc}}/\sqrt{2}$. This region of operation may be encountered, for example, when there is sustained error between the ac-

8.5. Overmodulation

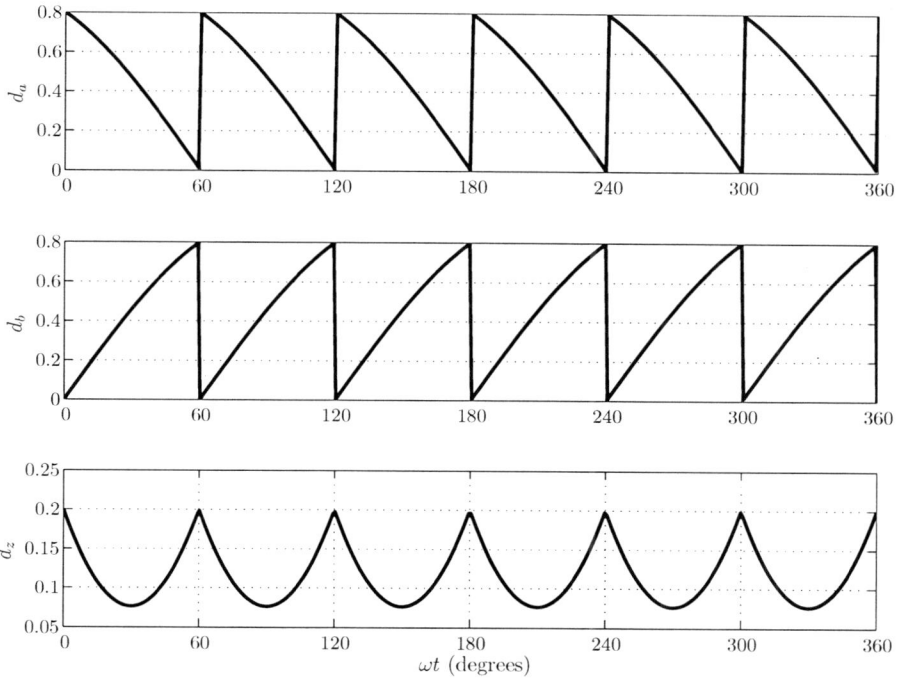

Figure 8.11: The duty ratios for the space vectors to support a commanded magnitude of 80% of $\sqrt{2/3}V_{\text{dc}}$ for $V_{\text{dc}} = 100$ V.

tual and commanded currents and the integral term within the control builds up to a high value. This can occur during start-up, or if there is a significant change in the operating point.

In overmodulation it is no longer possible for the inverter to synthesize voltages that sweep out a circular locus. The situation is depicted in Fig. 8.13, where the desired locus of $\left|\vec{V}^*\right|$ extends beyond the hexagon that bounds the available inverter output. Initially this occurs in the middle of the sector, but as $\left|\vec{V}^*\right|$ grows it pushes out toward the vertices of the hexagonal vector space. Overmodulation is usually considered to have two modes[4]. In the first mode, circular arcs in output voltage are followed when possible. The hexagon is followed when it is not possible. In the second mode, operation is beginning to transition toward six-step operation. In six-step operation the switching frequency essentially drops to one-sixth of the out-

---

[4] Additional discussion on overmodulation can be found, for example, in Chapter 4 of B. K. Bose, ed., *Power Electronics and Variable Frequency Drives*, IEEE Press, 1997, and Chapter 8 of D. G. Holmes and T. A. Lipo, *Pulse Width Modulation for Power Converters*, IEEE Press, 2003.

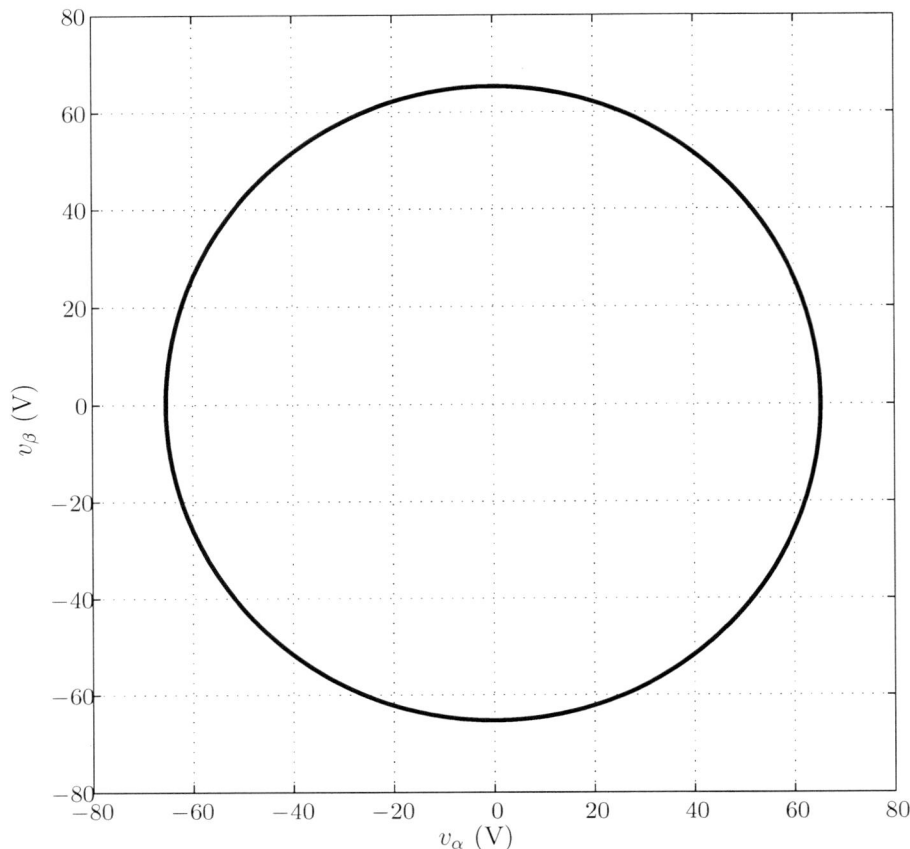

Figure 8.12: The commanded voltage trajectory for the duty ratios of Fig. 8.11. The commanded magnitude is 80% of $\sqrt{2/3}V_{\text{dc}}$ for $V_{\text{dc}} = 100\,\text{V}$.

put frequency, with the output voltage sitting at each non-zero space vector for one-sixth of the cycle.

In Fig. 8.13, the inverter is able to support the desired circular locus of $\vec{V}^*$ over the segment between points a and b and again over the segment between points c and d. Between points b and c, the best that can be accomplished is to follow the hexagonal boundary of the vector space. In this generic sector, the line describing the hexagon boundary is given by

$$v_v = -\sqrt{3}v_h + \sqrt{2}V_{\text{dc}} \quad , \tag{8.28}$$

where v_v and v_h are the vertical and horizontal components of voltage, re-

8.5. Overmodulation

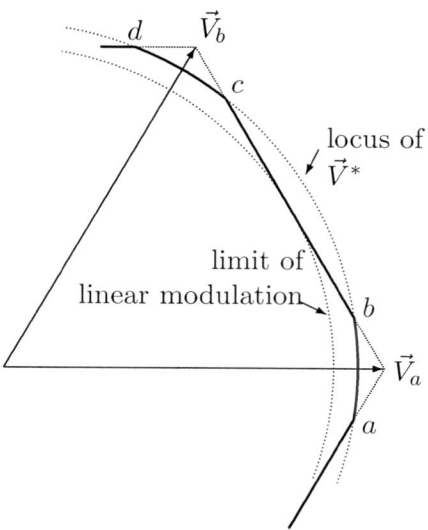

Figure 8.13: A depiction of operation in the overmodulation region, which occurs when $|\vec{V}^*| > V_{\text{dc}}/\sqrt{2}$.

spectively. To implement the correct angle of $\vec{V}^*$ we require

$$\frac{v_v}{v_h} = \tan \gamma' \quad . \tag{8.29}$$

Self-consistently satisfying Eqs. 8.28 and 8.29 gives

$$v_v = \frac{\sqrt{2} V_{\text{dc}} \sin \gamma'}{\sqrt{3} \cos \gamma' + \sin \gamma'} \quad . \tag{8.30}$$

We also know that during our switching cycle

$$v_v = \sqrt{\frac{2}{3}} V_{\text{dc}} d_b \sin 60° \quad . \tag{8.31}$$

It follows that

$$d_b = \frac{2 \sin \gamma'}{\sqrt{3} \cos \gamma' + \sin \gamma'} \quad . \tag{8.32}$$

Since $d_a + d_b = 1$ along the hexagon,

$$d_a = \frac{\sqrt{3} \cos \gamma' - \sin \gamma'}{\sqrt{3} \cos \gamma' + \sin \gamma'} \quad . \tag{8.33}$$

We know that we need to turn to Eqs. 8.32 and 8.33 when $d_z < 0$ from Eq. 8.24.

When the implemented voltage vectors fall on the perimeter of the hexagon, the commanded values of the voltages should be adjusted to reflect the reduced amplitude of the available voltage. On the hexagon we have

$$v_v = \frac{V_{\text{dc}}}{\sqrt{2}} d_b \quad, \tag{8.34}$$

and

$$v_h = v_v / \tan \gamma' \quad. \tag{8.35}$$

The commanded values of v_α and v_β then become

$$v^*_{\alpha_{\text{new}}} = \frac{\sqrt{v_h^2 + v_v^2}}{\sqrt{v_\alpha^2 + v_\beta^2}} v_\alpha \quad, \tag{8.36}$$

and

$$v^*_{\beta_{\text{new}}} = \frac{\sqrt{v_h^2 + v_v^2}}{\sqrt{v_\alpha^2 + v_\beta^2}} v_\beta \quad, \tag{8.37}$$

respectively, consistent with the implementation. The new values of v^*_α and v^*_β respect the angle of $\vec{V}^*$ but adjust the magnitude.

Figure 8.14 shows the duty ratios required to implement space vector modulation where the commanded voltage magnitude is partially within the linear modulation region and partially in the overmodulation region. That is, $\left|\vec{V}^*\right| > V_{\text{dc}}/\sqrt{2}$. Figure 8.15 shows the commanded voltage trajectory for the duty ratios of Fig. 8.14. It can be observed that the trajectory follows the desired circular locus near the vertices of the hexagon, but is limited to follow the hexagon perimeter in the region away from the vertices.

As the commanded voltage $\vec{V}^*$ continues to grow, it is usual practice to smoothly transition into six-step operation. Consider V_2 to be $\left|\vec{V}^*\right|$ when the transition toward six-step operation begins. Consider V_6 to be $\left|\vec{V}^*\right|$ when the transition toward six-step operation is complete. We have $V_{\text{dc}}/\sqrt{2} < V_2 < V_6 < \sqrt{2/3} V_{\text{dc}}$. (Be careful not to confuse V_2 and V_6 with space vectors $\vec{V}_2$ and $\vec{V}_6$.)

The effect of this transition into six-step operation is to dwell for successively longer intervals at each vertex of the hexagon. By the time $\left|\vec{V}^*\right| = V_6$, we only spend time at the vertices of the hexagonal vector space. We can accomplish this by requiring

$$d_a = 1 \quad \text{if} \quad \gamma' \leq \frac{\pi}{6} \left(\frac{\left|\vec{V}^*\right| - V_2}{V_6 - V_2} \right) \quad, \tag{8.38}$$

8.5. Overmodulation

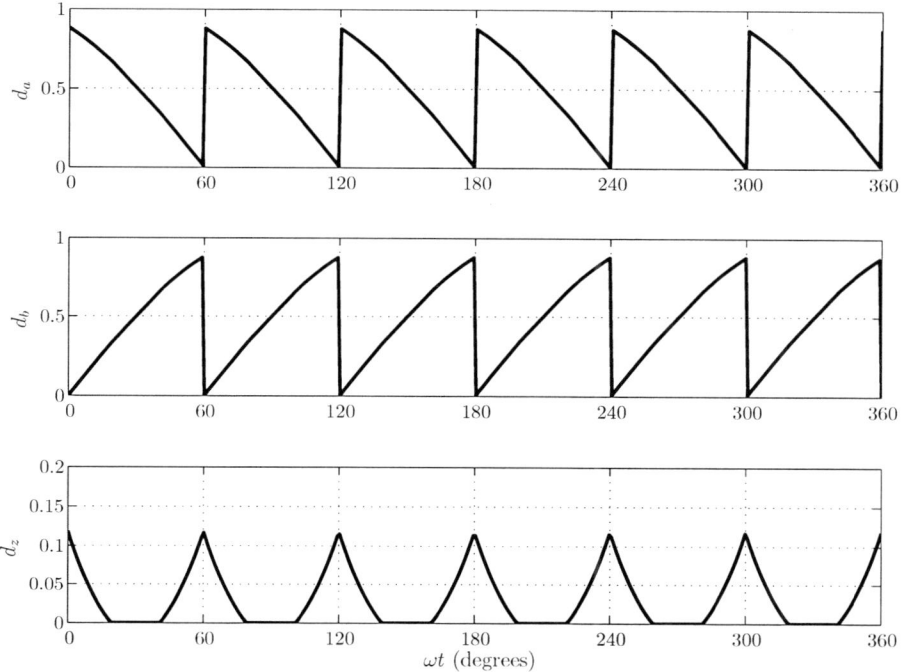

Figure 8.14: The duty ratios for the space vectors to support a commanded magnitude of 72 V for $V_{\text{dc}} = 100$ V.

and

$$d_b = 1 \quad \text{if} \quad \gamma' \geq \frac{\pi}{6} \left(\frac{2V_6 - \left|\vec{V}^*\right| - V_2}{V_6 - V_2} \right). \tag{8.39}$$

Equations 8.38 and 8.39 force $d_a = 1$ for the first half of the segment and $d_b = 1$ for the second half of the segment, respectively, for $\left|\vec{V}^*\right| = V_6$. In this mode of overmodulation, the switching frequency is effectively dropping since we are starting to string together multiple switching cycles where the same voltage vector is being implemented. In addition, dwelling at the vertices for an increasing amount of time will effectively reduce the transition time along the outside of the hexagon.

For any switching cycle in which $d_a = 1$, the implemented values of v_α and v_β are

$$v_{\alpha_{\text{implemented}}} = \sqrt{\frac{2}{3}} \cos\left[(\text{Segment} - 1) \times \frac{\pi}{3}\right], \tag{8.40}$$

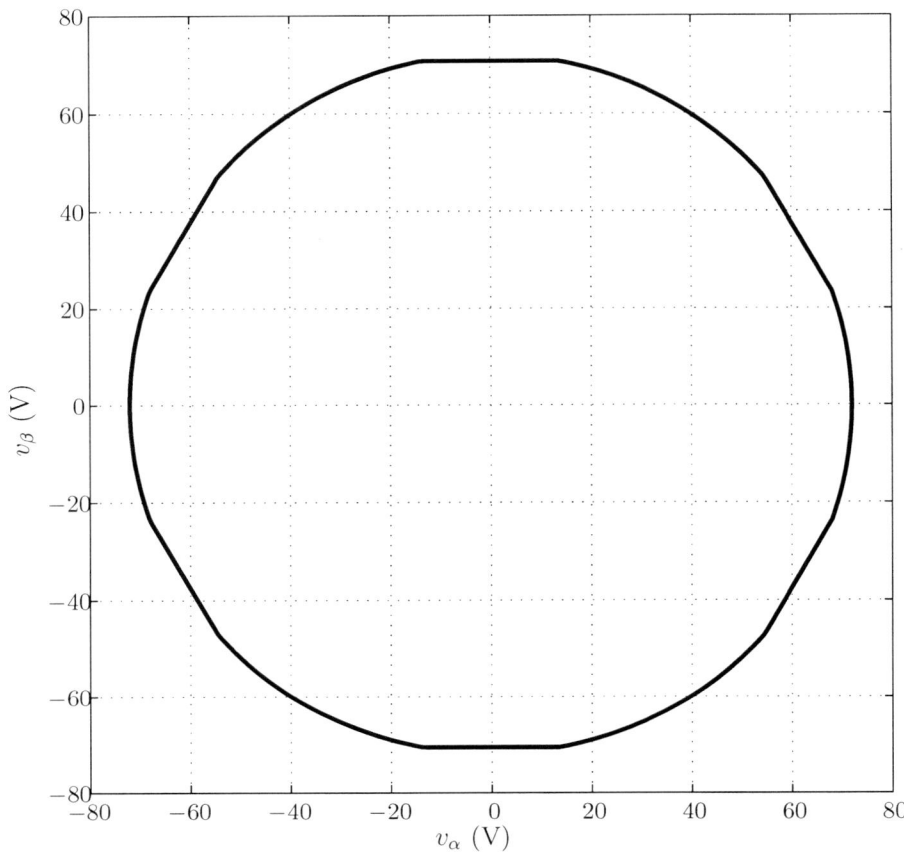

Figure 8.15: The commanded voltage trajectory for the duty ratios of Fig. 8.14. The commanded magnitude is 72 V for $V_{\text{dc}} = 100$ V.

and

$$v_{\beta_{\text{implemented}}} = \sqrt{\frac{2}{3}} \sin\left[(\text{Segment} - 1) \times \frac{\pi}{3}\right] \quad . \tag{8.41}$$

For any switching cycle in which $d_b = 1$, the implemented values of v_α and v_β become

$$v_{\alpha_{\text{implemented}}} = \sqrt{\frac{2}{3}} \cos\left[\text{Segment} \times \frac{\pi}{3}\right] \quad , \tag{8.42}$$

and

$$v_{\beta_{\text{implemented}}} = \sqrt{\frac{2}{3}} \sin\left[\text{Segment} \times \frac{\pi}{3}\right] \quad . \tag{8.43}$$

These modifications to v_α and v_β do not respect the commanded angle of $\vec{V}^*$. For purposes of modifying the commanded voltage to respect the limitations

8.5. Overmodulation

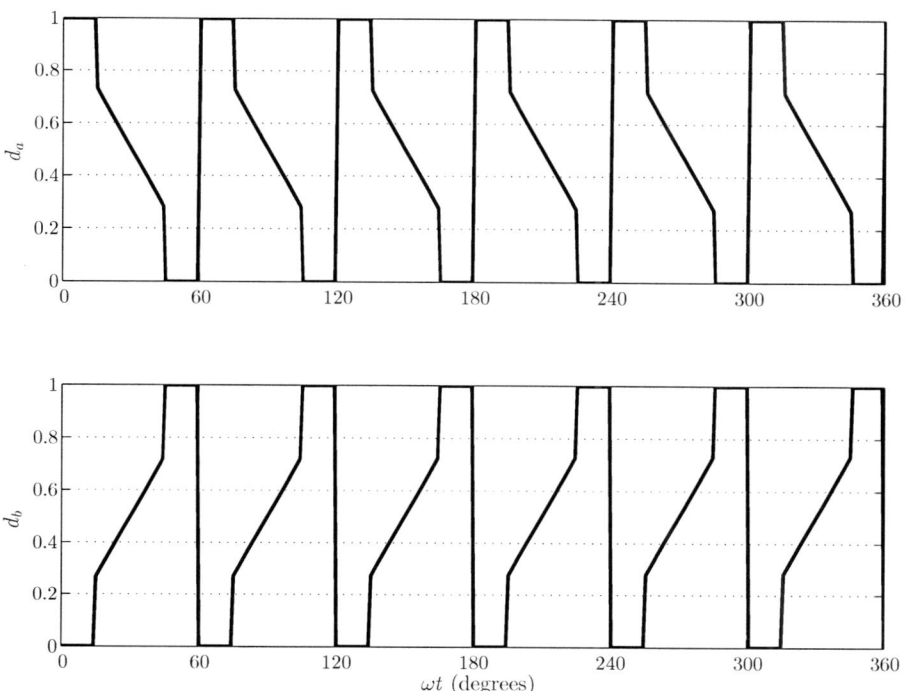

Figure 8.16: The duty ratios for the space vectors to support a commanded magnitude of 75.5 V for $V_{dc} = 100$ V; $V_2 = 73$ V and $V_6 = 78$ V.

being imposed by the modulator, we can again appeal to Eqs. 8.34 through 8.37 even though the implemented voltages may be substantially different. Preservation of the commanded angle of $\vec{V}^*$ will be valuable as the modulator comes out of saturation.

Figure 8.16 shows the duty ratios required to implement space vector modulation where the commanded voltage magnitude is $V_2 < \left|\vec{V}^*\right| < V_6$, so the modulator is transitioning into six-step operation. Figure 8.17 shows the commanded voltage trajectory for the duty ratios of Fig. 8.16. The gaps in the voltage trajectory are because of the dwell time at each hexagon vertex. For a portion of each segment the voltage trajectory follows the perimeter of the hexagon.

Figure 8.18 shows the duty ratios required to implement space vector modulation where the commanded voltage magnitude is $\left|\vec{V}^*\right| > V_6$, so the modulator has completely transitioned into six-step operation. In six-step operation, the duty ratios d_a and d_b are complementary and take on values of either one or zero. Figure 8.19 shows the commanded voltage trajectory

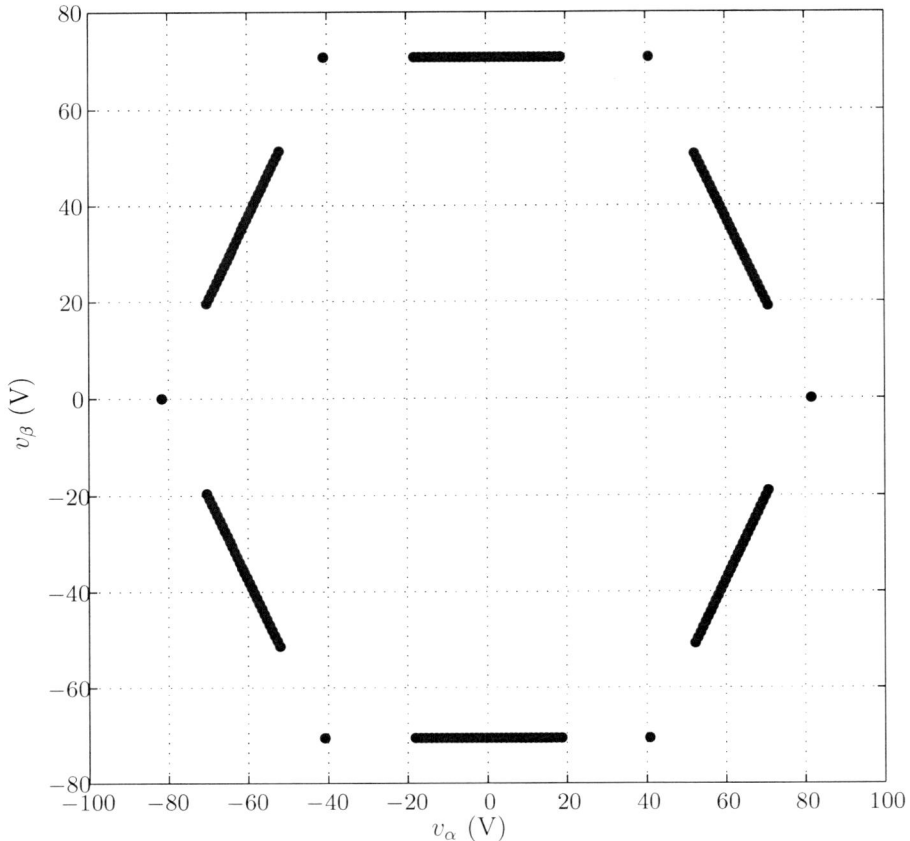

Figure 8.17: The commanded voltage trajectory for the duty ratios of Fig. 8.16. The commanded magnitude is 75.5 V for $V_{dc} = 100$ V; $V_2 = 73$ V and $V_6 = 78$ V.

for the duty ratios of Fig. 8.18. The only implemented space vectors are at the vertices of the hexagon.

When implementing the approach to overmodulation described here it is important to preserve the commanded angle for the desired voltage, even if the required voltage ends up being implemented with six-step operation. Preserving the commanded angle is needed to help bring the modulator out of saturation as soon as possible, and with the correct angle implementation as it comes out of saturation. This requires carefully distinguishing between commanded voltages and implemented voltages.

8.6. An Example Current Regulator

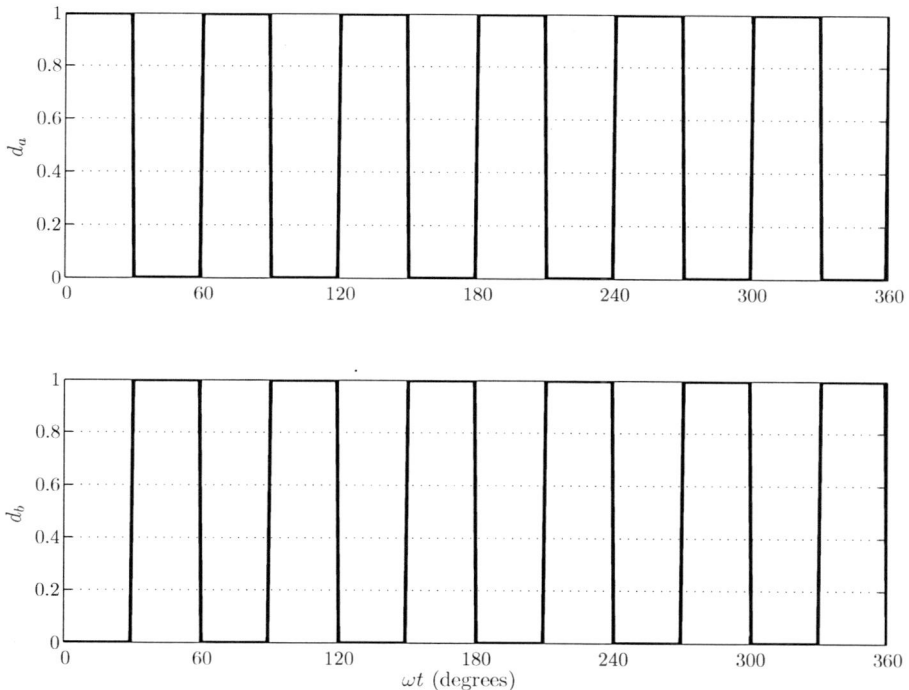

Figure 8.18: The duty ratios for the space vectors to support a commanded magnitude of 80 V for $V_{dc} = 100$ V; $V_6 = 78$ V.

8.6 An Example Current Regulator

Consider application of space-vector modulation in a current regulator as shown in Fig. 8.2. The objective of the current regulator is to force the phase currents to be in phase and of the same shape as the ac voltage sources. That is, we have

$$i_\alpha^* = \mu v_{\alpha n} \quad , \tag{8.44}$$

and

$$i_\beta^* = \mu v_{\beta n} \quad . \tag{8.45}$$

The proportionality factor μ sets the amplitude of the currents, and may be set by, for example, a voltage control loop trying to regulate the amplitude of V_{dc} in an application where dc energy is to be delivered to a fixed frequency utility. In another application, μ might be set by a speed control loop within a motor drive such that μ sets the shaft power output by the motor. In the motor drive application, the frequency ω of the voltage sources is tied to the rotational speed of the motor.

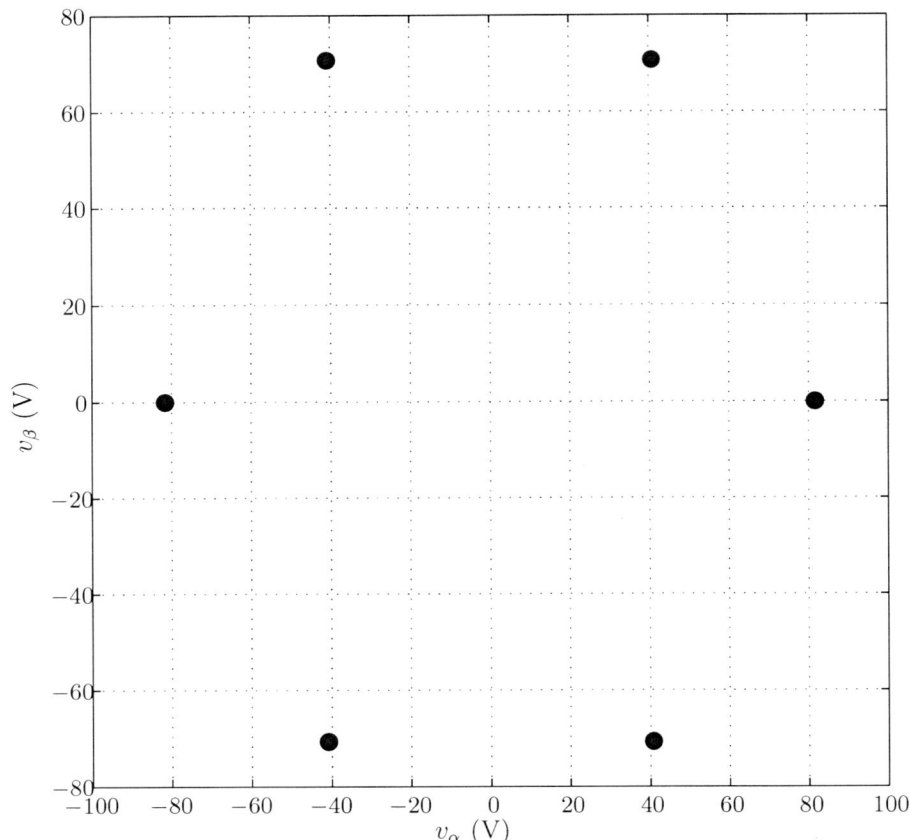

Figure 8.19: The commanded voltage trajectory for the duty ratios of Fig. 8.18. The commanded magnitude is 80 V for $V_{\text{dc}} = 100$ V; $V_6 = 78$ V.

For the balanced set of three phase voltages

$$v_{an} = V_s \sin(\omega t) \quad , \tag{8.46}$$

$$v_{bn} = V_s \sin\left(\omega t - \frac{2\pi}{3}\right) \quad , \tag{8.47}$$

and

$$v_{cn} = V_s \sin\left(\omega t + \frac{2\pi}{3}\right) \quad , \tag{8.48}$$

application of Eq. 6.8 gives

$$v_{\alpha n} = \sqrt{\frac{3}{2}} V_s \sin(\omega t) \quad , \tag{8.49}$$

8.6. An Example Current Regulator

Table 8.3: The parameters used in the sample current regulator.

Parameter	Value	Units
L	5	mH
R	0.1	Ω
V_{dc}	700	V
$|v_{an}|$	375.6	V
f_s	15	kHz
K_i	44,410	V/A·s
K_p	47.12	V/A
K_s	47.12	A/V
V_2	510.3	V
V_6	556.2	V
μ	0.0666	A/V

and
$$v_{\beta n} = -\sqrt{\frac{3}{2}} V_s \cos(\omega t) \quad . \tag{8.50}$$

Our state equations are given in Eq. 8.11:

$$\frac{d}{dt}\begin{bmatrix} i_\alpha \\ i_\beta \\ x_\alpha \\ x_\beta \end{bmatrix} = \begin{bmatrix} \frac{1}{L}(v'_\alpha - v_{\alpha n}) - \frac{R}{L} i_\alpha \\ \frac{1}{L}(v'_\beta - v_{\beta n}) - \frac{R}{L} i_\beta \\ (i^*_\alpha - i_\alpha) \\ (i^*_\beta - i_\beta) \end{bmatrix} \quad . \tag{8.51}$$

In addition, we use Eqs. 8.9 and 8.10 to provide the commanded inverter voltages:
$$v'^*_\alpha = K_i x_\alpha + K_p (i^*_\alpha - i_\alpha) \quad , \tag{8.52}$$
and
$$v'^*_\beta = K_i x_\beta + K_p (i^*_\beta - i_\beta) \quad , \tag{8.53}$$

respectively. These state equations form the core of the current regulator simulation. Voltages v'_α and v'_β are created by the space vector modulator using state variables v'^*_α and v'^*_β consistent with the discussion given in Secs. 8.4 and 8.5. Table 8.6 gives the load and controller parameters for the current regulator considered here. The simulation code for this example is given in Appendix C.

As a practical matter, it is sound practice to take steps to ensure that the integral terms within the controller do not have an opportunity to increase

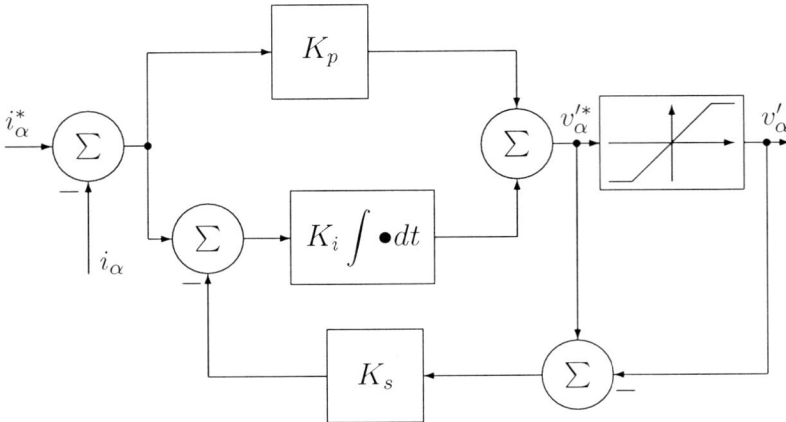

Figure 8.20: A practical implementation of a proportional-integral controller with integrator anti-windup.

without bound. Physical systems have natural limitations, and it is best to limit integrator output once these natural limitations are reached. In our current regulator, once the inverter output voltage has reached the limitations imposed by V_{dc}, it does us no good to let the integrator continue to accumulate error. In fact, continued accumulation of error will only result in drawing out the transient response, since any accumulated error is going to be compensated with error of the opposite sign before steady state behavior can be achieved.

Figure 8.20 shows the approach to preventing integrator windup used on the α axis in the current regulator example given here. An identical structure is used on the β axis. There are many other approaches to preventing integrator windup, but they generally share common features. In the approach taken here, a feedback loop is added that detects when the natural limits on inverter output voltage are reached. The saturation block in Fig. 8.20 represents the action of the modulator in detecting when the commanded voltage is being pushed outside of the space vector hexagon.

Once saturation occurs, the difference between the proportional-integral controller output and the saturated inverter voltage is fed back to the input of the integrator. This second control loop acts to force $v_\alpha'^*$ to the saturated value so that the controller will come out of saturation as quickly as possible. In the implementation considered here, $K_s = K_p$ gives acceptable performance.

Figure 8.21 shows the Bode plot for the forward loop gain of the example

8.6. An Example Current Regulator

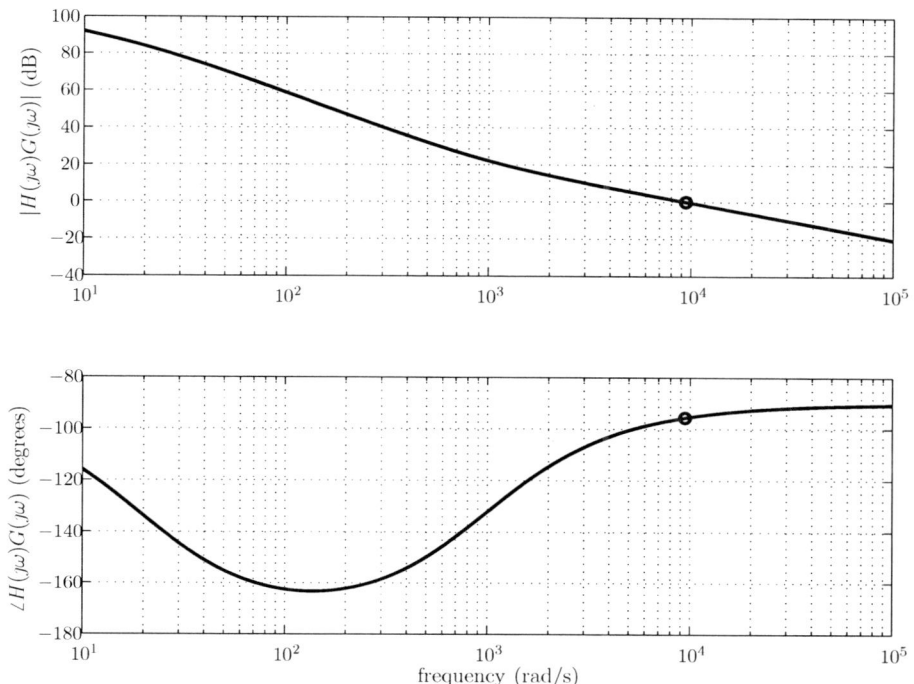

Figure 8.21: The Bode plots of the forward loop gain for the example current regulator using the parameters given in Table 8.6. The circles indicate values at crossover.

current regulator in the absence of saturation. From the parameters in Table 8.6 we see that $R/L = 20\,\text{rad/s}$, well below the frequency we are trying to synthesize. Accordingly, we use the controller to push crossover out to a frequency that provides an acceptable dynamic response. Because of the low corner frequency of the load, the corner frequency of the proportional-integral controller is placed well above that of the load.

Consistent with the corner frequencies provided in Fig. 8.8, the controller seeks to achieve crossover at $9.424\,\text{krad/s}$ by setting $K_p/L = 9.424\,\text{krad/s}$ giving $K_p = 47.12\,\Omega$. In addition, the corner frequency of the proportional-integral controller is selected to be one decade away from crossover, so $K_i/K_p = 942.4\,\text{rad/s}$ giving $K_i = 44.41\,\text{kV/A} \cdot \text{s}$.

The Bode plots show transition to a slope of -40 db/decade at $20\,\text{rad/s}$ as we pass through the pole associated with the load. At $942\,\text{rad/s}$ the slope reduces to -20db/decade as we pass through the corner frequency of the zero associated with the proportional-integral controller. Finally, we achieve crossover at $9.424\,\text{krad/s}$.

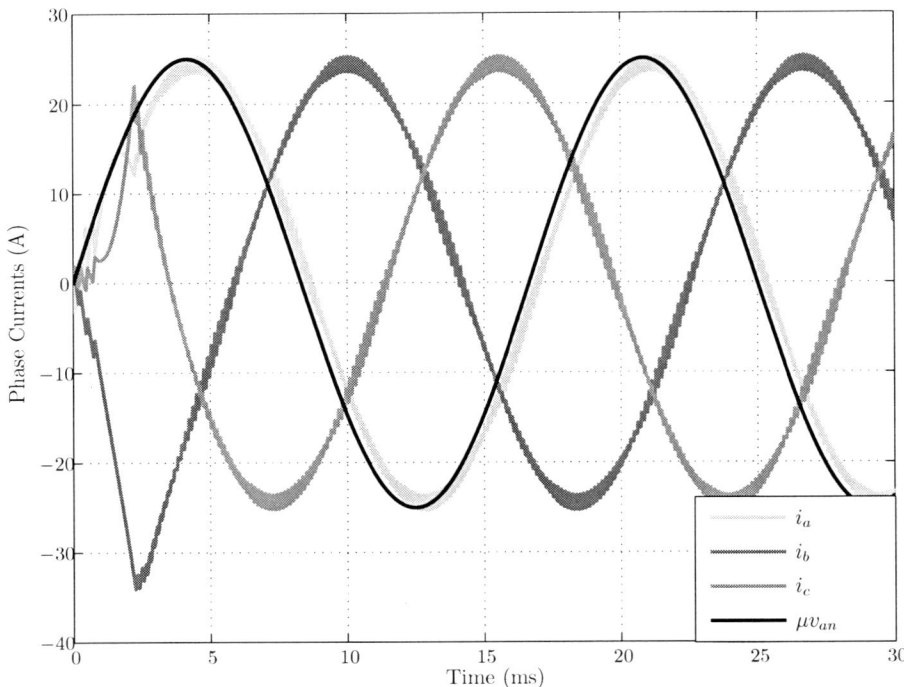

Figure 8.22: The simulated three-phase currents obtained for the example current regulator using the parameters given in Table 8.6.

Figure 8.22 shows the three phase currents generated by simulation of the current regulator. In addition to the three phase currents, the phase current command is shown for i_a. We see that the current regulator quickly locks onto the correct values of phase currents. It is worth noting that there is a slight phase delay between i_a^* and i_a. This is because all of the correction is being undertaken in the feedback loop. Increasing the bandwidth of the control loop would tend to reduce this phase delay.

Figure 8.23 shows the evolution of the currents in the $\alpha - \beta$ plane relative to the commanded current. Starting at the origin, the regulator quickly pushes the currents out to the commanded current locus. The ripple in the current due to switching the inverter states is more visible in this view than in Fig. 8.22.

Figure 8.24 shows the commanded voltages that are responsible for the current trajectory given in Fig. 8.23. The linear segments correspond to operation at the limit of the modulator, as the controller tries to quickly force the phase currents to track their commanded values. (The linear segments are an artifact of how the data is plotted. The plot connects all data points

8.6. An Example Current Regulator

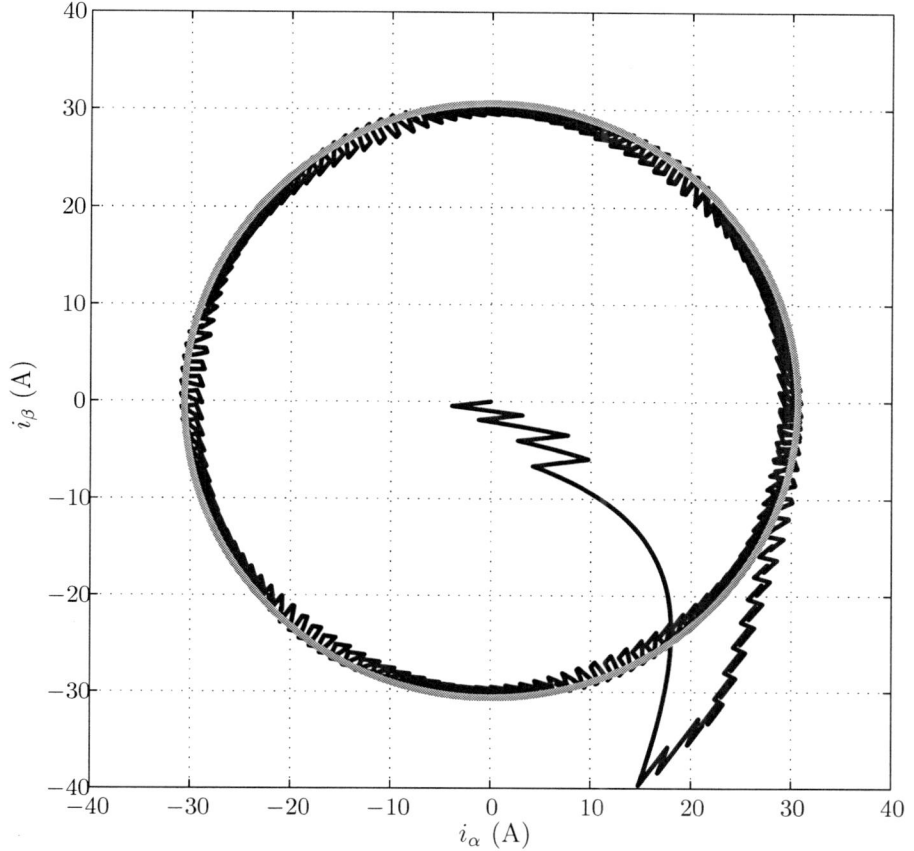

Figure 8.23: The simulated two-phase current trajectory obtained for the example current regulator using the parameters given in Table 8.6. The reference trajectory is also shown.

with lines, so while operation is actually at the vertices of the hexagon, it appears to be along the sides of the hexagon.) Once the phase currents catch up to their commands, the inverter output voltages fall back to a circular locus. It can be observed that the inverter output voltage locus has a magnitude of approximately 474 V, which corresponds to a line-neutral amplitude of $\sqrt{2/3} \times 474 = 387$ V. Since this is only slightly more than $|v_{an}|$, we conclude that the current regulator output voltages are producing the commanded phase currents largely through phase shift relative to the load voltages.

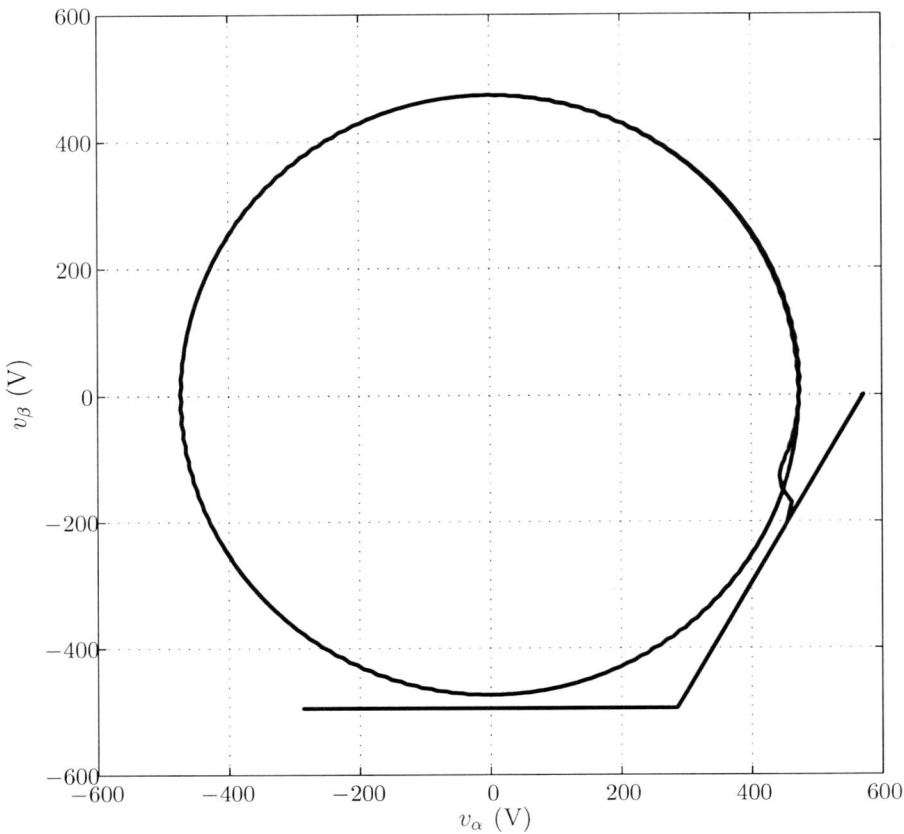

Figure 8.24: The simulated voltage commands applied to the space vector modulator for the example current regulator using the parameters given in Table 8.6.

8.7 A Space-Vector Predictive Current Regulator

The structure of the current regulator considered in the previous section relies on correcting errors in phase currents through the use of feedback. This always has the implemented voltages lagging behind what is necessary to produce the desired currents. If the sampling rate is high and the switching frequency is high, drive performance will not suffer appreciably. However, as the power level of the drive increases, the switching frequency and the bandwidth of the current regulator must be reduced.

In this case it makes sense to take advantage of knowing what we want the currents to do in the future. This forms the basis of a predictive current regulator in which we assume that the load voltages are predictable and we

8.7. A Space-Vector Predictive Current Regulator

use that information to determine what the inverter output voltages must be so that the phase currents have no error at the end of the current sampling interval. This approach forms a predictive current regulator.

Starting with our state equation for the load currents given by Eq. 8.2, we can approximate the current derivative over one sampling interval T as

$$\frac{d}{dt}\begin{bmatrix} i_\alpha \\ i_\beta \end{bmatrix} = \frac{di_{\alpha\beta}}{dt} \approx \frac{i_{\alpha\beta}(\omega(t+T)) - i_{\alpha\beta}(\omega t)}{T} \quad , \qquad (8.54)$$

where i is shorthand notation for our current vector. Using this in Eq. 8.2 and solving for the inverter output voltage gives

$$v'_{\alpha\beta}(\omega t) = v_{\alpha\beta n}(\omega t) + \frac{L}{T}\left\{ i_{\alpha\beta}(\omega(t+T)) + \left[\frac{RT}{L}I - I\right] i_{\alpha\beta}(\omega t) \right\} \quad , \qquad (8.55)$$

where $v'_{\alpha\beta}$ is the inverter output voltage vector and $v_{\alpha\beta n}$ is the load voltage vector. It follows that if we know the desired value of the current vector at the end of the current switching interval, then we know what voltage must be applied by the inverter to move the current vector there. In other words, if we can predict the desired current, then we can set our inverter output voltage during each switching cycle to simultaneously eliminate any residual error in the current and always drive toward the desired current vector.

Drawing on our experience with the example current regulator in the last section, producing load currents that are of the same shape and in phase with the load voltages leads to a circular locus in the $\alpha - \beta$ plane. We can use this to determine our desired current at the end of the current switching cycle. That is,

$$i(\omega(t+T)) = \mu v_{\alpha\beta n}(\omega(t+T)) = \mu e^{J\omega T} v_{\alpha\beta n}(\omega t) \quad , \qquad (8.56)$$

where μ is the proportionality factor between load voltage and desired load current, and $e^{J\omega T}$ is the rotation matrix (transformation) given by Eq. 6.13 that rotates the voltage vector through angle ωT during the switching period.

Using Eq. 8.56 in Eq. 8.55 gives

$$v'_{\alpha\beta}(\omega t) = \left[I + \frac{\mu L}{T} e^{J\omega T}\right] v_{\alpha\beta n}(\omega t) + \left[RI - \frac{L}{T}I\right] i(\omega t) \quad . \qquad (8.57)$$

From this we see that the measured values of load voltage and load current are used to develop the required inverter output voltages. Equation 8.57 forms the basis of a predictive current regulator. Naturally, a predictive current regulator is subject to the same limitations of inverter output voltage as any other current regulator.

8.8 Current Regulation in the dq Reference Frame

As we will see when we consider the control of induction and permanent magnet machines in Chapters 10 and 12, it is often convenient to view machine control from the perspective of an observer that is traveling with the air gap magnetic field. That is, control objectives are easier to see when viewed in the dq reference frame, where the direct axis is aligned with a particular field of interest, such as the magnet flux in the case of a permanent magnet machine. It is also possible to consider current regulation in the dq reference frame. A benefit of this is the suppression of the time varying quantities that are characteristic of the abc or $\alpha\beta 0$ reference frames.

The $\alpha\beta$ description of the inverter of Fig. 8.3 was given by Eq. 8.2 as

$$\frac{d}{dt}\begin{bmatrix} i_\alpha \\ i_\beta \end{bmatrix} = \frac{1}{L}\left\{\begin{bmatrix} v'_\alpha \\ v'_\beta \end{bmatrix} - \begin{bmatrix} v_{\alpha n} \\ v_{\beta n} \end{bmatrix}\right\} - \frac{R}{L}\begin{bmatrix} i_\alpha \\ i_\beta \end{bmatrix} . \qquad (8.58)$$

Using the rotation transformation given by Eq. 6.13, we can transform quantities from the $\alpha\beta$ reference frame into the dq reference frame using

$$\begin{bmatrix} x_\alpha \\ x_\beta \end{bmatrix} = e^{-J\phi}\begin{bmatrix} x_d \\ x_q \end{bmatrix} , \qquad (8.59)$$

where x is any voltage or current. Using Eq. 8.59 appropriately in Eq. 8.58 gives

$$\frac{d}{dt}\left\{e^{-J\phi}\begin{bmatrix} i_d \\ i_q \end{bmatrix}\right\} = \frac{1}{L}\left\{e^{-J\phi}\begin{bmatrix} v'_d \\ v'_q \end{bmatrix} - e^{-J\phi}\begin{bmatrix} v_{dn} \\ v_{qn} \end{bmatrix}\right\} - \frac{R}{L}e^{-J\phi}\begin{bmatrix} i_d \\ i_q \end{bmatrix} . \qquad (8.60)$$

Applying the product rule to the time derivative on the left gives

$$\frac{d}{dt}\left\{e^{-J\phi}\begin{bmatrix} i_d \\ i_q \end{bmatrix}\right\} = -e^{-J\phi}J\begin{bmatrix} i_d \\ i_q \end{bmatrix}\frac{d\phi}{dt} + e^{-J\phi}\frac{d}{dt}\begin{bmatrix} i_d \\ i_q \end{bmatrix} . \qquad (8.61)$$

Putting this into Eq. 8.60 and premultiplying by $e^{J\phi}$ gives

$$\frac{d}{dt}\begin{bmatrix} i_d \\ i_q \end{bmatrix} = \frac{1}{L}\left\{\begin{bmatrix} v'_d \\ v'_q \end{bmatrix} - \begin{bmatrix} v_{dn} \\ v_{qn} \end{bmatrix}\right\} - \frac{R}{L}\begin{bmatrix} i_d \\ i_q \end{bmatrix} + \omega J\begin{bmatrix} i_d \\ i_q \end{bmatrix} , \qquad (8.62)$$

where $d\phi/dt = \omega$. The state equations of Eq. 8.62 describe the equivalent circuits given in Figs. 8.25 and 8.26 for the d and q axes, respectively.

A block diagram that suggests the dq current regulator implementation is shown in Fig. 8.27. The structure is quite similar to that of the $\alpha\beta$ current regulator of Fig. 8.2, except that the rotation angle ϕ must be determined to

8.9. Summary

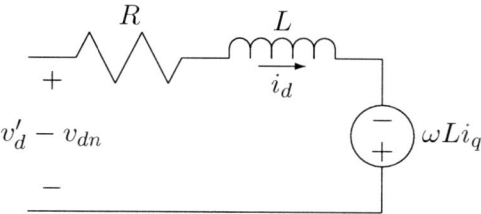

Figure 8.25: The equivalent circuit model for the d axis of the inverter circuit of Fig. 8.3.

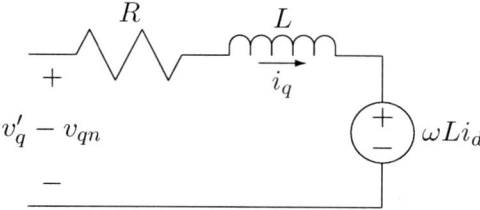

Figure 8.26: The equivalent circuit model for the q axis of the inverter circuit of Fig. 8.3.

facilitate transformation of $i_{\alpha\beta}$ into i_{dq}. A common approach to determining ϕ is to use a phase-locked loop that is driven by measurement of the line-line voltages of the ac sources shown in Fig. 8.3. In an electric machine system ϕ is often based on measurement of the rotor position, such as with an encoder or resolver.

8.9 Summary

This chapter has studied current regulators, in which voltage source inverters are used to regulate the currents supplied by the inverter. By closing the feedback loop around the voltage source inverter, we are able to effectively convert the voltage source into a programmable current source. The ability of the current regulator to accomplish its mission is tied to the switching frequency and the collection of available inverter output voltages. Multilevel inverters are able to offer an increased number of possible inverter voltage outputs. This becomes important particularly as the switching frequency is reduced at higher power levels.

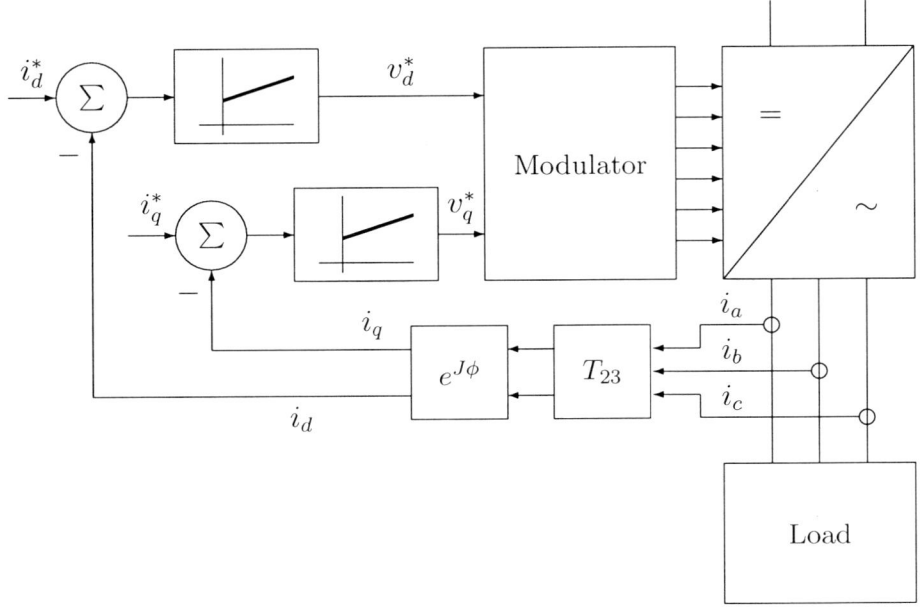

Figure 8.27: A block diagram of a current regulator based on the *dq* reference frame.

In the chapters that follow, it is often assumed that programmable currents can be applied to the electric machine under consideration. This is accomplished through the current regulator structures provided here. As the required precision of the machine output increases, so will the complexity of the current regulator structure. This will allow, for example, compensation for the ripple torques created by the interaction of current harmonics with space harmonics as discussed in Sec. 4.5.

Chapter 9

Induction Machine Models

9.1 Introduction

Induction machines have been the workhorse of industry for over a century. Their invention is credited to Nikola Tesla, as part of his development of polyphase ac electric systems. Tesla, with Westinghouse as his ally, challenged Edison and the proliferation of dc systems. It should be obvious who won, though there are certain applications where dc is making something of a resurgence.

Induction machines convert energy between electrical and mechanical forms through the interaction of rotor and stator magnetic fields. The rotor fields are induced by virtue of the air gap magnetic field moving at a different speed than the rotor. Accordingly, there are important similarities between the model of the induction machine and a transformer. In the induction machine, however, not only is voltage transformed, but also frequency. Because the rotor generally moves at a speed other than the speed of the air gap field, induction machines are also referred to as asynchronous machines.

It is important to note that although the rotor does not rotate at the same speed as the rotating magnetic field in the air gap, the field produced by the rotor does. That is, the rotor magnetic field travels relative to the rotor by a speed dictated by the frequency of the induced currents. The speed of the rotor field, superimposed on the rotor speed, equals the speed of the air gap magnetic field. That is, the condition for average power conversion discussed in Sec. 2.7, $\omega_m + \omega_r = \omega_s$, is satisfied.

This chapter develops a number of models for induction machines. These models begin with a physical model that is motivated by the structure of the machine. Each of the subsequent models are developed from the physical model through appropriate manipulation to put the model into a particular reference frame. Some of these models are used in the next chapter when we look at methods of controlling induction machines.

9.2 A Physical Model

An induction machine is constructed of a stationary cylindrical annulus, known as the stator, in which a cylinder is able to rotate. The cylinder that is free to rotate is known as the rotor. The stator is constructed of laminated soft ferromagnetic material, such as silicon steel. The stator laminations have slots into which the phase windings are placed. The windings for each phase are distributed approximately sinusoidally in space as discussed in Chapter 3. The phase windings are symmetrically displaced from one another. Most induction machines are built with three phases on the stator, though other numbers of phases have been considered. We will only consider three-phase machines here.

The rotor of the induction machine also carries windings. So-called wound rotor induction machines have physical windings which are accessible through slip rings thereby facilitating external interaction with the rotor circuit. The more common induction machines have rotor windings formed by a casting embedded within the silicon steel laminations. The casting takes the shape of circumferentially distributed bars that are connected together by circular rings on the ends of the rotor stack. This winding is often referred to as a squirrel cage, because it resembles the familiar ring toy used to entertain hamsters and other small rodents. The end rings often support axial fins that act as fans when the rotor spins. Regardless of the specifics of how the rotor windings are constructed, it is convenient for modeling purposes to consider the rotor windings to be formed by three sinusoidally distributed, symmetrically displaced windings.

Most construction details of the induction machine are beyond our scope here. The interested reader is directed to any of the multitude of electric machinery texts for additional details[1]. Figure 9.1 shows the construction of an induction machine that is adopted for modeling purposes. Each winding on the stator and rotor is assumed to be sinusoidally distributed, despite being shown by a concentrated coil in the figure. In practice, the winding distribution is not perfectly sinusoidal, but it is sufficiently close that we only need to consider the fundamental component in describing how windings

---

[1] See, for example, H. Majmudar, *Electromechanical Energy Converters*, Allyn and Bacon, 1965; G. Slemon and A. Straughen, *Electric Machines*, Addison-Wesley, 1980; and A. E. Fitzgerald, C. Kingsley, Jr., and S. D. Umans, *Electric Machinery*, 6th ed., McGraw-Hill, 2003.

9.2. A Physical Model

interact with one another.

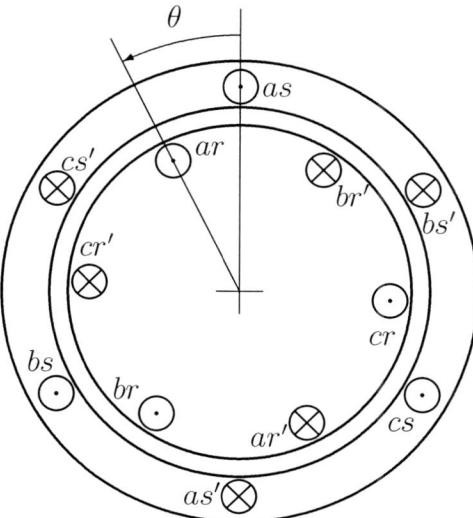

Figure 9.1: The physical model of the induction machine. The windings are assumed to be sinusoidally distributed. This machine has two poles on both the stator and rotor. That is, $N_p = 2$.

As discussed in Chapter 3, a typical design parameter for an electric machines is the number of magnetic poles. Since there must be the same number of north and south poles, the number of poles must be even. In an electric machine with two magnetic poles, one electrical cycle is completed in each revolution of the rotor. If there are four magnetic poles, there will be two electrical cycles completed in each revolution of the rotor. Accordingly, the number of magnetic poles prescribes the relationship between mechanical position and electrical position:

$$\theta_e = \frac{N_p}{2} \theta \quad , \tag{9.1}$$

where θ_e is the electrical position that corresponds to the mechanical position θ in a machine with N_p poles. In order for the electric machine to produce average torque, the stator and rotor must have the same number of poles. In a squirrel cage induction machine this is handled automatically since the induced currents will reflect the pole structure of the stator field.

The windings within the induction machine are coupled. That is, magnetic flux produced by current in one phase winding links one or more of the other phase windings. This coupling is described through the inductance matrix, which describes how current in any one winding contributes to the flux linking all of the other windings:

$$\begin{bmatrix} \lambda_{sa} \\ \lambda_{sb} \\ \lambda_{sc} \\ \lambda_{ra} \\ \lambda_{rb} \\ \lambda_{rc} \end{bmatrix} = \begin{bmatrix} L_s & -L_{ss} & -L_{ss} & L_{sr}\cos(\theta_e) & L_{sr}\cos\left(\theta_e + \frac{2\pi}{3}\right) & L_{sr}\cos\left(\theta_e - \frac{2\pi}{3}\right) \\ -L_{ss} & L_s & -L_{ss} & L_{sr}\cos\left(\theta_e - \frac{2\pi}{3}\right) & L_{sr}\cos(\theta_e) & L_{sr}\cos\left(\theta_e + \frac{2\pi}{3}\right) \\ -L_{ss} & -L_{ss} & L_s & L_{sr}\cos\left(\theta_e + \frac{2\pi}{3}\right) & L_{sr}\cos\left(\theta_e - \frac{2\pi}{3}\right) & L_{sr}\cos(\theta_e) \\ L_{rs}\cos(\theta_e) & L_{rs}\cos\left(\theta_e - \frac{2\pi}{3}\right) & L_{rs}\cos\left(\theta_e + \frac{2\pi}{3}\right) & L_r & -L_{rr} & -L_{rr} \\ L_{rs}\cos\left(\theta_e + \frac{2\pi}{3}\right) & L_{rs}\cos(\theta_e) & L_{rs}\cos\left(\theta_e - \frac{2\pi}{3}\right) & -L_{rr} & L_r & -L_{rr} \\ L_{rs}\cos\left(\theta_e - \frac{2\pi}{3}\right) & L_{rs}\cos\left(\theta_e + \frac{2\pi}{3}\right) & L_{rs}\cos(\theta_e) & -L_{rr} & -L_{rr} & L_r \end{bmatrix} \times \begin{bmatrix} i_{sa} \\ i_{sb} \\ i_{sc} \\ i_{ra} \\ i_{rb} \\ i_{rc} \end{bmatrix}, \quad (9.2)$$

where L_s is the self-inductance of each stator phase winding; $-L_{ss}$ is mutual inductance between stator windings; L_r is the self-inductance of each rotor phase winding; $-L_{rr}$ is the mutual inductance between rotor phase windings; L_{sr} is the mutual inductance between rotor currents and stator phase windings; L_{rs} is the mutual inductance between stator currents and rotor phase windings; and θ_e is the electrical rotor position based on the number of magnetic pole pairs of each winding on the stator and rotor. Note that $L_{sr} = L_{rs}$ by reciprocity.

In a machine with symmetrically displaced windings, $L_s \approx L_{s\ell} + 2L_{ss}$, where the factor of two is due to the phase displacement. The term $L_{s\ell}$ is a leakage term that accounts for flux linking a phase winding due to currents

9.2. A Physical Model

in that phase winding where that flux does not link either of the other stator windings. This leakage inductance is created by slotting effects and end turns as discussed in Chapter 4. Similar arguments hold for the rotor windings.

Equation 9.2 can be written more compactly as

$$\lambda_{abc} = L_{abc}(\theta_e) i_{abc} \quad , \tag{9.3}$$

where λ_{abc} and i_{abc} are vectors and $L_{abc}(\theta_e)$ is a matrix dependent on rotor position. The subscript abc is used to denote the abc reference frame. This representation is based on a physical description of the machine, though we will find it to be of limited value as we develop advanced control techniques for the induction machine.

The electrical dynamics for the induction machine can be written very succinctly using vector notation as

$$v_{abc} = \frac{d\lambda_{abc}}{dt} + R_{abc} i_{abc} \quad , \tag{9.4}$$

where the resistance matrix R_{abc} is

$$R_{abc} = \begin{bmatrix} R_s & 0 & 0 & 0 & 0 & 0 \\ 0 & R_s & 0 & 0 & 0 & 0 \\ 0 & 0 & R_s & 0 & 0 & 0 \\ 0 & 0 & 0 & R_r & 0 & 0 \\ 0 & 0 & 0 & 0 & R_r & 0 \\ 0 & 0 & 0 & 0 & 0 & R_r \end{bmatrix} . \tag{9.5}$$

The electrical dynamics depend on the mechanical dynamics through the dependence on rotor position within L_{abc}. Further, the mechanical dynamics are coupled to the electrical dynamics through torque production. Torque is produced by virtue of the interaction of the magnetic fields established by the stator and rotor windings.

The description of force and torque through conservation of energy (often termed the method of virtual work) was discussed in Chapter 2, using explicit representation of every winding. Making use of vector notation, we have

$$dW_m = i^T d\lambda - T^e d\theta \quad , \tag{9.6}$$

from which we conclude

$$dW_m = \frac{\partial W_m}{\partial \lambda} d\lambda + \frac{\partial W_m}{\partial \theta} d\theta \quad , \tag{9.7}$$

so that

$$i^T = \frac{\partial W_m}{\partial \lambda} \quad ; \tag{9.8}$$

$$T^e = -\frac{\partial W_m}{\partial \theta} \quad . \tag{9.9}$$

Assuming that $T^e = 0$ when the machine is unexcited ($\lambda = 0$), the energy is found to be

$$W_m = \int_0^\lambda \tilde{\lambda}^T L^{-1}(\theta) d\tilde{\lambda} = \frac{1}{2}\lambda^T L^{-1}(\theta)\lambda \quad , \tag{9.10}$$

from which we compute the electromagnetic torque to be

$$T^e = -\frac{1}{2}\lambda^T \frac{dL^{-1}(\theta)}{d\theta}\lambda \quad . \tag{9.11}$$

This is entirely consistent with Eq. 2.63. Equation 9.11 can also be written as

$$T^e = \frac{1}{2}i^T \frac{dL(\theta)}{d\theta} i \quad , \tag{9.12}$$

consistent with the coenergy formulation from Eq. 2.88 noting that the cross terms combine because $L_{xy} = L_{yx}$.

The mechanical dynamics are

$$H\frac{d\omega}{dt} = T^e - \tau_l \quad , \tag{9.13}$$

where the electromagnetic torque is given by Eq. 9.12 and the load torque τ_l includes windage and friction in addition to the shaft load. The moment of inertia H is assumed to include the inertia of the induction machine and whatever is connected to the induction machine through its shaft.

Taken together, Eqs. 9.3, 9.4, 9.12, and 9.13 summarize the electromechanical dynamics of the induction machine. The description of these dynamics has been motivated by the physical construction of the machine. This description, however, is inconvenient for studying dynamics and control for three reasons. First, the order of the system is large. Second, the dependence on θ_e gives rise to a time-varying model. Third, the model developed so far mixes the stator and rotor reference frames. The sections that follow convert the physical model into a number of different forms through the use of the transformations developed in Chapter 6.

9.3 The Phase Equivalent Model

9.3.1 Model Development

The phase equivalent model of the induction machine is the most commonly encountered model in electric machinery texts. It is a sinusoidal steady

9.3. The Phase Equivalent Model

state model of the machine appropriate for phasor analysis. Despite being a steady state model, it can provide useful insights into the operation of the induction machine.

To get the phase equivalent model, we must move rotor quantities from the rotor reference frame into the stator reference frame. To be explicit as we do this, we break apart Eq. 9.3 into

$$\lambda^s_{s,abc} = L_s i^s_{s,abc} + L_{sr} i^r_{r,abc} \quad , \tag{9.14}$$

and

$$\lambda^r_{r,abc} = L_{sr}^T i^s_{s,abc} + L_r i^r_{r,abc} \quad , \tag{9.15}$$

where

$$L_s = \begin{bmatrix} L_s & -L_{ss} & -L_{ss} \\ -L_{ss} & L_s & -L_{ss} \\ -L_{ss} & -L_{ss} & L_s \end{bmatrix} \quad , \tag{9.16}$$

$$L_{sr} = \begin{bmatrix} L_{sr}\cos(\theta_e) & L_{sr}\cos\left(\theta_e + \frac{2\pi}{3}\right) & L_{sr}\cos\left(\theta_e - \frac{2\pi}{3}\right) \\ L_{sr}\cos\left(\theta_e - \frac{2\pi}{3}\right) & L_{sr}\cos(\theta_e) & L_{sr}\cos\left(\theta_e + \frac{2\pi}{3}\right) \\ L_{sr}\cos\left(\theta_e + \frac{2\pi}{3}\right) & L_{sr}\cos\left(\theta_e - \frac{2\pi}{3}\right) & L_{sr}\cos(\theta_e) \end{bmatrix} \quad , \tag{9.17}$$

and

$$L_r = \begin{bmatrix} L_r & -L_{rr} & -L_{rr} \\ -L_{rr} & L_r & -L_{rr} \\ -L_{rr} & -L_{rr} & L_r \end{bmatrix} \quad . \tag{9.18}$$

The subscript notation used for the flux linkages and currents tracks whether the quantities are on the stator or rotor and the type of coordinate system. The superscript notation indicates whether the quantities are in the stator or rotor reference frame. For example, $i^s_{r,abc}$ denotes rotor abc currents in the stator reference frame. Our objective is to get everything into the stator reference frame so that we have a description of machine behavior based on stator excitation.

To move rotor flux and current from the rotor reference frame to the stator reference frame, we use the transformation introduced in Chapter 6, Eq. 6.17:

$$T^T R_{\theta_e} T = \frac{2}{3} \times$$
$$\begin{bmatrix} \cos\theta_e + \frac{1}{2} & \cos\left(\theta_e - \frac{2\pi}{3}\right) + \frac{1}{2} & \cos\left(\theta_e + \frac{2\pi}{3}\right) + \frac{1}{2} \\ \cos\left(\theta_e + \frac{2\pi}{3}\right) + \frac{1}{2} & \cos\theta_e + \frac{1}{2} & \cos\left(\theta_e - \frac{2\pi}{3}\right) + \frac{1}{2} \\ \cos\left(\theta_e - \frac{2\pi}{3}\right) + \frac{1}{2} & \cos\left(\theta_e + \frac{2\pi}{3}\right) + \frac{1}{2} & \cos\theta_e + \frac{1}{2} \end{bmatrix} \tag{9.19}$$

Through this transformation, we have

$$\lambda^r_{r,abc} = T^T R_{\theta_e} T \lambda^s_{r,abc} \quad , \tag{9.20}$$

and

$$i^r_{r,abc} = T^T R_{\theta_e} T i^s_{r,abc} \quad . \tag{9.21}$$

Putting these expressions into Eqs. 9.14 and 9.15 gives

$$\lambda^s_{s,abc} = L_s i^s_{s,abc} + L_{sr} T^T R_{\theta_e} T i^s_{r,abc} \quad , \tag{9.22}$$

and

$$T^T R_{\theta_e} T \lambda^s_{r,abc} = L^T_{sr} i^s_{s,abc} + L_r T^T R_{\theta_e} T i^s_{r,abc} \quad , \tag{9.23}$$

Performing the indicated operations in Eq. 9.22 gives

$$\lambda^s_{s,abc} = L_s i^s_{s,abc} + L'_{sr} i^s_{r,abc} \quad , \tag{9.24}$$

where

$$L'_{sr} = \begin{bmatrix} L_{sr} & -\frac{1}{2}L_{sr} & -\frac{1}{2}L_{sr} \\ -\frac{1}{2}L_{sr} & L_{sr} & -\frac{1}{2}L_{sr} \\ -\frac{1}{2}L_{sr} & -\frac{1}{2}L_{sr} & L_{sr} \end{bmatrix} \quad . \tag{9.25}$$

Moving the rotor flux and current from the rotor reference frame to the stator reference frame has removed the dependence on rotor position. To simplify Eq. 9.23, we premultiply both sides by $T^T R_{-\theta_e} T$ and reduce to get[2]

$$\lambda^s_{r,abc} = L'^T_{sr} i^s_{s,abc} + L_r i^s_{r,abc} \quad . \tag{9.26}$$

While these manipulations have removed the dependence on rotor position, there is still coupling among the phases.

In a balanced three phase machine, we have

$$i_{sa} + i_{sb} + i_{sc} = 0 \quad , \tag{9.27}$$

and

$$i_{ra} + i_{rb} + i_{rc} = 0 \quad , \tag{9.28}$$

which allows us to conclude that

$$\lambda^s_{sa} = (L_s + L_{ss}) i^s_{sa} + \frac{3}{2} L_{sr} i^s_{ra} \quad , \tag{9.29}$$

and

$$\lambda^s_{ra} = \frac{3}{2} L_{sr} i^s_{sa} + (L_r + L_{rr}) i^s_{ra} \quad . \tag{9.30}$$

---

[2]First, note that $(T^T R_{-\theta_e} T)(T^T R_{\theta_e} T) = I$. Second, note that $T^T R_{-\theta_e} T L^T_{sr} = (L_{sr} T^T R_{\theta_e} T)^T$. Third, the symmetry of L_r provides for $T^T R_{-\theta_e} T L_r T^T R_{\theta_e} T = L_r$.

9.3. The Phase Equivalent Model

Together, Eqs. 9.29 and 9.30 show how the stator and rotor flux linkages depend on stator and rotor currents, respectively. Accordingly, these two expressions provide the coupling between the stator and rotor. To complete our model, we must also address the stator and rotor terminal equations.

On the stator, the voltage applied to the terminals must either go into resistive voltage drop or changing the stator flux linkage:

$$v_{sa}^s = R_s i_{sa}^s + \frac{d\lambda_{sa}^s}{dt} \quad . \tag{9.31}$$

On the rotor, we have a similar situation:

$$v_{ra}^r = R_r i_{ra}^r + \frac{d\lambda_{ra}^r}{dt} \quad . \tag{9.32}$$

In a squirrel cage induction machine the rotor windings are closed on themselves so $v_{ra}^r = 0$. Accordingly, in the sinusoidal steady state we have

$$\hat{V}_{sa}^s = R_s \hat{I}_{sa}^s + \jmath\omega_s \hat{\lambda}_{sa}^s \quad , \tag{9.33}$$

and

$$\hat{V}_{ra}^r = 0 = R_r \hat{I}_{ra}^r + \jmath\omega_r \hat{\lambda}_{ra}^r \quad , \tag{9.34}$$

where $\hat{X}$ denotes the phasor of $x(t)$.

In the rotor reference frame, the time rate of change in the flux linkage is tied to the difference between the speed of the air gap magnetic field and the speed of the rotor. The speed of the air gap magnetic field is tied to the frequency of the stator currents and the number of poles in the machine. The synchronous speed is denoted as ω_{syn}, where $\omega_{\text{syn}} = 2\omega_s/N_p$. As discussed in Sec. 2.7, average power conversion takes place when

$$\frac{N_p}{2}\omega_m = \omega_s - \omega_r \quad . \tag{9.35}$$

In induction machines, it is common to refer to the frequency of the rotor currents as the slip frequency, where the slip is defined as

$$s = \frac{\omega_r}{\omega_s} = \frac{\omega_{\text{syn}} - \omega_m}{\omega_{\text{syn}}} \quad . \tag{9.36}$$

Using the slip, Eq. 9.33 becomes

$$0 = R_r \hat{I}_{ra}^r + \jmath s\omega_s \hat{\lambda}_{ra}^r \quad . \tag{9.37}$$

To move this to the stator frame at frequency ω_s, we divide through by s:

$$\jmath\omega_s \hat{\lambda}_{ra}^r = -\frac{R_r}{s}\hat{I}_{ra}^r \quad . \tag{9.38}$$

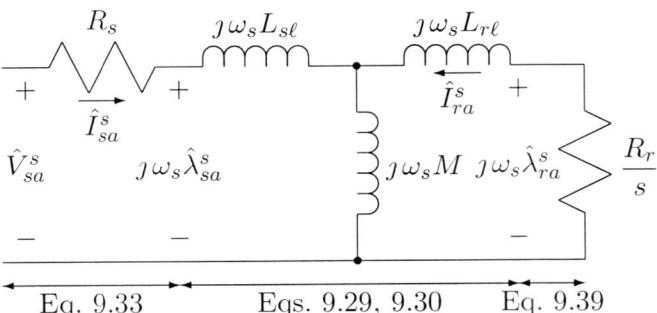

Figure 9.2: The single phase equivalent circuit model of an induction machine in sinusoidal steady state.

Because $\hat{\lambda}_{ra}^r$ and $\hat{I}_{ra}^r$ will transform to the stator frame in exactly the same way, we have

$$j\omega_s \hat{\lambda}_{ra}^s = -\frac{R_r}{s} \hat{I}_{ra}^s \quad . \tag{9.39}$$

We are now in a position to construct our single phase equivalent circuit model using Eqs. 9.29, 9.30, 9.33, and 9.39. Equations 9.29 and 9.30 are used to connect the stator and rotor flux linkages to the stator and rotor currents. These two equations can be interpreted as a T-connection of three inductors. Equation 9.33 shows how stator terminal voltage either goes into stator resistance drop or stator flux linkage. Equation 9.39 shows the relationship between rotor flux linkage and rotor current.

Putting all of these pieces together gives the single phase equivalent circuit model shown in Fig. 9.2. It will be noted that

$$L_{s\ell} = (L_s + L_{ss}) - \frac{3}{2} L_{sr} \quad , \tag{9.40}$$

$$M = \frac{3}{2} L_{sr} \quad , \tag{9.41}$$

and

$$L_{r\ell} = (L_r + L_{rr}) - \frac{3}{2} L_{sr} \quad . \tag{9.42}$$

In this model, the leakage inductances represent flux linking the stator winding that does not link the rotor winding and vice versa. This stator-rotor leakage (Eq. 9.40) is different than the stator-stator and rotor-rotor leakage discussed earlier. The stator-stator and rotor-rotor leakage inductances contribute to the stator-rotor leakage inductances, but do not form the entire inductance.

9.3.2 Model Analysis

With our single phase equivalent model, we can actually determine quite a lot about induction machine behavior. To begin, we see that the applied stator voltage is the source of all fields in the machine. That is, it is the applied stator voltage that is responsible for establishing stator and rotor currents.

The representation of the stator and rotor inductances makes clear the contribution of the stator and rotor currents to the air gap field. The air gap flux is the flux associated with the mutual inductance M. The air gap flux is the basis of torque production because it is what couples the stator and rotor.

In the rotor circuit we have a speed dependent resistance: R_r/s. When the rotor is moving at synchronous speed $s = 0$, making the effective rotor resistance infinite. In response, there is no rotor current and the air gap field is due entirely to stator currents. This makes complete sense based on our physical understanding of machine operation. In order for currents to be induced in the rotor, the rotor must see a time-varying magnetic field. However, if the rotor is moving at the same speed as the applied field, there is no time variation, hence no current. We conclude from this that the induction machine can only produce torque when the rotor is not moving at synchronous speed.

The speed dependent resistance takes the sign of the slip. When $\omega_m < \omega_{\text{syn}}$, $s > 0$ and the resistance is positive suggesting power is being dissipated in the resistance. When $\omega_m > \omega_{\text{syn}}$, $s < 0$ and the resistance is negative suggesting the resistance is actually supplying power. These regions will perhaps make more sense after some discussion of torque production.

The power being delivered to the rotor is $\left|\hat{I}_{ra}^s\right|^2 R_r/s$.[3] Of this power, some is being dissipated in the rotor resistance. The remaining power is that which is available for output through the shaft. That is, the power being converted is

$$P_{\text{shaft}} = \omega_m T^e = N_\phi \frac{1-s}{s} R_r \left|\hat{I}_{ra}^s\right|^2 . \qquad (9.43)$$

Since $\omega_m = (1-s)\omega_{\text{syn}}$ the torque is

$$T^e = N_\phi \frac{R_r}{s\omega_{\text{syn}}} \left|\hat{I}_{ra}^s\right|^2 . \qquad (9.44)$$

We see that the sign of the torque is dictated by the sign of the slip. When the slip is positive, torque is positive. When the slip is negative, torque is

---

[3] Recall that for phasor quantities $\left|\hat{I}_{ra}^s\right|^2 = \hat{I}_{ra}^s \hat{I}_{ra}^{s*}$, where * denotes complex conjugation.

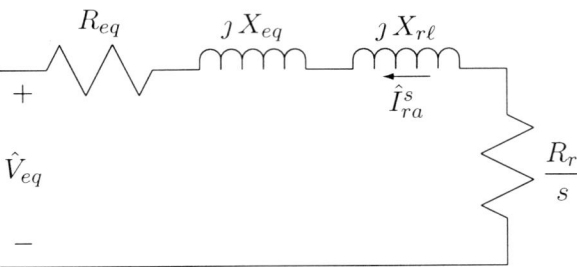

Figure 9.3: The reduced single phase equivalent circuit model of an induction machine in sinusoidal steady state.

negative. The sign of the torque is independent of the direction of rotation, so the induction machine is capable of four-quadrant operation.

Further analysis of the torque speed characteristic of the induction machine is aided if we recast the circuit model of Fig. 9.2 into the equivalent circuit of Fig. 9.3 which replaces everything left of the rotor leakage inductance as a Thevenin equivalent. The open circuit (Thevenin) voltage is:

$$\hat{V}_{eq} = \hat{V}_{sa}^{s} \frac{\jmath X_m}{R_s + \jmath(X_{s\ell} + X_m)} \quad . \tag{9.45}$$

The Thevenin impedance is:

$$R_{eq} + \jmath X_{eq} = \frac{(R_s + \jmath X_{s\ell})\jmath X_m}{R_s + \jmath(X_{s\ell} + X_m)} \quad . \tag{9.46}$$

All reactances are based on the stator frequency ω_s. For example, $X_m = \omega_s M$.

Using the circuit of Fig. 9.3, our torque expression becomes

$$T^e = N_\phi \frac{R_r}{s\omega_{\text{syn}}} \frac{\hat{V}_{eq}^2}{\left(R_{eq} + \frac{R_r}{s}\right)^2 + (X_{eq} + X_{r\ell})^2} \quad . \tag{9.47}$$

Equation 9.47 shows us that for small values of slip (so R_r/s dominates in the denominator) the torque is proportional to slip. For large values of slip (so R_r/s can be neglected in the denominator) the torque is inversely proportional to slip. Figure 9.4 shows a qualitative torque-speed curve with important points and regions of operation noted. The peak torque is known as the breakdown torque. The slip associated with peak torque is referred to as $s_{\max}$.

9.3. The Phase Equivalent Model

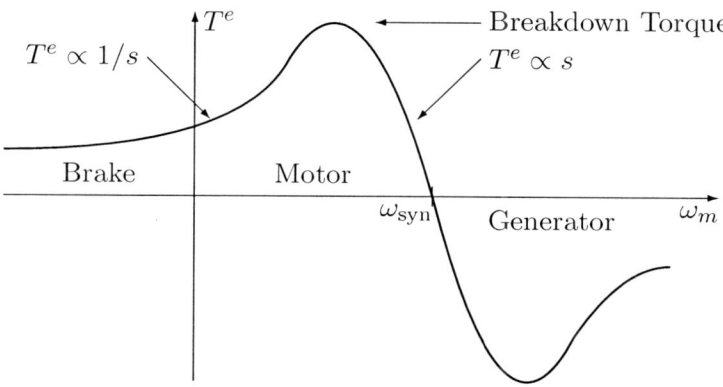

Figure 9.4: The general nature of a torque-speed curve for an induction machine.

The character of the torque-speed curve can be made to vary substantially based on the design objectives for the machine. There are four NEMA[4] standard designs of induction machines, labeled design classes A through D; there are also comparable IEC design classes. These designs provide different values of starting torque and slip at peak torque. Designs for highest efficiency are generally design Class A where the rotor resistance is small and full-load torque is produced with the smallest slip, usually just a few percent. Accordingly, the torque speed characteristic is very steep near synchronous speed. Figure 9.5 shows the torque-speed characteristic for a Class A induction machine with the parameters given in Table 9.1.

From an analysis of Eq. 9.47[5], we determine that the peak torque occurs when

$$\frac{R_r}{s_{\max}} = \pm\sqrt{R_{eq}^2 + (X_{eq} + X_{r\ell})^2} \quad . \tag{9.48}$$

The peak torque is also known as the breakdown torque. This should not be a surprise since it is predicted by the maximum power transfer theorem for the circuit of Fig. 9.4! The peak torque in motor and generator mode are symmetrically displaced from synchronous speed. Note that as R_r gets smaller, the slip at peak torque also gets smaller and the torque-speed curve gets steeper. We also note that the peak torque is independent of R_r and that peak torque is larger for generator operation than for motor operation.

---

[4]National Electrical Manufacturers Association, Standard MG-1.
[5]For inflection points in the torque expression we set $\partial T^e / \partial (R_r/s) = 0$.

Table 9.1: The parameters for a three-phase induction machine. The torque speed curve of this machine is given in Fig. 9.5.

Quantity	Symbol	Value
Number of Phases	N_ϕ	3
Rated Power	P_{rated}	7.46 kW
Rated Stator Voltage	V_{rated}	220 V
Rated Frequency	f_{rated}	60 Hz
Rated Speed	n_{rated}	1164 rpm
Number of Poles	N_p	6
Stator Resistance	R_s	0.294 Ω/ph
Stator Leakage Reactance	$X_{s\ell}$	0.524 Ω/ph
Rotor Resistance	R_r	0.156 Ω/ph
Rotor Leakage Reactance	$X_{r\ell}$	0.279 Ω/ph
Magnetizing Reactance	X_m	15.457 Ω/ph
Moment of Inertia	H	0.4 kg m²

9.4 Two Phase Equivalent Models

There are two, two-phase equivalent models that are of interest to us. One model is in the stator reference frame. We will refer to this model as the $\alpha\beta$ model. The second model is based on projecting both stator and rotor quantities onto a common set of axes that are free to move as a function of time. We will refer to this model as the dq model. We will develop these models in two steps, starting with the $\alpha\beta$ model.

Technically the models we are developing are three-phase models. However, we will see that only two of the phases contribute to torque production. We will take this as an opportunity to drop the unproductive phase from further consideration, thereby reducing the order of the model.

9.4.1 The $\alpha\beta$ Model

The development of the two phase equivalent models of the induction machine starts with the physical model represented by Eqs. 9.3, 9.4, 9.13, and

9.4. Two Phase Equivalent Models

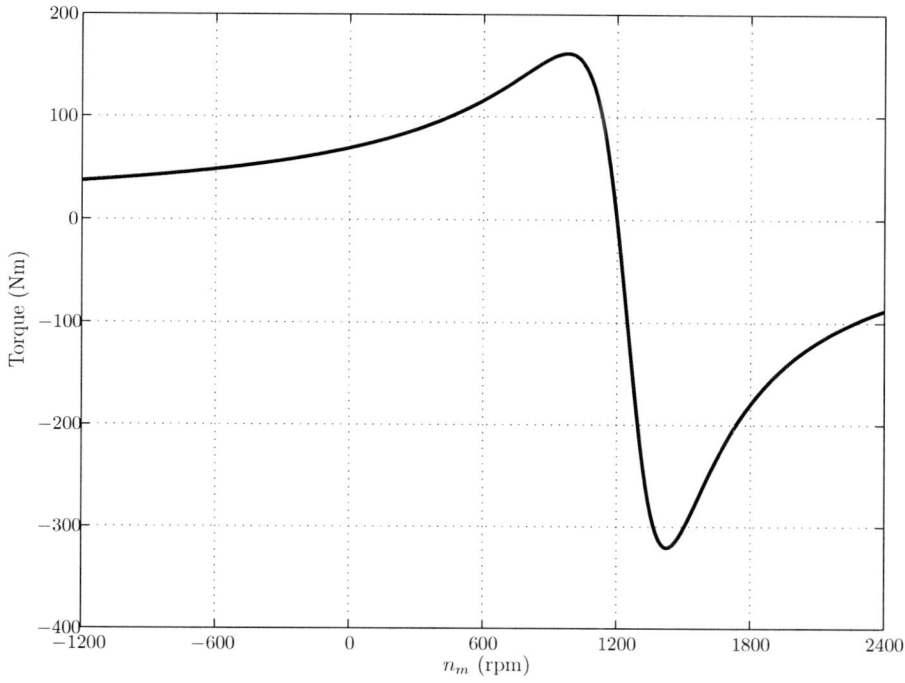

Figure 9.5: The torque speed characteristic for the three-phase induction machine having the parameters given in Table 9.1.

9.12. If we apply our transformation T (Eq. 6.8) to Eq. 9.3 we have

$$\begin{bmatrix} \lambda_{s\alpha} \\ \lambda_{s\beta} \\ \lambda_{s0} \\ \lambda_{r\alpha} \\ \lambda_{r\beta} \\ \lambda_{r0} \end{bmatrix} = \begin{bmatrix} T & 0 \\ 0 & T \end{bmatrix} L_{abc}(\theta_e) \begin{bmatrix} T^{-1} & 0 \\ 0 & T^{-1} \end{bmatrix} \begin{bmatrix} i_{s\alpha} \\ i_{s\beta} \\ i_{s0} \\ i_{r\alpha} \\ i_{r\beta} \\ i_{r0} \end{bmatrix} . \tag{9.49}$$

Carrying out the indicated multiplication gives

$$\lambda_{\alpha\beta 0} = L_{\alpha\beta 0}(\theta_e) i_{\alpha\beta 0} , \tag{9.50}$$

where

$$L_{\alpha\beta 0}(\theta_e) = \begin{bmatrix} L_s + L_{ss} & 0 & 0 \\ 0 & L_s + L_{ss} & 0 \\ 0 & 0 & L_s - 2L_{ss} \\ \frac{3}{2}L_{rs}\cos(\theta_e) & \frac{3}{2}L_{rs}\sin(\theta_e) & 0 \\ -\frac{3}{2}L_{rs}\sin(\theta_e) & \frac{3}{2}L_{rs}\cos(\theta_e) & 0 \\ 0 & 0 & 0 \end{bmatrix}$$

$$\left[\begin{array}{ccc} \frac{3}{2}L_{sr}\cos(\theta_e) & -\frac{3}{2}L_{sr}\sin(\theta_e) & 0 \\ \frac{3}{2}L_{sr}\sin(\theta_e) & \frac{3}{2}L_{sr}\cos(\theta_e) & 0 \\ 0 & 0 & 0 \\ L_r + L_{rr} & 0 & 0 \\ 0 & L_r + L_{rr} & 0 \\ 0 & 0 & L_r - 2L_{rr} \end{array}\right]. \quad (9.51)$$

It is important to note that the 0 components for both stator and rotor do not couple to either the α or β phases. Further, the 0 components do not contribute to torque production by virtue of their independence of rotor position. It follows that it is best not to excite these 0 components, implying that they can be left out of the model, thereby reducing the order of the model from six states to four states.

With the suppression of the 0 components, the electrical dynamics become

$$\left[\begin{array}{c} \lambda_{s\alpha} \\ \lambda_{s\beta} \\ \lambda_{r\alpha} \\ \lambda_{r\beta} \end{array}\right] = L_{\alpha\beta}(\theta_e) \left[\begin{array}{c} i_{s\alpha} \\ i_{s\beta} \\ i_{r\alpha} \\ i_{r\beta} \end{array}\right] \quad ; \quad (9.52)$$

$$v_{\alpha\beta} = \frac{d\lambda_{\alpha\beta}}{dt} + R_{\alpha\beta} i_{\alpha\beta} \quad , \quad (9.53)$$

where

$$\begin{aligned} L_{\alpha\beta}(\theta_e) &= \left[\begin{array}{cccc} L_s + L_{ss} & 0 & \frac{3}{2}L_{sr}\cos(\theta_e) & -\frac{3}{2}L_{sr}\sin(\theta_e) \\ 0 & L_s + L_{ss} & \frac{3}{2}L_{sr}\sin(\theta_e) & \frac{3}{2}L_{sr}\cos(\theta_e) \\ \frac{3}{2}L_{rs}\cos(\theta_e) & \frac{3}{2}L_{rs}\sin(\theta_e) & L_r + L_{rr} & 0 \\ -\frac{3}{2}L_{rs}\sin(\theta_e) & \frac{3}{2}L_{rs}\cos(\theta_e) & 0 & L_r + L_{rr} \end{array}\right] \\ &= \left[\begin{array}{cc} (L_s + L_{ss})I & \frac{3}{2}L_{sr}e^{-J\theta_e} \\ \frac{3}{2}L_{rs}e^{J\theta_e} & (L_r + L_{rr})I \end{array}\right] \quad , \quad (9.54) \end{aligned}$$

The result in Eq. 9.54 recognizes the mutual inductance submatrix can be expressed in terms of the rotation transformation given in Eq. 6.13. This form of the induction machine model was developed by applying T_{23} to the original physical model, thereby projecting stator abc quantities onto a stator $\alpha\beta$ coordinate system. Similarly, rotor abc quantities are projected onto a rotor $\alpha\beta$ coordinate system.

Just like the sinusoidal steady state model, we need all of the dynamics projected into the same reference frame. To accomplish this, we can recognize that the rotor $\alpha\beta$ quantities can be projected onto the stator $\alpha\beta$ frame by using the rotary transformation of Eq. 6.13, where the (electrical) angle between the two reference frames is θ_e. To be clear about whether quantities are relative to the stator or rotor $\alpha\beta$ reference frame, we will again use a

9.4. Two Phase Equivalent Models

superscript s to denote the stator reference frame and a superscript r to denote the rotor reference frame. That is,

$$\lambda^r_{r,\alpha\beta} = e^{J\theta_e} \lambda^s_{r,\alpha\beta} \quad , \tag{9.55}$$

and

$$i^r_{r,\alpha\beta} = e^{J\theta_e} i^s_{r,\alpha\beta} \quad . \tag{9.56}$$

Substituting these two expressions into Eq. 9.52 gives

$$\begin{bmatrix} \lambda^s_{s,\alpha\beta} \\ e^{J\theta_e} \lambda^s_{r,\alpha\beta} \end{bmatrix} = \begin{bmatrix} L_S I & M e^{-J\theta_e} \\ M e^{J\theta_e} & L_R I \end{bmatrix} \begin{bmatrix} i^s_{s,\alpha\beta} \\ e^{J\theta_e} i^s_{r,\alpha\beta} \end{bmatrix} \quad , \tag{9.57}$$

which reduces to

$$\begin{bmatrix} \lambda^s_{s,\alpha\beta} \\ \lambda^s_{r,\alpha\beta} \end{bmatrix} = \begin{bmatrix} L_S I & M I \\ M I & L_R I \end{bmatrix} \begin{bmatrix} i^s_{s,\alpha\beta} \\ i^s_{r,\alpha\beta} \end{bmatrix} \quad , \tag{9.58}$$

where

$$L_S = L_s + L_{ss} \quad , \tag{9.59}$$

$$M = \frac{3}{2} L_{sr} \quad , \tag{9.60}$$

and

$$L_R = L_r + L_{rr} \quad . \tag{9.61}$$

The stator voltage equations are as given in Eq. 9.53:

$$v^s_{s,\alpha\beta} = R_s i^s_{s,\alpha\beta} + \frac{d\lambda^s_{s,\alpha\beta}}{dt} \quad . \tag{9.62}$$

The rotor voltage equations need to be transformed to the stator reference frame, again by using the rotary transformation:

$$e^{J\theta_e} v^s_{r,\alpha\beta} = R_r e^{J\theta_e} i^s_{r,\alpha\beta} + \frac{d}{dt} \left(e^{J\theta_e} \lambda^s_{r,\alpha\beta} \right) \quad . \tag{9.63}$$

Carrying out the indicated differentiation gives

$$e^{J\theta_e} v^s_{r,\alpha\beta} = R_r e^{J\theta_e} i^s_{r,\alpha\beta} + \frac{d\theta_e}{dt} J e^{J\theta_e} \lambda^s_{r,\alpha\beta} + e^{JP\theta} \frac{d\lambda^s_{r,\alpha\beta}}{dt} \quad . \tag{9.64}$$

Recognizing that $J e^{J\theta_e} = e^{J\theta_e} J$, this reduces to

$$v^s_{r,\alpha\beta} = R_r i^s_{r,\alpha\beta} + \frac{d\theta_e}{dt} J \lambda^s_{r,\alpha\beta} + \frac{d\lambda^s_{r,\alpha\beta}}{dt} \quad . \tag{9.65}$$

We are now in a position to develop equivalent circuits for the α and β axis dynamics. These equivalent circuits are shown in Figs. 9.6 and 9.7 for

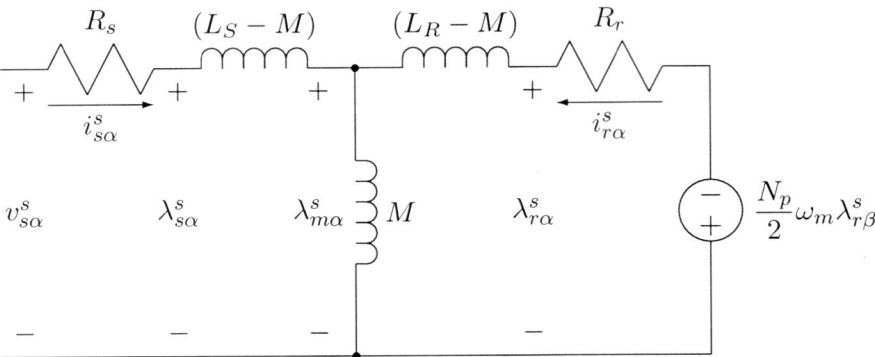

Figure 9.6: The equivalent α axis circuit for the induction machine, with all quantities referred to the stator frame.

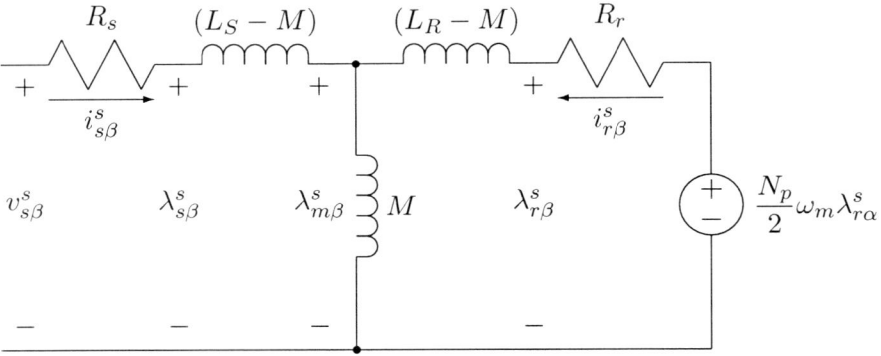

Figure 9.7: The equivalent β axis circuit for the induction machine, with all quantities referred to the stator frame.

the α and β axes, respectively. Each equivalent circuit model is developed by using the stator voltage equation on the input side (Eq. 9.62), the flux linkage (Eq. 9.58) expression to develop the T inductance network, and the rotor voltage equation (Eq. 9.65) to complete the circuit. The process of constructing the $\alpha\beta$ circuit description has a lot in common with the approach used to develop the phase equivalent model.

In the $\alpha\beta$ model, the mechanical dynamics are again given by Eq. 9.13

9.4. Two Phase Equivalent Models

where the electromagnetic torque is given by

$$T^e = \frac{N_p M}{2} \left(i^s_{s\beta} i^s_{r\alpha} - i^s_{s\alpha} i^s_{r\beta} \right) \quad . \tag{9.66}$$

Development of Eq. 9.66 is straightforward when starting with the general torque expression of Eq. 9.12 and using Eq. 9.52 and the stator and rotor currents in their natural reference frames. The rotor currents are subsequently transformed to their stator frame equivalents.

9.4.2 The dq Model

The development of the $\alpha\beta$ model started with the physical model and decoupled the three phases. The dq model starts with the $\alpha\beta$ model and projects quantities onto a common reference frame. Figure 9.8 shows the relationship among the three coordinate systems, where the superscript s and r indicate stator and rotor reference frames, respectively. Accordingly, we can transform the $\alpha\beta$ dynamics of the stator by rotation through an angle ϕ_e, thereby projecting them on the dq reference frame. The $\alpha\beta$ dynamics of the rotor are transformed to the dq frame through rotation of an angle $(\phi_e - \theta_e)$. It follows that

$$\begin{bmatrix} \lambda_{sd} \\ \lambda_{sq} \\ \lambda_{rd} \\ \lambda_{rq} \end{bmatrix} = \begin{bmatrix} e^{J\phi_e} & 0 \\ 0 & e^{J(\phi_e - \theta_e)} \end{bmatrix} \begin{bmatrix} \lambda_{s\alpha} \\ \lambda_{s\beta} \\ \lambda_{r\alpha} \\ \lambda_{r\beta} \end{bmatrix} \quad . \tag{9.67}$$

With

$$L_{\alpha\beta}(\theta_e) = \begin{bmatrix} L_S I & M e^{-J\theta_e} \\ M e^{J\theta_e} & L_R I \end{bmatrix} , \tag{9.68}$$

then our projection of all quantities into the dq reference frame becomes

$$\begin{bmatrix} \lambda_{sd} \\ \lambda_{sq} \\ \lambda_{rd} \\ \lambda_{rq} \end{bmatrix} = \begin{bmatrix} e^{J\phi_e} & 0 \\ 0 & e^{J(\phi_e - \theta_e)} \end{bmatrix} \begin{bmatrix} L_S I & M e^{-J\theta_e} \\ M e^{J\theta_e} & L_R I \end{bmatrix} \times$$

$$\begin{bmatrix} e^{-J\phi_e} & 0 \\ 0 & e^{-J(\phi_e - \theta_e)} \end{bmatrix} \begin{bmatrix} i_{sd} \\ i_{sq} \\ i_{rd} \\ i_{rq} \end{bmatrix} \quad . \tag{9.69}$$

Completing the indicated multiplication gives

$$\begin{bmatrix} \lambda_{sd} \\ \lambda_{sq} \\ \lambda_{rd} \\ \lambda_{rq} \end{bmatrix} = \begin{bmatrix} L_S I & M I \\ M I & L_R I \end{bmatrix} \begin{bmatrix} i_{sd} \\ i_{sq} \\ i_{rd} \\ i_{rq} \end{bmatrix} \quad . \tag{9.70}$$

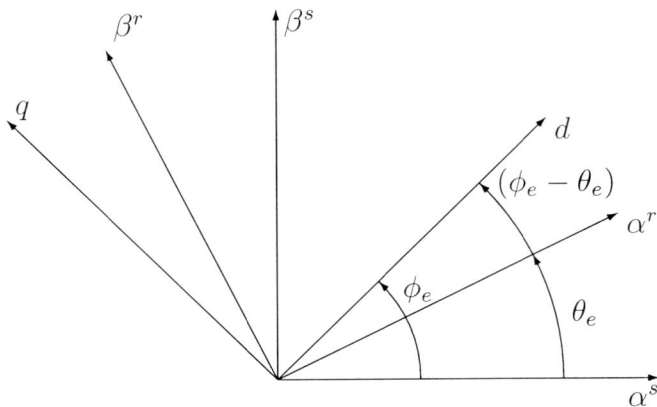

Figure 9.8: The relationship among the stator $\alpha\beta$ axes, the rotor $\alpha\beta$ axes, and the axes of the dq reference frame.

In a similar manner,

$$v_{\alpha\beta} = \frac{d\lambda_{\alpha\beta}}{dt} + R_{\alpha\beta} i_{\alpha\beta} \qquad (9.71)$$

is transformed into

$$v_{dq} = \frac{d\lambda_{dq}}{dt} + R_{dq} i_{dq} - \begin{bmatrix} \frac{d\phi_e}{dt} J & 0 \\ 0 & \left(\frac{d\phi_e}{dt} - \frac{d\theta_e}{dt}\right) J \end{bmatrix} \lambda_{dq} \quad . \qquad (9.72)$$

The last term indicates that the electrical and mechanical dynamics are still coupled, though they have been simplified significantly by eliminating the spatial coupling between current and flux. Through the matrix J, we see that direct axis flux produces voltage in the quadrature axis, and flux in the quadrature axis produces negative voltage in the direct axis. The quantity $d\phi_e/dt = \omega_s$ is the electrical angular velocity of the air gap magnetic field and is dictated by the electrical frequency of the stator currents. The quantity $(d\phi_e/dt - d\theta_e/dt) = \omega_{sl} = s\omega_{\text{syn}}$ is the slip frequency, and indicates the electrical velocity of the air gap magnetic field relative to the rotor.

In the dq reference frame the dynamics are

$$\lambda_{dq} = \begin{bmatrix} L_S I & M I \\ M I & L_R I \end{bmatrix} i_{dq} \quad ; \qquad (9.73)$$

$$v_{dq} = \frac{d\lambda_{dq}}{dt} + R_{dq} i_{dq} - \begin{bmatrix} \omega_s J & 0 \\ 0 & \omega_{sl} J \end{bmatrix} \lambda_{dq} \quad . \qquad (9.74)$$

9.4. Two Phase Equivalent Models

The torque transforms into

$$T^e = \frac{3N_p L_{sr}}{8} i_{dq}^T \begin{bmatrix} 0 & -J \\ J & 0 \end{bmatrix} i_{dq} \quad , \tag{9.75}$$

which reduces to

$$T^e = \frac{3N_p L_{sr}}{4} (i_{sq}i_{rd} - i_{sd}i_{rq}) = \frac{N_p}{2} M (i_{sq}i_{rd} - i_{sd}i_{rq}) \quad . \tag{9.76}$$

Equation 9.76 shows that the torque can be viewed as the vector cross product between stator and rotor currents (or fluxes) along the direct and quadrature axes.

Equations 9.74, 9.73 and 9.76 can be used to develop two descriptions of the induction machine that are useful for simulation. The first description is an equivalent circuit model of the induction machine in the dq reference frame. The second description is a block diagram of the induction machine.

Construction of the equivalent circuit model focuses on Eqs. 9.74 and 9.73. To construct the equivalent circuits, we begin with the stator voltage equations that show that each terminal voltage (v_{sd} and v_{sq}) is equal to the superposition of a resistive voltage drop, a motional voltage, and an induced voltage. The next step in the construction is to make use of the flux linkage relationships, Eq. 9.73, and recognizing that these relationships can be cast as an inductor T network comprised of three inductances: $(L_S - M)$, M, and $(L_R - M)$. The final step is to make use of the rotor voltage equations in Eq. 9.74.

Figure 9.9 shows the direct axis circuit for the induction machine. Proceeding from the left, the stator voltage equation motivates the circuit up to the stator flux linkage. The T network of inductors comes from the flux linkage relationships. The balance of the circuit comes from the rotor voltage equation. Figure 9.10 shows the quadrature axis circuit for the induction machine. In both of these circuits, the quantities λ_{md} and λ_{mq} represent the mutual fluxes in the direct and quadrature axes, respectively. The mutual fluxes are those that are seen by both the stator and the rotor. Physically these fluxes are in the air gap of the machine.

To construct a block diagram model of the induction machine, we begin with the voltage equations for the rotor. Inserting the rotor flux linkage relationships, solved for the rotor current, allows formulation of dynamic equations for the rotor fluxes in terms of rotor flux and stator current. The final stage of the block diagram goes from rotor fluxes to electromagnetic torque. The expression shown in the block diagram can be found by taking Eq. 9.76 and substituting the rotor fluxes for the rotor currents and reducing. Figure 9.11 shows the block diagram for the induction machine. In the block

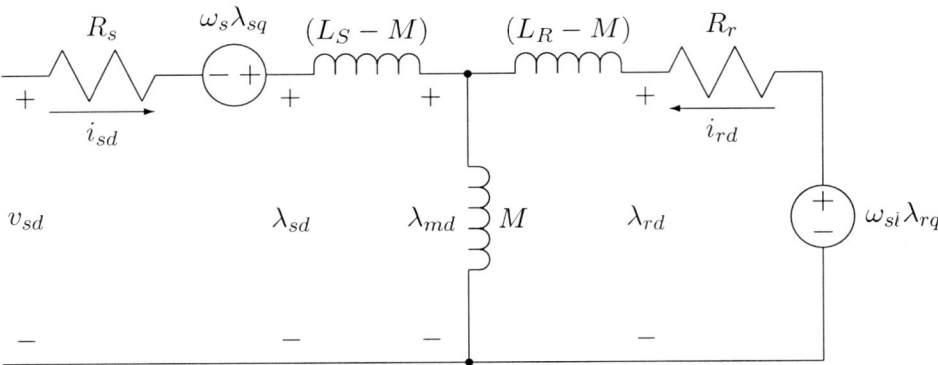

Figure 9.9: The equivalent direct axis circuit for the induction machine.

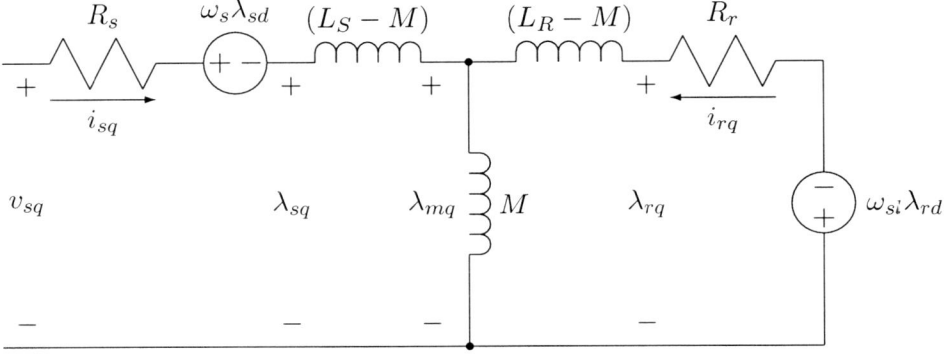

Figure 9.10: The equivalent quadrature axis circuit for the induction machine.

diagram, it should be noted that the rotor electrical time constant is used, defined as

$$\tau_r = \frac{L_R}{R_r} .$$ (9.77)

The block diagram shows the significant coupling between the direct and quadrature axes of the induction machine, in both the electrical dynamics and in the production of electromagnetic torque.

9.5. Using the dq Induction Machine Model

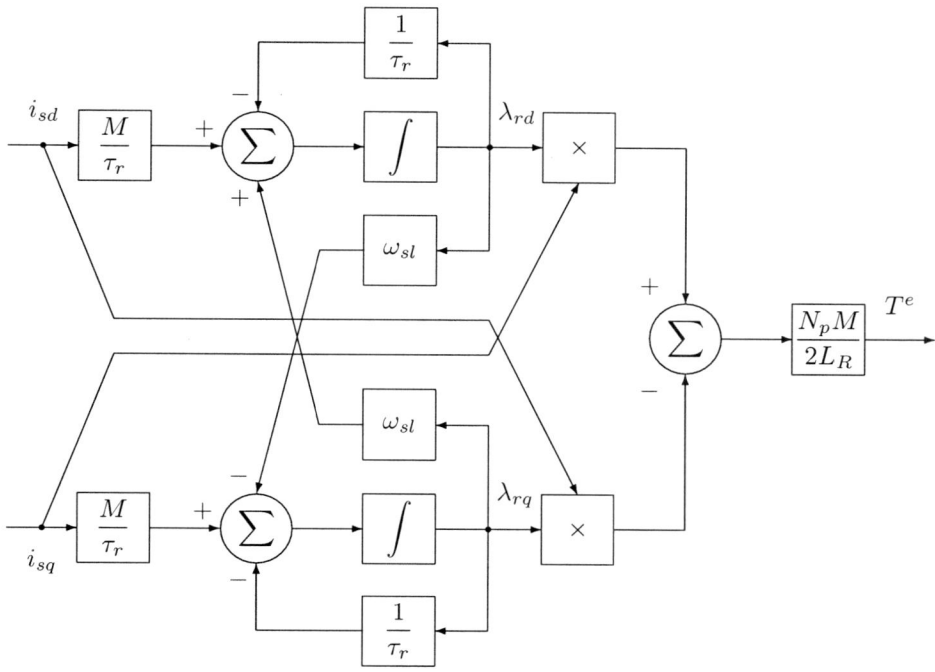

Figure 9.11: A block diagram representation for the induction machine.

9.5 Using the dq Induction Machine Model

The induction machine model developed in the last section really works! To demonstrate the model, it is applied to the induction machine described in Table 9.2; this is the same induction machine described in Table 9.1. The induction machine is assumed to be line-started from a constant amplitude, constant frequency source while driving a fan load, so the load torque is taken to be

$$\tau_l = k_t \omega_m^2 \quad . \tag{9.78}$$

The induction machine is simulated in Matlab; the source code is given in Appendix C. The system could just as easily have been simulated using Simulink and the block diagram of Fig. 9.11, augmented to include the stator dynamics when excited by a voltage source.

The simulation results from Matlab are given in Figs. 9.12 through 9.19 and reflect system variables in both the stationary and rotating reference frames. Figure 9.12 shows the rotor speed of the induction machine. The oscillations are caused by the torque ripple that is created by the transient response of the stator and rotor magnetic circuits. Figure 9.13 shows the

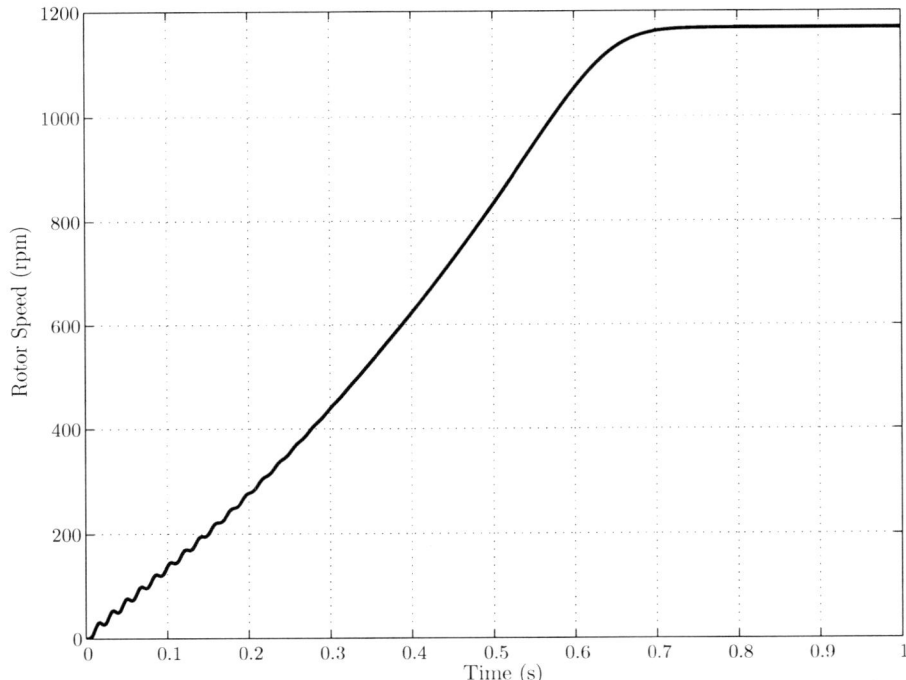

Figure 9.12: The rotor speed during the line-start transient.

electromagnetic and load torques. Again, the torque ripple is caused by the transient response of the magnetic circuits. Figure 9.14 gives the direct and quadrature stator fluxes. The tendency of λ_{sq} to drop to zero in steady state suggests a natural tendency of the induction machine to adopt a field-oriented state; this will be discussed further in Chapter 10. Figure 9.15 gives the direct and quadrature rotor fluxes. Figure 9.16 gives the direct and quadrature stator currents. Figure 9.17 gives the direct and quadrature rotor currents. Figures 9.18 and 9.19 give the stator and rotor currents in phase a in the stationary abc frame, respectively. The frequency of the rotor currents drops as the rotor speed increases and the slip decreases.

9.6 Summary

This chapter has focused on the induction machine and the development of models that support not only a physical understanding of the machine, but can be used for understanding how to control the machine. The physical model of the induction machine was motivated by capturing the interaction between the stator and rotor circuits that is embodied in the inductance

9.6. Summary

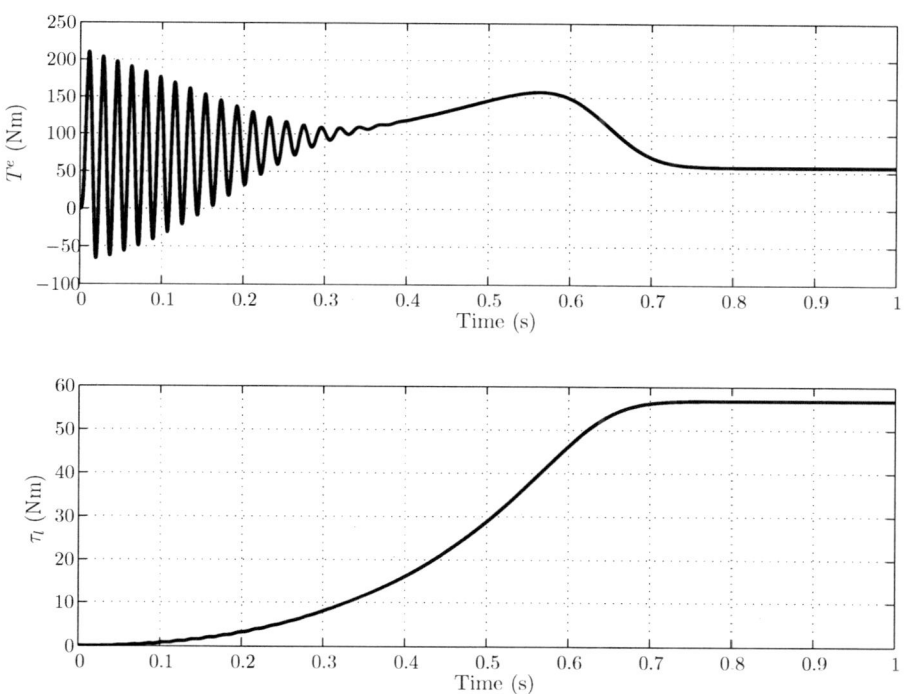

Figure 9.13: The electromagnetic and load torques during the line-start transient.

matrix. The physical model started by modeling rotor quantities relative to the rotor and stator quantities relative to the stator.

To move toward a consistent representation of the machine, we employed the transformations developed in Chapter 6 to move currents and flux linkages from one reference frame to another. In the single phase equivalent model, rotor quantities were moved to the stator while preserving the coupled three-phase nature of the machine.

In the $\alpha\beta$ model, stator and rotor quantities were taken from a coupled abc representation to a decoupled $\alpha\beta$ representation. The 0 phase components were suppressed because they do not contribute to torque production. For the $\alpha\beta$ model rotor quantities were again put into the stator reference frame.

The dq model was developed by projecting stator and rotor quantities of the $\alpha\beta$ model into a common reference frame that is allowed to rotate. It is quite natural to tie this dq reference frame to the air gap magnetic fields. As we will see in Chapter 10, the direct axis is usually tied to either the rotor, air gap, or stator flux, establishing the basis for a technique known as field

Table 9.2: The parameters used to simulate an induction machine using Matlab.

Parameter	Value	Units
Rated Power	7.46	kW
Rated Line Voltage	220	V
Rated Frequency	60	Hz
Rated Speed	1164	rpm
N_p	6	
R_s	0.249	Ω/phase
R_r	0.156	Ω/phase
$L_s - 2L_{ss}$	1.39	mH/phase
$L_r - 2L_{rr}$	0.74	mH/phase
L_{sr}	41.0	mH/phase
H	0.5	kg·m^2
k_t	0.0038	kg·m^2

orientation.

9.6. Summary

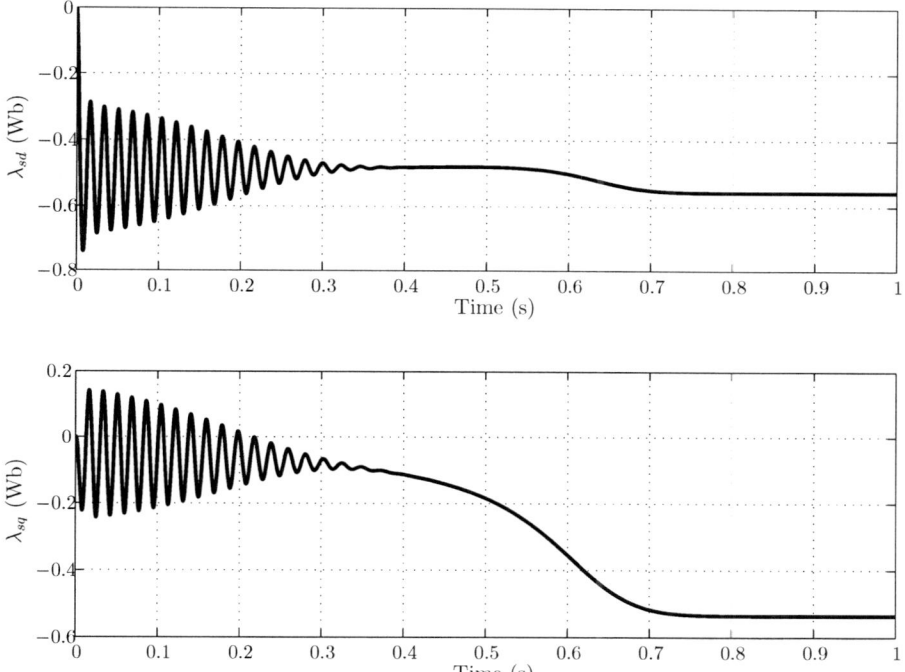

Figure 9.14: The direct and quadrature stator fluxes during the line-start transient.

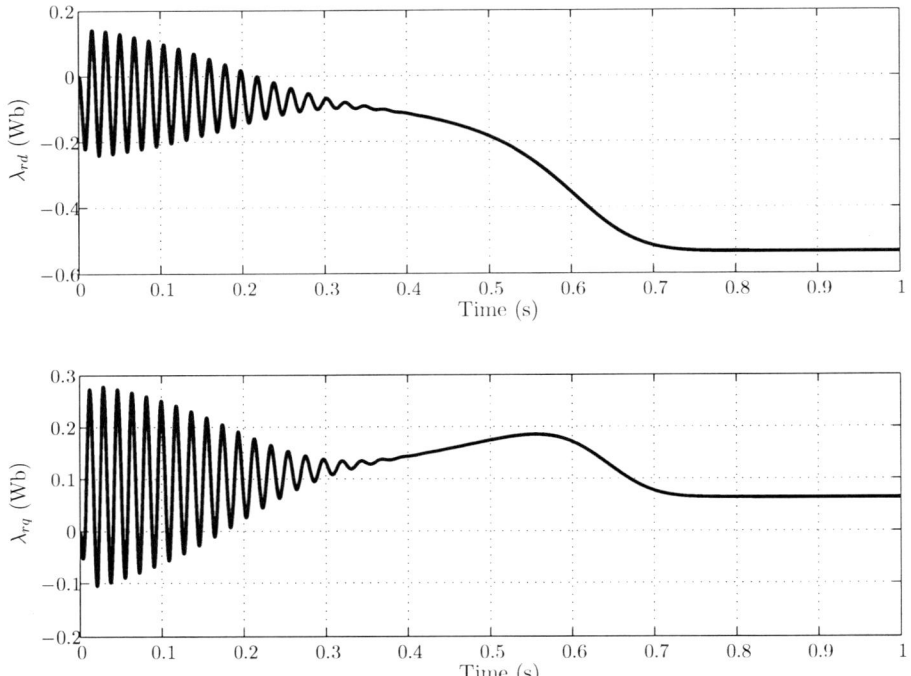

Figure 9.15: The direct and quadrature rotor fluxes during the line-start transient.

9.6. Summary

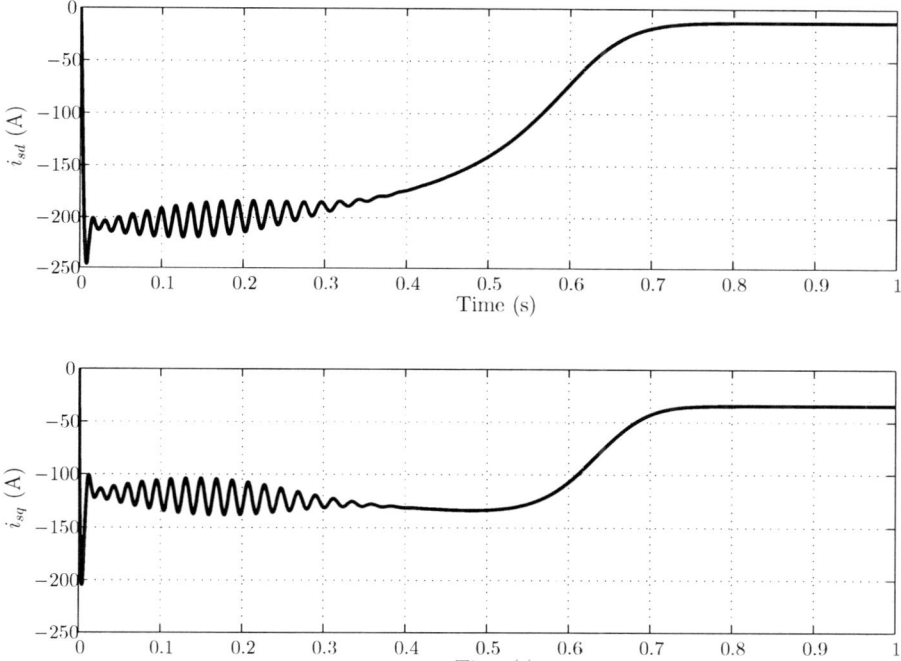

Figure 9.16: The direct and quadrature stator currents during the line-start transient.

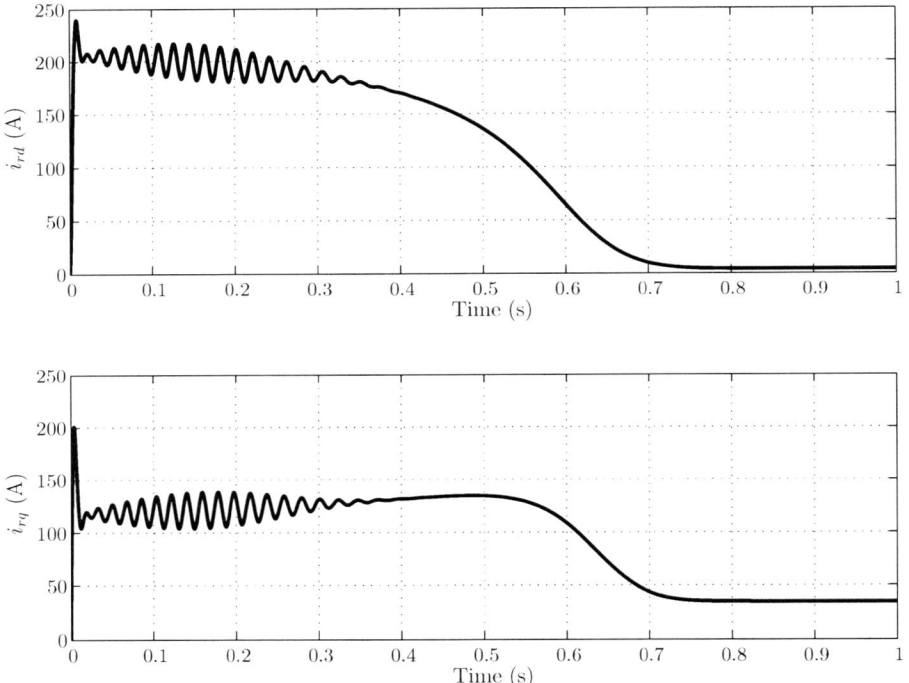

Figure 9.17: The direct and quadrature rotor currents during the line-start transient.

9.6. Summary

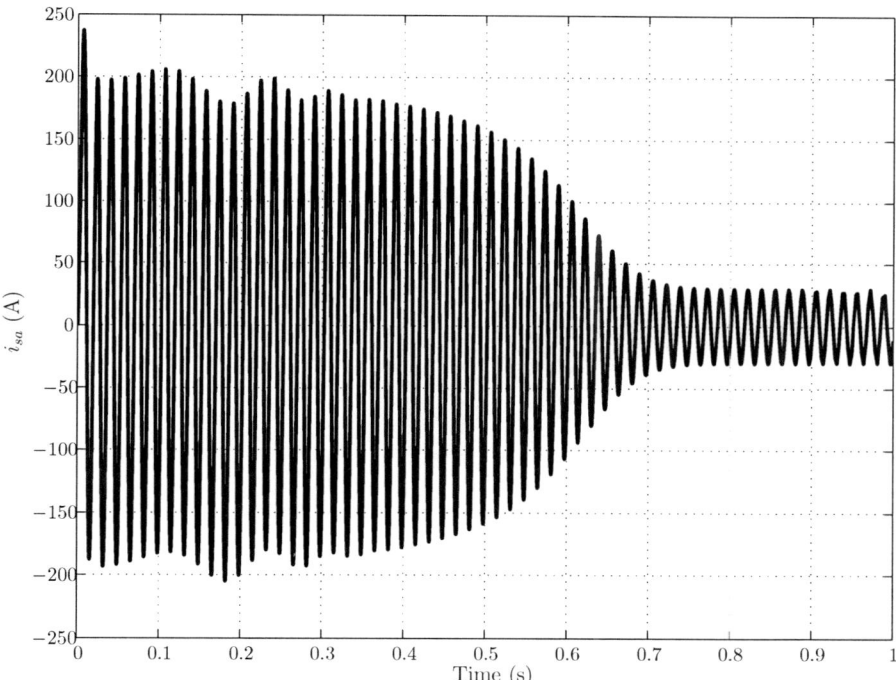

Figure 9.18: The phase a stator current during the line-start transient in the stationary reference frame.

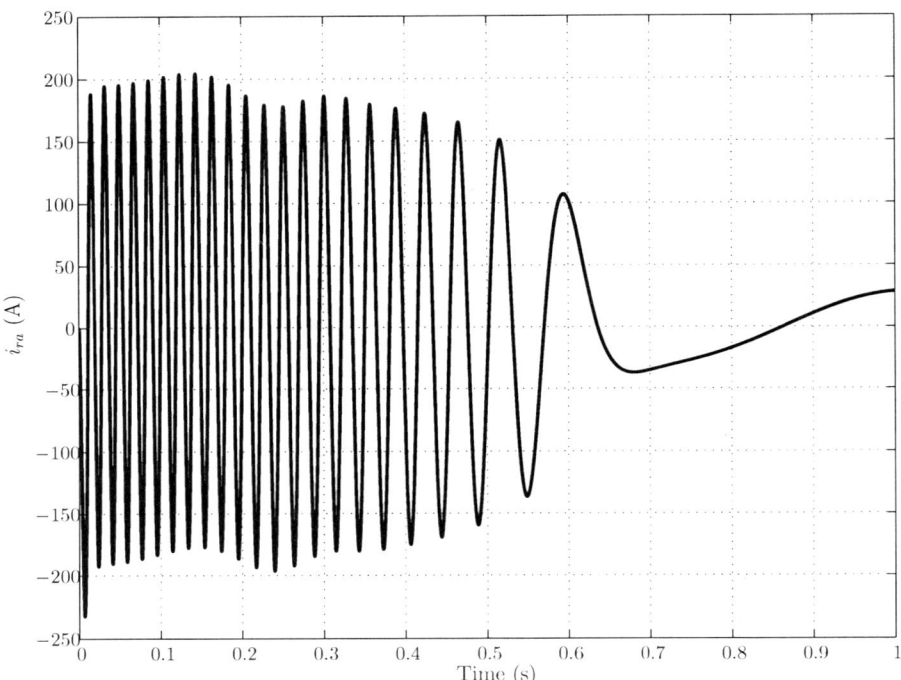

Figure 9.19: The phase a rotor current during the line-start transient.

Chapter 10

Control of Induction Machines

10.1 Introduction

The analysis of the single phase equivalent circuit model undertaken in Chapter 9 shows that the induction machine is nearly a constant-speed device. This begs the question of how to adjust the speed. Again our analysis of the single phase equivalent circuit suggests that varying the voltage adjusts the peak torque, but the steepness of the torque speed curve in the vicinity of the synchronous speed suggests that varying the voltage by itself is going to be quite ineffective as a control method.

From Fig. 9.5, we see that operation is always going to be in the vicinity of ω_{syn}, which is tied to ω_s through the pole structure of the machine. This suggests that adjusting the frequency of the voltage applied to the stator could be quite an effective way to control the speed of the induction machine. Adjusting the speed of an induction machine through adjusting the stator frequency is often called scalar control. We will see that the voltage amplitude needs to be adjusted with the frequency. While effective, there are limitations on its performance, particularly as the speed reduces toward zero.

A more sophisticated form of control is the method of field orientation. This method was developed in the early 1970's, and has become widely adopted throughout industry[1]. Under field orientation, control is facilitated

---

[1] F. Blaschke, "Das Verfahren der Feldorientierung zur Regelung der Drehfelmachine" ("The method of field orientation for control of three phase machines"), Ph.D. Dissertation, TU Braunschweig, 1973; and F. Blaschke, "The principle of field orientation as applied to the new transvektor closed-loop control system for rotating-field machines," *Siemens*

from the perspective of an observer that is riding through the air gap of the machine at the same speed as the rotating magnetic fields. That is, we are in the dq reference frame. A particular magnetic field is used to define the direct axis; this magnetic field could be that of the rotor, stator, or the air gap.

Direct torque control of the induction machine bears some similarity to field orientation, but it is trying to directly select the stator voltages that are required to simultaneously keep torque and speed within acceptable windows.

10.2 Scalar Control

The simplest form of speed control for an induction machine is known as scalar control, and consists of adjusting the frequency applied to the stator. As we will see, however, it is not sufficient to simply adjust the frequency. Typically the voltage is also adjusted as a function of frequency.

Figure 10.1 shows two torque speed characteristics for the induction machine of Table 9.1. The characteristic passing through zero torque at 1200 rpm is based on a line-line voltage of 220 V and a frequency of 60 Hz. The characteristic passing through zero torque at 600 rpm is based on a line-line voltage of 220 V and a frequency of 30 Hz. It is a safe assumption that the magnetically linear model of the induction has been pushed beyond its reasonable limits. The actual torque would not reach the values shown because magnetic saturation will limit the torque.

Figure 10.1 shows that it is insufficient to vary speed by adjusting only the frequency. To do so increases the flux to an unreasonable level. To maintain a reasonable level of flux in the machine, the voltage must be adjusted with the frequency. To understand why, consider that the flux is proportional to the integral of the applied voltage. If the applied voltage is

$$v_s = V_s \cos \omega_s t \quad , \tag{10.1}$$

then the flux linkage, neglecting stator resistance, is

$$\lambda_s = \int V_s \cos \omega_s t \, dt \quad , \tag{10.2}$$

or

$$\lambda_s = -\frac{V_s}{\omega_s} \sin \omega_s t \quad . \tag{10.3}$$

To first order, the amplitude of the flux linkage is maintained constant if the ratio of voltage to frequency is maintained constant.

Review, Vol. 34, pp. 217-220, May 1972.

10.2. Scalar Control

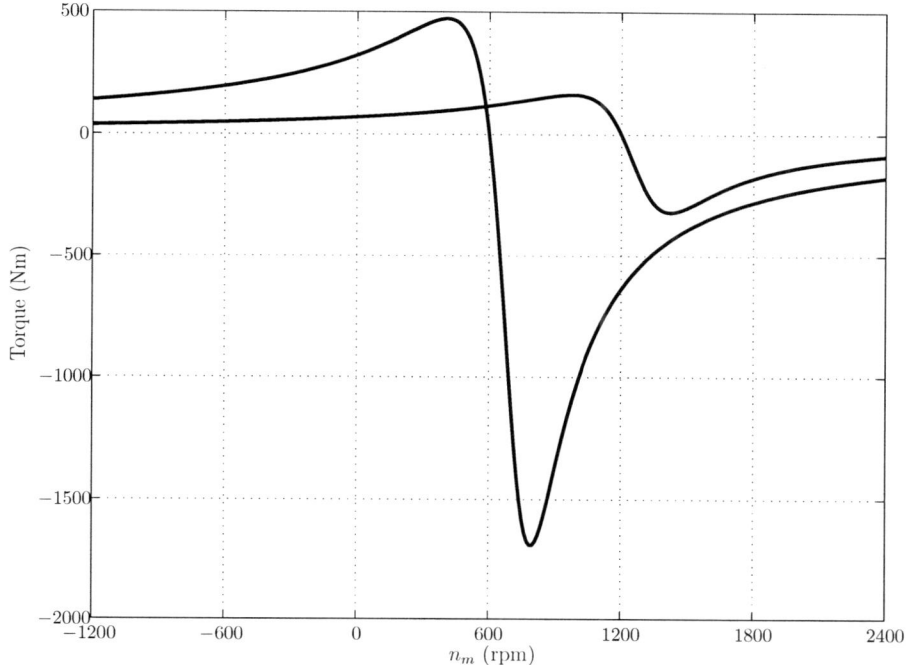

Figure 10.1: Two torque speed characteristics for the induction machine of Table 9.1 when the voltage amplitude is kept constant and the frequency is changed.

Figure 10.2 shows a family of torque speed curves for the induction machine of Table 9.1 when the applied voltage is scaled with frequency. Figure 10.3 shows the same family focusing in on the region of positive torque production. The applied frequencies are 60 Hz, 45 Hz, 30 Hz, and 15 Hz. Tracking the breakdown torque as a function of frequency, we see that maintaining constant breakdown torque requires more than maintaining a constant ratio between applied voltage and frequency. The reason for this is the influence of the stator resistance. As the frequency drops to 15 Hz (synchronous speed of 300 rpm) the breakdown torque has begun to roll off significantly. To compensate for this roll off, it is common to boost the applied voltage at low frequencies, such as shown in Fig. 10.4.

Scalar control provides a modest level of performance over a fairly broad speed range of 8 : 1. For a four-pole machine, this would correspond to a speed range of 225 rpm to 1800 rpm. Performance begins to deteriorate when the voltage necessary to maintain constant flux exceeds the supply voltage; this typically happens shortly above rated synchronous speed. Performance also begins to deteriorate as the commanded speed gets small, due to the

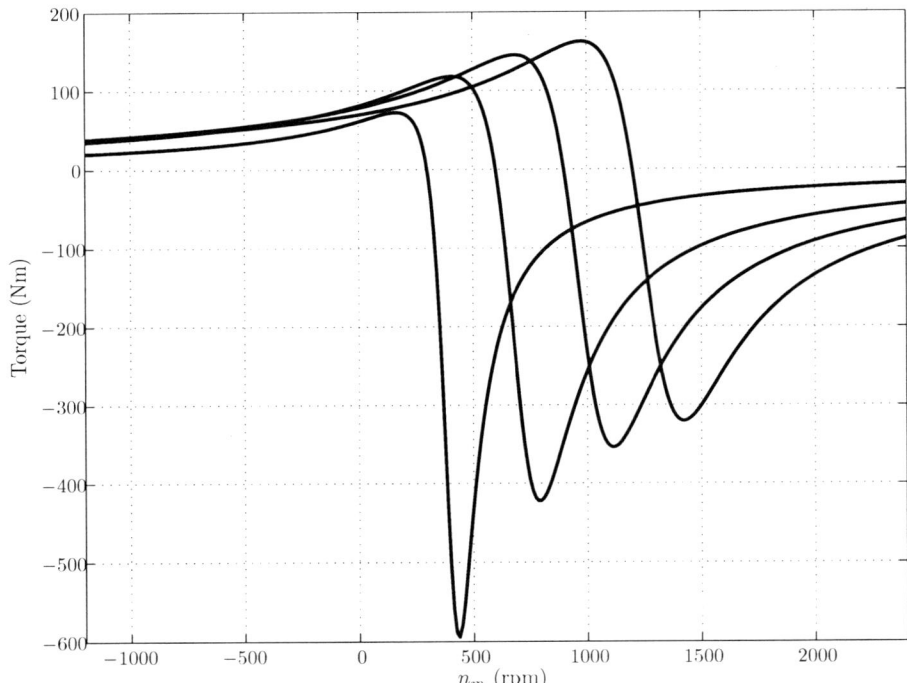

Figure 10.2: A family of torque speed characteristics for the induction machine of Table 9.1 when the voltage amplitude is scaled with frequency. Frequencies of 60, 45, 30, and 15 Hz are shown.

increasing influence of stator resistance and difficulty in inducing significant rotor currents at lower frequencies. If a higher level of performance is required, either in terms of dynamic response or wider dynamic range, it is necessary to consider a more sophisticated approach to control. One such approach is field oriented control.

10.3 Field Oriented Control

Field orientation is a technique that structures the control of an induction to be entirely parallel to that of a separately excited dc machine. That is, the field flux is oriented to be orthogonal to the torque-producing current. This section outlines how this is accomplished. There are some excellent texts on field oriented control[2].

---

[2]See, for example, A. M. Trzynadlowski, *The Field Orientation Principle in Control of Induction Motors*, Kluwer, 1994; and D. W. Novotny and T. A. Lipo, *Vector Control and*

10.3. Field Oriented Control

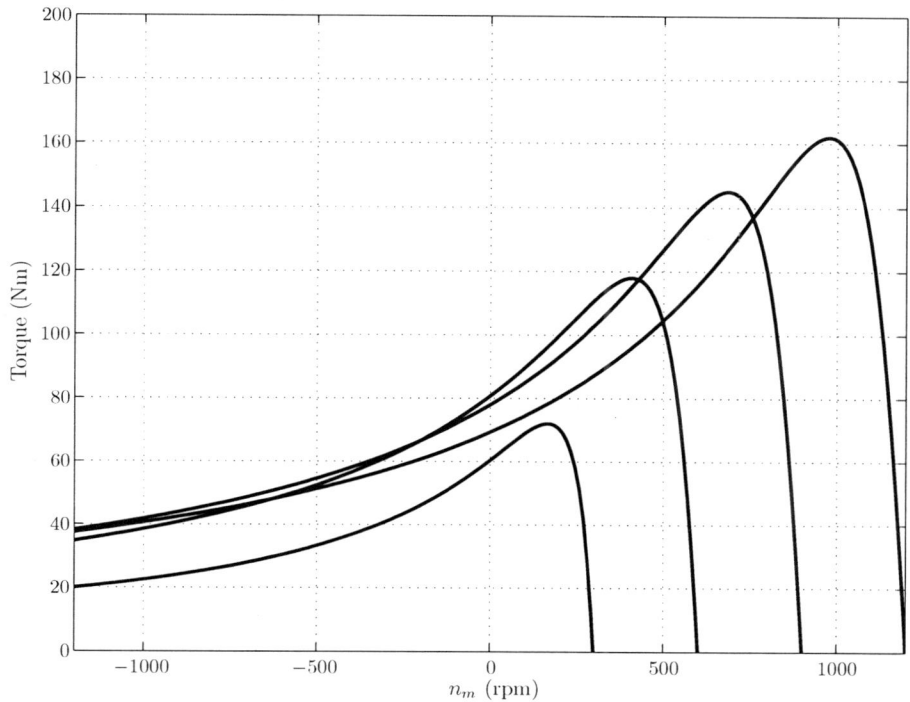

Figure 10.3: The family of torque speed characteristics from Fig. 10.2, homing in on the positive torque region of the characteristic.

10.3.1 Torque Expressions

The starting point with field orientation is the torque expression, Eq. 9.76, which is repeated here for convenience.

$$T^e = \frac{N_p}{2} M \left(i_{sq} i_{rd} - i_{sd} i_{rq} \right) \quad . \tag{10.4}$$

Equation 10.4 gives the electromagnetic torque in terms of a vector cross product between stator and rotor currents. We can, however, substitute stator or rotor flux linkages for currents by using the flux linkage expressions, Eq. 9.73. We can also substitute the mutual (air gap) flux for currents by using

$$\begin{bmatrix} \lambda_{md} \\ \lambda_{mq} \end{bmatrix} = M \begin{bmatrix} 1 & 0 & 1 & 0 \\ 0 & 1 & 0 & 1 \end{bmatrix} \begin{bmatrix} i_{sd} \\ i_{sq} \\ i_{rd} \\ i_{rq} \end{bmatrix} \quad . \tag{10.5}$$

Dynamics of AC Drives, Oxford University Press, 1997.

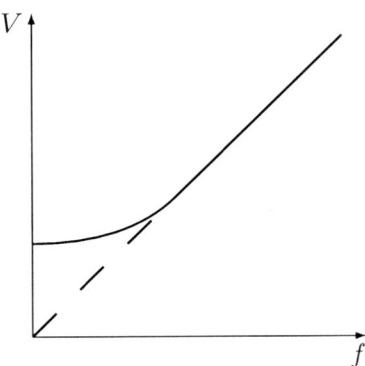

Figure 10.4: The voltage-frequency relationship typically used for scalar control of induction machine speed, showing the voltage boost at low frequencies to compensate for the increasing influence of stator resistance.

The mutual fluxes are indicated in Figs. 9.9 and 9.10. The mutual flux is taken to be the flux in the air gap of the machine that links both the stator and rotor windings.

There are three commonly discussed versions of field orientation: rotor, stator and air gap. In each, the torque is given by the vector cross product between fluxes and stators currents. The fluxes involved are tied to the type of orientation. That is, rotor orientation uses rotor fluxes. Stator orientation uses stator fluxes and air gap orientation uses mutual fluxes. The following subsections develop the appropriate torque expressions.

Rotor Flux Orientation

Here we want the torque in terms of the stator currents and the rotor fluxes. From Eq. 9.73 we have

$$i_{rd} = \frac{1}{L_R}\lambda_{rd} - \frac{M}{L_R}i_{sd} \quad , \tag{10.6}$$

and

$$i_{rq} = \frac{1}{L_R}\lambda_{rq} - \frac{M}{L_R}i_{sq} \quad . \tag{10.7}$$

Putting these into Eq. 10.4 gives

$$T^e = \frac{N_p}{2}\frac{M}{L_R}(i_{sq}\lambda_{rd} - i_{sd}\lambda_{rq}) \quad . \tag{10.8}$$

10.3. Field Oriented Control

Stator Flux Orientation

Here we want the torque in terms of the stator currents and the stator fluxes. From Eq. 9.73 we have

$$i_{rd} = \frac{1}{M}\lambda_{sd} - \frac{L_S}{M}i_{sd} , \qquad (10.9)$$

and

$$i_{rq} = \frac{1}{M}\lambda_{sq} - \frac{L_S}{M}i_{sq} . \qquad (10.10)$$

Putting these into Eq. 10.4 gives

$$T^e = \frac{N_p}{2}(i_{sq}\lambda_{sd} - i_{sd}\lambda_{sq}) . \qquad (10.11)$$

Air Gap Flux Orientation

Here we want the torque in terms of the stator currents and the air gap fluxes. From Eq. 10.5 we have

$$i_{rd} = \frac{1}{M}\lambda_{md} - \frac{1}{M}i_{sd} , \qquad (10.12)$$

and

$$i_{rq} = \frac{1}{M}\lambda_{mq} - \frac{1}{M}i_{sq} . \qquad (10.13)$$

Putting these into Eq. 10.4 gives

$$T^e = \frac{N_p}{2}(i_{sq}\lambda_{md} - i_{sd}\lambda_{mq}) . \qquad (10.14)$$

10.3.2 Field Orientation

The starting point for a discussion of field orientation is with the rotor; consideration of stator and air gap orientation are given below. Consider Eq. 10.8. If we are able to force $\lambda_{rq} = 0$ the torque is given by

$$T^e = \frac{N_p}{2}\frac{M}{L_R}(i_{sq}\lambda_{rd}) . \qquad (10.15)$$

This implies that currents fed to the machine are used for two things: establishing λ_{rd} and producing torque. As we will see, a current component in phase with λ_{rd} is responsible for establishing λ_{rd}. The current component in quadrature with λ_{rd} is responsible for producing torque. That is, currents along the direct axis are responsible for producing flux, while currents along the quadrature axis are responsible for producing torque.

Assuming the (rotor) field orientation condition ($\lambda_{rq} = 0$) is satisfied and the machine is supplied from a current-regulated source, our rotor electrical dynamics become

$$\frac{d\lambda_{rd}}{dt} = -R_r i_{rd} \tag{10.16}$$

on the direct axis and

$$\frac{d\lambda_{rq}}{dt} = -R_r i_{rq} - \omega_{sl}\lambda_{rd} \quad (=0) \tag{10.17}$$

on the quadrature axis. Using the relationship between quadrature rotor flux and the currents gives

$$\lambda_{rq} = Mi_{sq} + L_R i_{rq} = 0 \implies i_{rq} = -\frac{M}{L_R}i_{sq} \quad . \tag{10.18}$$

Combining these last two expressions gives the so-called slip condition:

$$\omega_{sl} = \frac{R_r}{L_R}M\frac{i_{sq}}{\lambda_{rd}} = \frac{M}{\tau_r}\frac{i_{sq}}{\lambda_{rd}} \quad . \tag{10.19}$$

(There are different slip conditions for rotor, stator, and air gap field orientation.)

In steady state $i_{rd} = 0$; we have seen this in Fig. 9.17. During transients we have

$$i_{rd} = -\frac{1}{R_r}\frac{d\lambda_{rd}}{dt} = \frac{\lambda_{rd} - Mi_{sd}}{L_R} \quad , \tag{10.20}$$

from which we get

$$\frac{d\lambda_{rd}}{dt} + \frac{1}{\tau_r}\lambda_{rd} = \frac{M}{\tau_r}i_{sd} \quad . \tag{10.21}$$

This shows that direct axis stator current only drives direct axis rotor flux, implying a decoupling between the field and torque producing components of stator current.

With this we can draw an analogy between an induction machine and a separately excited dc machine. The direct axis stator current is responsible for setting up the direct axis rotor flux, making this current analogous to the field current in the dc machine. It also follows that the direct axis rotor flux is analogous to the field flux in the dc machine. The quadrature axis stator current is responsible for creating torque in the presence of the field on the direct axis, thereby making the quadrature stator current analogous to the armature current in the dc machine. In summary, the analogy of the induction machine under field orientation and the dc machine can be summarized as:

$$i_{sd} \iff i_f$$

10.3. Field Oriented Control

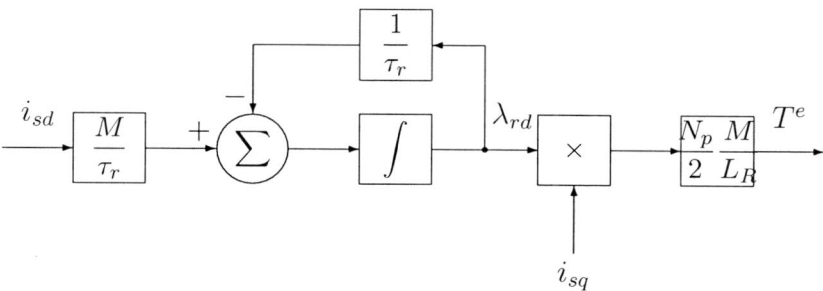

Figure 10.5: A block diagram for the induction machine when operated under rotor field orientation.

$$\lambda_{rd} \iff \phi_f$$
$$i_{sq} \iff i_a$$
$$T^e = \frac{N_p}{2}\frac{M}{L_R}\lambda_{rd}i_{sq} \iff T^e = k\phi_f i_a$$

Under field orientation, we are forcing the induction machine to maintain orthogonality of λ_{rd} and i_{sq} through active control. The commutator accomplishes this function in a dc machine. Figure 10.5 shows a block diagram for the induction machine operating under field orientation. The simplifications over Fig. 9.11 are striking, and reflect the decoupling accomplished by forcing $\lambda_{rq} = 0$.

We can now go back and see that the induction machine naturally gravitates toward field orientation. In Figs. 9.16 and 9.17, we see that the direct axis currents tend toward zero as the machine reaches steady state. The quadrature currents, on the other hand, have to remain at a value that will support the load torque. We also note that the quadrature stator flux is seen to tend toward zero in steady state as well, further suggesting a natural predisposition toward field orientation. These observations lend credence to the field orientation approach, suggesting that if the machine has a natural tendency toward field orientation that it should not require excessive control effort to accomplish it.

10.3.3 Implementation of Field Oriented Control

Field oriented control is implemented by using a controller to enforce the direct and quadrature currents that are consistent with supporting the commanded flux and torque, respectively. The commands for flux and torque are generated by a higher-level controller that is not included in this discussion. The higher-level controller may have the objective, for example, of regulating the speed of the induction machine with high efficiency while having fast dynamic response.

Based on the commanded flux and torque, we generate desired values for i_{sd}^* and i_{sq}^*, respectively. The superscript * indicates a commanded, or reference, value to distinguish it from an actual value. By knowing the rotor position, we can determine the corresponding values of $i_{s\alpha}^*$ and $i_{s\beta}^*$. These currents can be converted into i_{sa}^*, i_{sb}^* and i_{sc}^*. These stator current commands are given to the current-regulated switching inverter responsible for their enforcement. It is common to use a voltage source inverter to feed the induction machine with closed-loop current control that effectively turns the inverter into a three-phase programmed current source. Pulse-width modulation (PWM) is routinely employed to control the inverter switches. An introduction to inverters and current regulators is given in Chapters 7 and 8, respectively.

Figure 10.6 shows a block diagram that gives the general implementation of the direct method for field oriented control of an induction machine. We can tell Fig. 10.6 is a direct method by virtue of the flux sensors that feed air gap flux information to the flux calculator. Generally the flux and torque commands are generated by higher-level controllers. The torque command could be generated, for example, by a controller that is trying to regulate machine speed. The flux command could be generated, for example, by a controller that monitors machine efficiency. Weakening the field could substantially increase the machine efficiency at light loads. The following subsections give more specific details about the direct method and the indirect method of field orientation, respectively. Essential to both approaches is the determination of the excitation angle.

Direct Field Orientation

The direct method of field orientation makes use of the air gap flux to determine the magnitude and orientation of the rotor flux vector. This information subsequently allows calculation of the torque produced by the machine. The outputs of the flux and torque calculators are used to close the flux and torque feedback loops, respectively. The air gap flux is measured directly, perhaps by Hall sensors.

Orientation of the rotor flux is determined as follows:

1. The currents $i_{s\alpha}$ and $i_{s\beta}$ are calculated from the measured currents i_{sa}, i_{sb} and i_{sc} using T_{23}.

10.3. Field Oriented Control

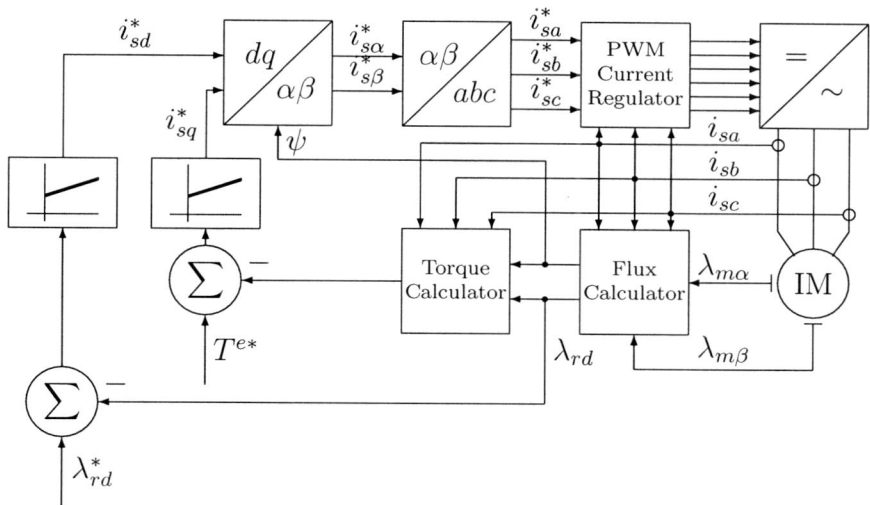

Figure 10.6: A general block diagram of how field oriented control is implemented.

2. The fluxes $\lambda_{r\alpha}$ and $\lambda_{r\beta}$ are calculated using

$$\lambda_{r\alpha} = \frac{L_R}{M}\lambda_{m\alpha} - (L_R - M)\,i_{s\alpha}$$

$$\lambda_{r\beta} = \frac{L_R}{M}\lambda_{m\beta} - (L_R - M)\,i_{s\beta} \quad .$$

The first expression makes use of

$$\lambda_{r\alpha} = \lambda_{m\alpha} + (L_R - M)\,i_{r\alpha}$$

and

$$i_{r\alpha} = \frac{1}{M}\lambda_{m\alpha} - i_{s\alpha} \quad ,$$

with similar reasoning for the second expression. These relationships are consistent with the direct and quadrature axis equivalent circuits, given in Figs. 9.9 and 9.10, respectively.

3. The magnitude and orientation of the rotor flux are determined using the rectangular to polar coordinate transformation:

$$|\lambda_r| = \lambda_{rd} = \sqrt{\lambda_{r\alpha}^2 + \lambda_{r\beta}^2} \quad ;$$

$$\psi = \tan^{-1}\frac{\lambda_{r\beta}}{\lambda_{r\alpha}} \quad .$$

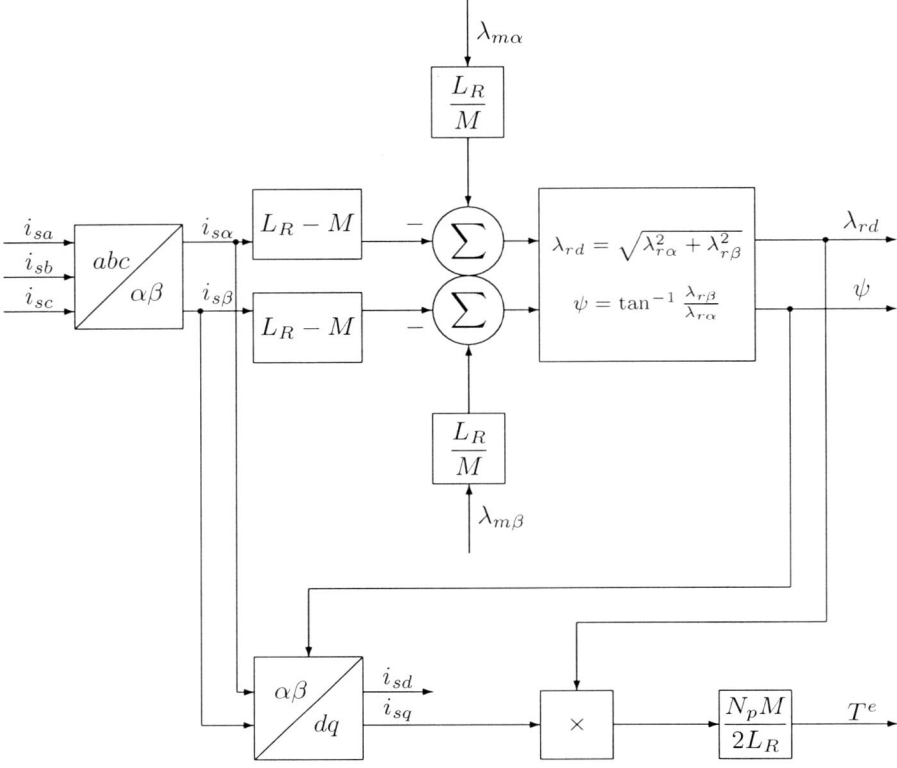

Figure 10.7: A block diagram of how rotor flux and torque are determined under direct field oriented control.

4. The rotor fluxes in the dq frame can now be computed through the transformation
$$\begin{bmatrix} \lambda_{rd} \\ \lambda_{rq} \end{bmatrix} = e^{J\psi} \begin{bmatrix} \lambda_{r\alpha} \\ \lambda_{r\beta} \end{bmatrix}.$$

Once the rotor flux is determined, it is a straightforward matter to calculate the torque. Figure 10.7 shows the integrated calculation of flux and torque for the direct field orientation method.

In practice, the ratio L_R/M and the rotor leakage inductance $(L_R - M)$ are not significantly affected by changes in operating conditions, such as temperature or saturation. However, Hall sensors are fragile, prone to noise, and require modification of the machine. Another approach would be to use measurement of the stator flux linkages through integration of the stator voltage (less resistive drop). The problem here is that the stator voltages are pulse-width modulated, making the integration more complicated. Inte-

10.3. Field Oriented Control

gration is further complicated at low rotational speeds where the period of integration becomes long, and integrator error has more time to accumulate.

Indirect Field Orientation

The indirect method of field orientation does not have the benefit of the measured air gap flux in order to determine the orientation of the rotor flux. The indirect approach is based on calculating the slip speed required for proper field orientation, and the imposition of this speed on the machine. In this sense, indirect field oriented control is a feed-forward approach. Here, we take advantage of the slip relation given in Eq. 10.19:

$$\omega_{sl} = \omega_s - \omega_r = \frac{R_r}{L_R} M \frac{i_{sq}}{\lambda_{rd}} = \frac{M}{T_r} \frac{i_{sq}}{\lambda_{rd}} \quad . \tag{10.22}$$

If the synchronous speed necessary to maintain the orthogonal orientation of λ_{rd} and i_{sq} under a set of operating conditions is ω_s^*, the angle ψ is given by

$$\psi = \int_0^t \omega_s^* d\tau = \int_0^t \omega_{sl}^* d\tau + \int_0^t \omega_r^* d\tau = \int_0^t \omega_{sl}^* d\tau + \psi_0 \quad , \tag{10.23}$$

where ψ_0 is the angular displacement of the rotor, typically measured with an incremental encoder.

The required value of the slip speed, ω_{sl}^*, can be computed from the equations of the machine under field orientation conditions. From the slip relation we have

$$\omega_{sl}^* = \frac{M}{T_r} \frac{i_{sq}^*}{\lambda_{rd}^*} \quad . \tag{10.24}$$

Signal i_{sd}^*, corresponding to a given reference rotor flux can be determined from the rotor flux dynamics:

$$\frac{d\lambda_{rd}^*}{dt} + \frac{1}{T_r}\lambda_{rd}^* = \frac{M}{T_r} i_{sd}^* \quad . \tag{10.25}$$

Signal i_{sq}^* can be determined from the required torque:

$$i_{sq}^* = \frac{2L_R}{N_p M} \frac{T^{e*}}{\lambda_{rd}^*} \quad . \tag{10.26}$$

A block diagram of indirect field orientation is given in Fig. 10.8. The indirect method of field orientation is relatively sensitive to estimation of the rotor time constant. Error in this parameter causes error in ω_{sl}^* that creates error in ψ that detunes the orientation. For this reason a lot of effort has gone into accurately estimating the rotor time constant.

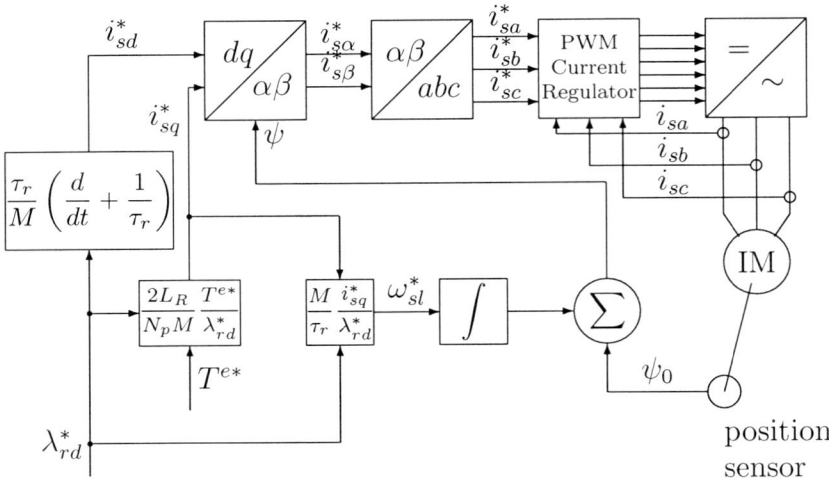

Figure 10.8: A block diagram of how indirect field oriented control is implemented.

10.3.4 Stator Field Orientation

The objective of stator field orientation is to use the direct axis stator current to control the flux in the machine and the quadrature axis stator current to control the torque. That is, we want $\lambda_{sq} = 0$ so that the torque expression, Eq. 10.11 reduces to

$$T^e = \frac{N_p}{2} i_{sq} \lambda_{sd} \quad . \tag{10.27}$$

Unlike rotor field orientation, there is coupling between the direct and quadrature stator axes even with $\lambda_{sq} = 0$. This coupling can be seen through the rotor dynamics and their influence on the stator dynamics. This coupling suggests that we are going to need i_{sd} to depend dynamically on not only the λ_{sd} command, but also on i_{sq}. Further, the slip condition under stator field orientation is going to be different than it was under rotor field orientation.

To develop the decoupling network used to achieve stator field orientation, we are going to carry out the following steps:

1. The rotor dynamics are put into terms of stator fluxes and currents. We will use the equivalent circuit models of Figs. 9.9 and 9.10 to do this, along with the flux linkage relations. Our focus is the rotor dynamics because it is here that the slip frequency enters into the picture.

2. The field orientation condition will be applied ($\lambda_{sq} = d\lambda_{sq}/dt = 0$) to eliminate the quadrature stator flux.

10.3. Field Oriented Control

3. The slip condition will be deduced from the quadrature axis dynamics.

4. The dynamic dependence of i_{sd} on λ_{sd}, i_{sq} and ω_{sl} will be deduced from the direct axis dynamics.

To begin, we can put the rotor fluxes in terms of stator fluxes and currents using the d- and q-axis equivalent circuits:

$$\lambda_{rd} = \lambda_{sd} - (L_S - M)i_{sd} + (L_R - M)i_{rd} \quad,$$

$$\lambda_{rq} = \lambda_{sq} - (L_S - M)i_{sq} + (L_R - M)i_{rq} \quad.$$

Further, the rotor currents can be put in terms of stator flux and stator current through the flux linkage expressions:

$$i_{rd} = \frac{\lambda_{sd} - L_S i_{sd}}{M} \quad, \tag{10.28}$$

$$i_{rq} = \frac{\lambda_{sq} - L_S i_{sq}}{M} \quad. \tag{10.29}$$

Combining these expressions gives

$$\lambda_{rd} = \frac{L_R}{M}\lambda_{sd} - \frac{L_S L_R}{M}\sigma i_{sd} \quad, \tag{10.30}$$

$$\lambda_{rq} = \frac{L_R}{M}\lambda_{sq} - \frac{L_S L_R}{M}\sigma i_{sq} \quad. \tag{10.31}$$

The quantity σ represents the coupling between the rotor and stator circuits, and is given by

$$\sigma = 1 - \frac{M^2}{L_S L_R} \quad. \tag{10.32}$$

We now drop Eqs. 10.28 through 10.31 into the rotor voltage equations, replacing rotor quantities with stator quantities. Carrying out this process on the direct axis we get

$$0 = \left(\frac{d}{dt} + \frac{1}{\tau_r}\right)\lambda_{sd} - \sigma L_S \left(\frac{d}{dt} + \frac{1}{\sigma\tau_r}\right)i_{sd} - \omega_{sl}\left(\lambda_{sq} - \sigma L_S i_{sq}\right) \quad. \tag{10.33}$$

On the quadrature axis we get

$$0 = \left(\frac{d}{dt} + \frac{1}{\tau_r}\right)\lambda_{sq} - \sigma L_S \left(\frac{d}{dt} + \frac{1}{\sigma\tau_r}\right)i_{sq} + \omega_{sl}\left(\lambda_{sd} - \sigma L_S i_{sd}\right) \quad. \tag{10.34}$$

Equations 10.33 and 10.34 reduce under conditions of field orientation. The direct axis expression is

$$0 = \frac{1}{\sigma L_S}\left(\frac{d}{dt} + \frac{1}{\tau_r}\right)\lambda_{sd} - \left(\frac{d}{dt} + \frac{1}{\sigma\tau_r}\right)i_{sd} + \omega_{sl}i_{sq} \quad, \tag{10.35}$$

and the quadrature axis expression is

$$0 = \left(\frac{d}{dt} + \frac{1}{\sigma\tau_r}\right) i_{sq} - \omega_{sl}\left(\frac{1}{\sigma L_S}\lambda_{sd} - i_{sd}\right) \quad . \tag{10.36}$$

Solving the quadrature axis expression gives the slip condition for stator field orientation to be

$$\omega_{sl} = \frac{\left(\frac{d}{dt} + \frac{1}{\sigma\tau_r}\right) i_{sq}}{\frac{1}{\sigma L_S}\lambda_{sd} - i_{sd}} \quad . \tag{10.37}$$

When Eq. 10.37 is satisfied, $\lambda_{sq} = 0$. That is, stator field orientation is achieved when the slip is as prescribed in Eq. 10.37.

Rearranging Eq. 10.35 shows us how the direct axis current must depend dynamically on quadrature stator current, direct axis flux, and slip frequency.

$$\left(\frac{d}{dt} + \frac{1}{\sigma\tau_r}\right) i_{sd} = \frac{1}{\sigma L_S}\left(\frac{d}{dt} + \frac{1}{\tau_r}\right)\lambda_{sd} + \omega_{sl} i_{sq} \quad . \tag{10.38}$$

The interpretation of Eq. 10.38 is that i_{sd} must depend dynamically on i_{sq} to prevent changes in torque command from inadvertently changing the flux level within the machine. It is only through decoupling this dynamic dependence of i_{sd} on λ_{sd}, i_{sq} and ω_{sl} that the field and torque can be independently controlled.

Figure 10.9 shows a block diagram for the decoupling network associated with stator field orientation. The block diagram is based on Eqs. 10.37 and 10.38. Commanded quantities are shown in Fig. 10.9 to reflect the location of the decoupling network in the overall structure of the field oriented controller, whether it is a direct or indirect implementation.

Operation under stator field orientation is more restrictive than under rotor field orientation. To understand why, consider steady state conditions, from which Eqs. 10.37 and 10.38 give, respectively,

$$\omega_{sl}^* = \frac{\frac{1}{\sigma\tau_r} i_{sq}^*}{\frac{1}{\sigma L_S}\lambda_{sd}^* - i_{sd}^*} \quad ; \tag{10.39}$$

$$i_{sd}^* = \frac{1}{L_S}\lambda_{sd}^* + \sigma\tau_r \omega_{sl}^* i_{sq}^* \quad . \tag{10.40}$$

Eliminating i_{sd}^* between these two equations gives a quadratic equation for ω_{sl}^*:

$$\omega_{sl}^{*2} - \frac{1}{\sigma^2 L_S \tau_r}(1-\sigma)\frac{\lambda_{sd}^*}{i_{sq}^*}\omega_{sl}^* + \frac{1}{(\sigma\tau_r)^2} = 0 \quad . \tag{10.41}$$

10.3. Field Oriented Control

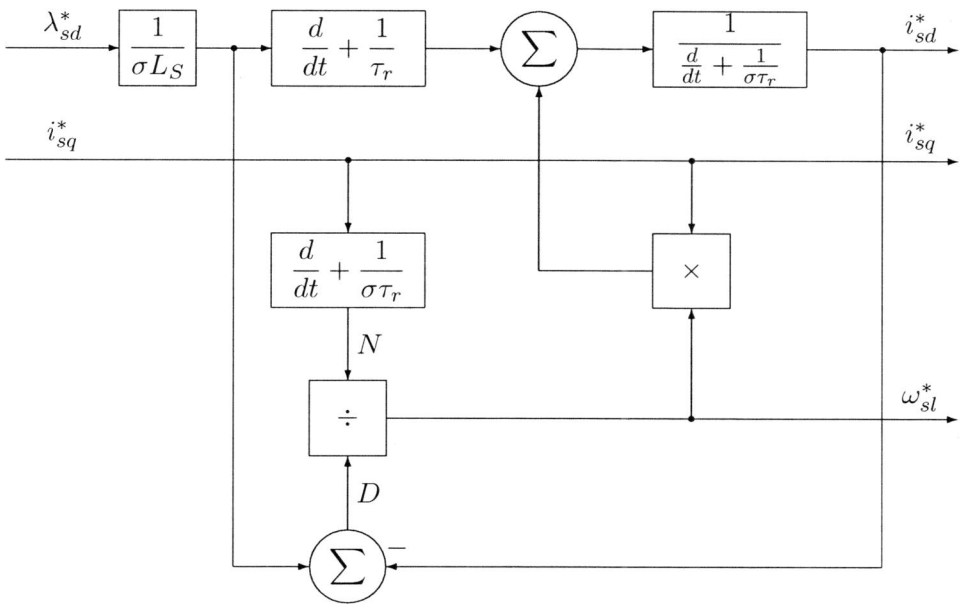

Figure 10.9: A block diagram of the decoupling network needed to implement stator field orientation.

Equation 10.41 only has real roots when the discriminant is greater than or equal to zero. Thus, we require

$$i_{sq}^{*\,2} \leq \left[\frac{1-\sigma}{2\sigma L_S}\lambda_{sd}^*\right]^2 \quad . \tag{10.42}$$

Since

$$T^{e*} = \frac{N_p}{2}\lambda_{sd}^* i_{sq}^* \quad , \tag{10.43}$$

we require

$$\frac{|T^{e*}|}{\lambda_{sd}^{*\,2}} \leq \frac{N_p}{2L_S}\left(\frac{1}{\sigma}-1\right) \quad . \tag{10.44}$$

Equation 10.44 indicates that the leakage factor imposes a torque limit under stator field orientation. Note that the torque limitation is eased as the coupling between the stator and rotor improves.

10.3.5 Air Gap Field Orientation

The objective of air gap field orientation is to use the direct axis stator current to control the flux in the machine and the quadrature axis stator current to control the torque. That is, we want $\lambda_{mq} = 0$ so that the torque expression, Eq. 10.14, reduces to

$$T^e = \frac{N_p}{2} i_{sq} \lambda_{md} \quad . \tag{10.45}$$

As with stator field orientation, there is coupling between the direct and quadrature stator axes, even with $\lambda_{mq} = 0$. This coupling can be seen through the rotor dynamics and their influence on the air gap dynamics. This coupling suggests that we are going to need i_{sd} to depend dynamically on not only the λ_{md} command, but also on i_{sq}. Further, the slip condition under air gap field orientation is going to be different than it was under either rotor field orientation or stator field orientation.

To develop the decoupling network used to achieve stator field orientation, we are going to carry out the same steps used to determine the decoupling network for stator field orientation, except we are going to work with air gap fluxes and stator currents in the rotor dynamical equations. Again, we will make use of Figs. 9.9 and 9.10 to do this, along with the flux linkage relations.

To begin, we can put the rotor fluxes in terms of air gap fluxes and currents using the d- and q-axis equivalent circuits:

$$\lambda_{rd} = \lambda_{md} + (L_R - M) i_{rd} \quad ,$$

$$\lambda_{rq} = \lambda_{mq} + (L_R - M) i_{rq} \quad .$$

Similarly, we can put stator fluxes in terms of the air gap fluxes and stator currents:

$$\lambda_{sd} = \lambda_{md} + (L_S - M) i_{sd} \quad ,$$

$$\lambda_{sq} = \lambda_{mq} + (L_S - M) i_{sq} \quad .$$

Further, the rotor currents can be put in terms of air gap flux and stator current through the flux linkage expressions:

$$i_{rd} = \frac{1}{M} \lambda_{md} - i_{sd} \quad , \tag{10.46}$$

$$i_{rq} = \frac{1}{M} \lambda_{mq} - i_{sq} \quad . \tag{10.47}$$

Combining these expressions gives

$$\lambda_{rd} = \frac{L_R}{M} \lambda_{md} - \sigma_r L_R i_{sd} \quad , \tag{10.48}$$

$$\lambda_{rq} = \frac{L_R}{M} \lambda_{mq} - \sigma_r L_R i_{sq} \quad . \tag{10.49}$$

10.3. Field Oriented Control

The quantity σ_r defines the rotor leakage, and is given by

$$\sigma_r = \frac{L_R - M}{L_R} \quad . \tag{10.50}$$

We now drop Eqs. 10.46 through 10.49 into the rotor voltage equations, replacing rotor fluxes with air gap fluxes and rotor currents with stator currents. Carrying out this process on the direct axis we get

$$0 = \frac{1}{\sigma_r M}\left(\frac{d}{dt} + \frac{1}{\tau_r}\right)\lambda_{md} - \left(\frac{d}{dt} - \frac{1}{\sigma_r \tau_r}\right)i_{sd}$$
$$- \omega_{sl}\left(\frac{1}{\sigma_r M}\lambda_{mq} - i_{sq}\right) \quad . \tag{10.51}$$

On the quadrature axis we get

$$0 = \frac{1}{\sigma_r M}\left(\frac{d}{dt} + \frac{1}{\tau_r}\right)\lambda_{mq} - \left(\frac{d}{dt} - \frac{1}{\sigma_r \tau_r}\right)i_{sq}$$
$$+ \omega_{sl}\left(\frac{1}{\sigma_r M}\lambda_{md} - i_{sd}\right) \quad . \tag{10.52}$$

Equations 10.51 and 10.52 reduce under conditions of air gap field orientation. The direct axis expression is

$$0 = \frac{1}{\sigma_r M}\left(\frac{d}{dt} + \frac{1}{\tau_r}\right)\lambda_{md} - \left(\frac{d}{dt} + \frac{1}{\sigma_r \tau_r}\right)i_{sd} + \omega_{sl}i_{sq} \quad , \tag{10.53}$$

and the quadrature axis expression is

$$0 = \left(\frac{d}{dt} + \frac{1}{\sigma_r \tau_r}\right)i_{sq} - \omega_{sl}\left(\frac{1}{\sigma_r M}\lambda_{md} - i_{sd}\right) \quad . \tag{10.54}$$

Solving the quadrature axis expression gives the slip condition for air gap field orientation to be

$$\omega_{sl} = \frac{\left(\frac{d}{dt} + \frac{1}{\sigma_r \tau_r}\right)i_{sq}}{\frac{1}{\sigma_r M}\lambda_{md} - i_{sd}} \quad . \tag{10.55}$$

When Eq. 10.55 is satisfied, $\lambda_{mq} = 0$. That is, air gap field orientation is achieved when the slip is as prescribed in Eq. 10.55.

Rearranging Eq. 10.53 shows us how the direct axis stator current must depend dynamically on quadrature stator current, direct axis flux, and slip frequency.

$$\left(\frac{d}{dt} + \frac{1}{\sigma_r \tau_r}\right)i_{sd} = \frac{1}{\sigma_r M}\left(\frac{d}{dt} + \frac{1}{\tau_r}\right)\lambda_{md} + \omega_{sl}i_{sq} \quad . \tag{10.56}$$

The interpretation of Eq. 10.56 is that i_{sd} must depend dynamically on i_{sq} to prevent changes in torque command from inadvertently changing the flux level within the machine. It is only through this dynamic dependence of i_{sd} on λ_{md}, i_{sq} and ω_{sl} that the field and torque can be independently controlled.

Figure 10.10 shows a block diagram for the decoupling network associated with air gap field orientation. The block diagram is based on Eqs. 10.55 and 10.56. Commanded quantities are shown in Fig. 10.10 to reflect the location of the decoupling network in the overall structure of the field oriented controller, whether it be a direct or indirect implementation.

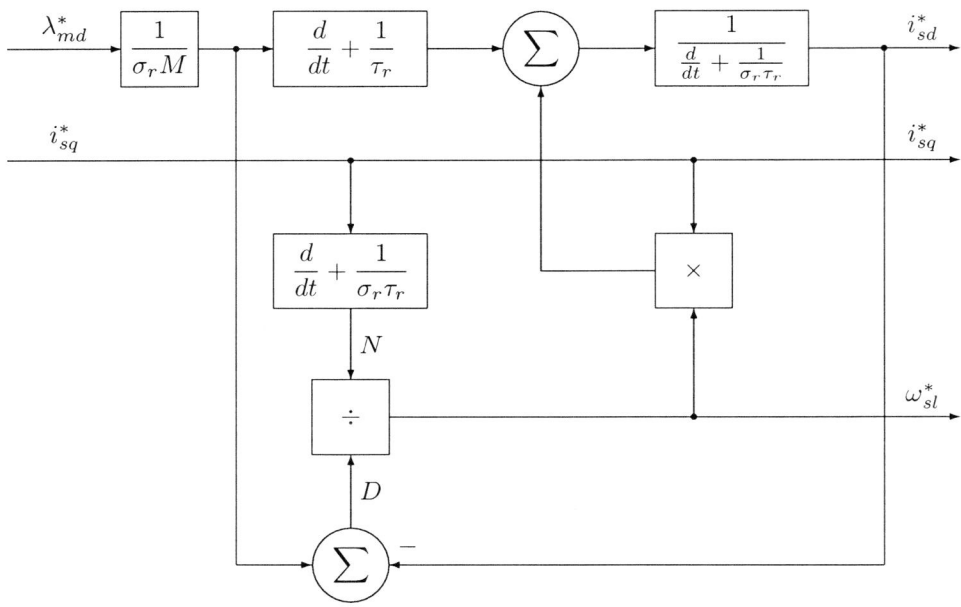

Figure 10.10: A block diagram of the decoupling network needed to implement air gap field orientation.

Similar to stator field orientation, operation under air gap field orientation is more restrictive than under rotor field orientation. To evaluate this restriction, consider steady state conditions, from which Eqs. 10.55 and 10.56 give, respectively,

$$\omega_{sl}^* = \frac{\frac{1}{\sigma_r \tau_r} i_{sq}^*}{\frac{1}{\sigma_r M} \lambda_{md}^* - i_{sd}^*} \quad ; \tag{10.57}$$

$$i_{sd}^* = \frac{1}{M} \lambda_{md}^* + \sigma_r \tau_r \omega_{sl}^* i_{sq}^* \quad . \tag{10.58}$$

10.4. Direct Torque Control

Eliminating i^*_{sd} between these two equations gives a quadratic equation for ω^*_{sl}:

$$\omega^{*\,2}_{sl} - \frac{1}{\sigma_r^2 M \tau_r}(1-\sigma_r)\frac{\lambda^*_{md}}{i^*_{sq}}\omega^*_{sl} + \frac{1}{(\sigma_r \tau_r)^2} = 0 \quad . \tag{10.59}$$

Equation 10.59 only has real roots when the discriminant is greater than or equal to zero. Thus, we require

$$i^{*\,2}_{sq} \leq \left[\frac{1-\sigma_r}{2\sigma_r M}\lambda^*_{md}\right]^2 \quad . \tag{10.60}$$

Since

$$T^{e*} = \frac{N_p}{2}\lambda^*_{md}i^*_{sq} \quad , \tag{10.61}$$

we require

$$\frac{|T^{e*}|}{\lambda^{*\,2}_{md}} \leq \frac{N_p}{4M}\left(\frac{1}{\sigma_r} - 1\right) \quad . \tag{10.62}$$

Equation 10.62 indicates that the rotor leakage factor imposes a torque limit under air gap field orientation. Since $\sigma_r < \sigma$, this condition is less restrictive than its stator field orientation counterpart.

10.4 Direct Torque Control

Under field oriented control, our torque expression was always in terms of a product of a flux and a current. This helped maintain the analogy between torque production in the induction machine with torque production in a dc machine. Under direct torque control, we work with a torque expression that depends only on flux linkage, thereby setting the stage for directly controlling torque by directly controlling flux.

Our torque expression used as the starting point for stator field oriented control was given in Eq. 10.11:

$$T^e = \frac{N_p}{2}(i_{sq}\lambda_{sd} - i_{sd}\lambda_{sq}) \quad . \tag{10.63}$$

Under field orientation, the λ_{sq} term was forced to zero. For direct torque control, we want to replace the stator currents by the rotor fluxes.

Starting with the flux linkage expressions embodied in Fig. 9.9 we have

$$\lambda_{sd} = \lambda_{rd} + (L_S - M)i_{sd} - (L_R - M)i_{rd} \quad . \tag{10.64}$$

In addition, from the stator flux linkage expression we have

$$i_{rd} = \frac{1}{M}(\lambda_{sd} - L_S i_{sd}) \quad . \tag{10.65}$$

Substituting Eq. 10.65 into Eq. 10.64 and reducing gives

$$\frac{L_R}{M}\lambda_{sd} = \lambda_{rd} + \frac{L_S L_R}{M}\sigma i_{sd} \quad , \tag{10.66}$$

where

$$\sigma = 1 - \frac{M^2}{L_S L_R} \quad . \tag{10.67}$$

Solving for the stator current gives

$$i_{sd} = \frac{M}{L_S L_R \sigma}\left(\frac{L_R}{M}\lambda_{sd} - \lambda_{rd}\right) \quad . \tag{10.68}$$

Similarly, on the quadrature axis we have

$$i_{sq} = \frac{M}{L_S L_R \sigma}\left(\frac{L_R}{M}\lambda_{sq} - \lambda_{rq}\right) \quad . \tag{10.69}$$

Putting Eqs. 10.68 and 10.69 into the torque expression, Eq. 10.63, gives

$$T^e = \frac{N_p M}{2 L_S L_R \sigma}(\lambda_{sq}\lambda_{rd} - \lambda_{sd}\lambda_{rq}) \quad . \tag{10.70}$$

We can view this result as the cross product between the stator and rotor fluxes:

$$T^e = \frac{N_p M}{2 L_S L_R \sigma}\left(\vec{\lambda}_r \times \vec{\lambda}_s\right) = \frac{N_p M}{2 L_S L_R \sigma}\left|\vec{\lambda}_r\right|\left|\vec{\lambda}_s\right|\sin\gamma \quad . \tag{10.71}$$

If we assume that the rotor flux is constant and the stator flux is changed incrementally through the application of stator voltage for an increment of time, the incremental change in torque is

$$\Delta T^e = \frac{N_p M}{2 L_S L_R \sigma}\left|\vec{\lambda}_r\right|\left|\vec{\lambda}_s + \Delta\vec{\lambda}_s\right|\sin\Delta\gamma \quad . \tag{10.72}$$

Direct torque control is based on changing (controlling) torque through changing (controlling) stator flux. A graphical depiction of Eq. 10.72 is shown in Fig. 10.11. The assumption of constant rotor flux is valid if the stator flux is updated much more frequently than the rotor time constant, $\tau_r = L_R/R_r$. The rotor time constant dictates the rate of rotor flux decay.

Figure 10.12 shows the structure of how direct torque control is implemented. The flux and torque commands are compared with calculated values of flux and torque. As the error in torque and/or current gets sufficiently large, the voltage vector command to the inverter is updated. Figure 10.13 shows the details within the "Signal Calculator" block of Fig. 10.12. In

10.4. Direct Torque Control

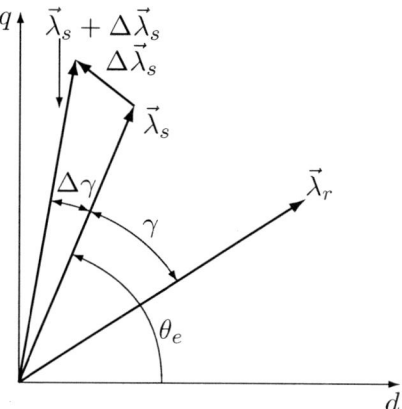

Figure 10.11: The flux vectors showing how changes in torque and stator flux are generated.

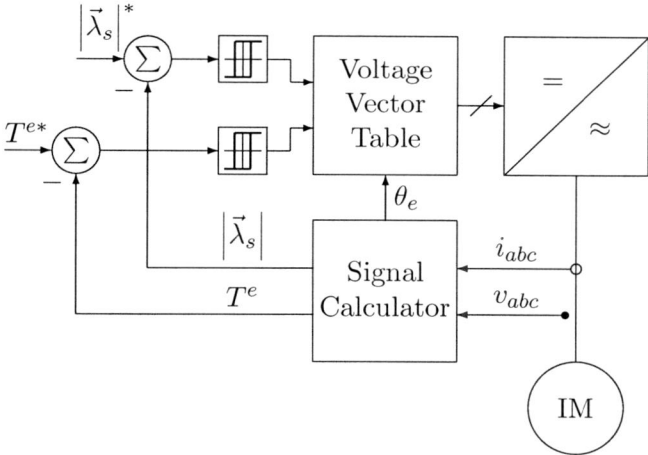

Figure 10.12: The structure for direct torque control implementation.

many respects, Fig. 10.13 has a lot in common with the calculation of flux and torque used in support of direct field orientation as shown in Fig. 10.7.

The incremental changes in flux linkage are created by applying a particular stator voltage for an increment of time. That is,

$$\Delta \vec{\lambda}_s \approx \vec{V}_s \Delta t \quad , \qquad (10.73)$$

if the voltage drop across the stator resistance is small. The vector repre-

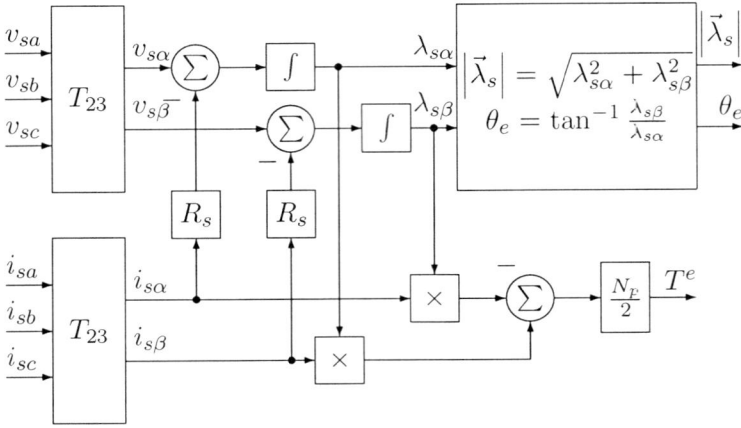

Figure 10.13: The contents of the "Signal Calculator" box in Fig. 10.12.

sentation of the voltage follows directly from the discrete combinations of voltage that can be created by the inverter. These voltage combinations can be considered to represent a vector in the $\alpha\beta$ coordinate system. For a two-level voltage-source inverter, there are eight valid combinations of switches that give rise to seven distinct output voltages. (The combinations of all upper switches conducting and all lower switches conducting are electrically equivalent and therefore redundant.) See Chapters 7 and 8 for a more detailed discussion of voltage space vectors produced by voltage source inverters.

Figure 10.14 shows the six non-zero incremental changes in flux linkage generated by the voltage space vectors, superimposed on the stator flux vector. Consistent with Fig. 10.11, we see at this operating point that vectors $\Delta\vec{\lambda}_1$, $\Delta\vec{\lambda}_2$, and $\Delta\vec{\lambda}_3$ increase $\left|\vec{\lambda}_s\right|$. Vectors $\Delta\vec{\lambda}_2$, $\Delta\vec{\lambda}_3$, and $\Delta\vec{\lambda}_4$ increase T^e. It is important to note that the role of the voltage vectors changes as the flux vector rotates through the six different segments of the vector space.

10.5 Summary

This chapter has examined three methods of dynamically controlling the speed of induction machines. Scalar control represents the most basic form of speed control by changing the synchronous speed through adjustment of the stator frequency. To maintain nearly constant flux in the machine, voltage is varied proportionately with frequency. At low speeds stator resistance requires some compensation through boosting the applied stator voltage.

10.5. Summary

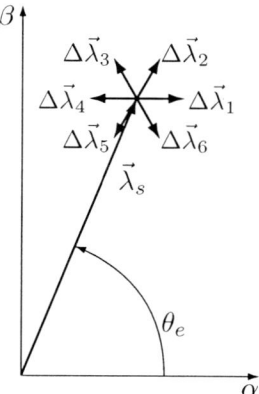

Figure 10.14: The possible changes in stator flux due to six symmetrically-displaced non-zero voltage vectors.

Field oriented control is based on viewing the control of the induction machine from the perspective of an observer that is riding in the air gap on a reference frame that is traveling at the same speed as the air gap magnetic field. From this perspective, the control of the induction machine looks remarkably similar to the control of a separately excited dc machine, with one component of current being responsible for maintaining the field and the second component of current being responsible for producing torque. There are six different forms of field oriented control based on the flux vector (rotor, stator, air gap) that is used to define the direct axis, and whether the implementation is direct or indirect. Direct field orientation uses feedback to actively control the flux and torque. Indirect field orientation is a feedforward technique that uses measurement of the rotor position to determine the slip frequency that must be imposed to maintain field orientation.

Direct torque control bears some similarity to direct field oriented control, but it is simpler in structure. Combining hysteretic control of flux and torque with the selection of the voltage vector that will drive the stator flux in the proper direction, direct torque control is able to close both flux and torque loops without relying heavily on coordinate transformations.

There are other alternatives to controlling induction machines that have not been discussed here. Adaptive control techniques can give dynamic performance that rivals field oriented control or direct torque control. In situations where the machine is driven heavily into saturation, adaptive techniques may actually provide superior performance because they are not as susceptible to detuning as field oriented control.

Chapter 11

Permanent Magnet Machine Models

11.1 Introduction

This chapter parallels Chapter 9, focusing on permanent magnet machines. With sustained improvements in permanent magnet materials over the last twenty years, permanent magnet machines are being considered for a wide range of applications as both motors and generators. The models that will be developed in this chapter are applicable to both axial- and radial-field motor designs. However, the development will be undertaken from the perspective of a radial-field design.

Windings within permanent magnet machines are often referred to as either trapezoidal or sinusoidal. These references are directed at the shape of the back emf waveform. If the windings are simple concentrated coils, the back emf waveform will tend to be trapezoidal in shape as discussed in Sec. 4.4 and shown in Figs. 4.10 and 4.11. If two or more trapezoidal back emfs are shifted relative to one another, the shape of the resultant waveform tends toward becoming a sinusoid, regardless of whether a fractional-slot or integral-slot winding is used.

In machines with trapezoidal windings, the usual intent is to produce torque by feeding the phases with constant current during the regions when the back emf is approximately constant. Under this mode of operation, torque is produced by only two phases at a time with torque production rotating through all combinations of phases. The ability of the inverter to support this type of current waveform degrades as the speed increases since there is less voltage available to drive changes in the phase current.

At higher speeds, machine performance is often improved by feeding the machine balanced three-phase sinusoidal currents.

In machines with sinusoidal windings, the intent is to produce torque simultaneously from all three phases by feeding the phases with currents that are symmetrically displaced in time and sinusoidal in shape. Simultaneous regulation of all three phase currents allows for increased control over the instantaneous torque production. High performance applications that demand low torque ripple will typically use sinusoidally wound machines.

11.2 A Physical Model

There are two fundamental variants in the design and construction of a permanent magnet machine. The first has to do with the windings; the second has to do with the design of the rotor. In this section we will develop a sufficiently general model that all variants are covered.

There are a wide range of windings used for permanent magnet machines. Permanent magnet machines can be found with at least the following kinds of windings:

- Distributed, integral slot coils that are formed into either single- or double-layer windings. This is the same type of winding structure found in most induction machines.

- Fractional slot coils that are used to form double-layer windings. The spatial shifting of the coils is used to control the harmonics in the back emf waveform.

- Concentric windings that give the same effect as a number of integral slot windings that are connected in series. Concentric windings are common in designs that do not use slots for holding the windings. Slotless machines and ironless machines fall into this category. (Note that ironless machines actually do contain ferromagnetic material. However, this material is used to entrain the magnet field, not to guide flux through the stator windings.)

Common characteristics of all machine windings include:

1. Self inductance.

2. Mutual inductance among stator phase windings.

3. Coupling between the windings and the permanent magnets.

11.2. A Physical Model

Accordingly, our model must include these characteristics. Because the way the machine is wound dictates the shape of the back emf, we will separately develop models that reflect sinusoidal and trapezoidal back emf waveforms.

There are a number of different rotor constructions that are used in permanent magnet machines. There are many considerations that go into the selection of the rotor structure, including:

- The speed range of the machine. Some rotor constructions more naturally hold the permanent magnets to the rotor without banding or other means of mechanical retention.

- The desired torque speed characteristic of the machine. For machine applications that require a range of speeds over which the machine supports constant power, some rotor structures are more amenable to weakening the field of the permanent magnets over speed. The concept of field weakening control is discussed in Chapter 12.

- Making optimal use of the magnet volume, however optimal may be defined. Some rotor structures act to focus magnet flux into the air gap, thereby allowing less expensive magnets to create fields that are larger than the magnet would otherwise be able to create.

- Optimizing the manufacture of the rotor, however optimal may be defined. Some rotor structures may be more amenable to re-use of existing plant equipment that would reduce overall manufactured cost.

Figures 11.1 through 11.3 show three relatively common rotor cross-sections illustrating some of the possibilities for a four pole machine. Two of these rotor construction possibilities give rise to magnetic saliency, making the stator winding inductances a function of rotor position. This is another feature that must be included in our model. Magnetic saliency gives rise to reluctance torque.

Figures 11.2 and 11.3 show rotor constructions that embed the permanent magnets into the rotor laminations. In Fig. 11.2 the magnets are oriented so that the magnet flux is directed in a nominal radial direction. An advantage in this type of construction is that the laminations hold the magnets in place. A consequence of this construction is that the inductance seen by the stator windings becomes a function of rotor position. This saliency can be exploited in torque production, as we will see in Chapter 12. One can imagine a number of more sophisticated ways of embellishing this concept at the expense of a more complicated rotor lamination and more magnet segments.

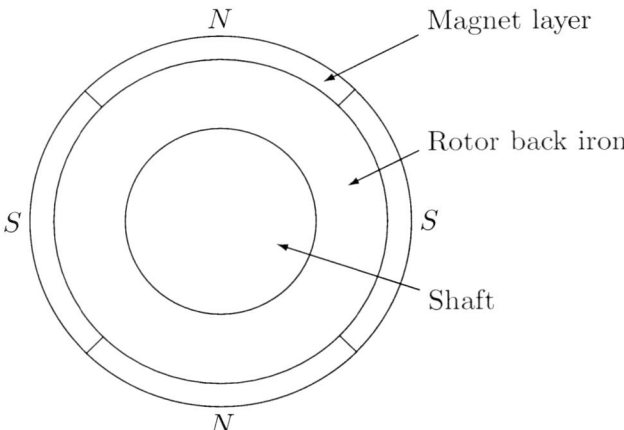

Figure 11.1: A permanent magnet machine rotor with the permanent magnets mounted to the surface of the rotor.

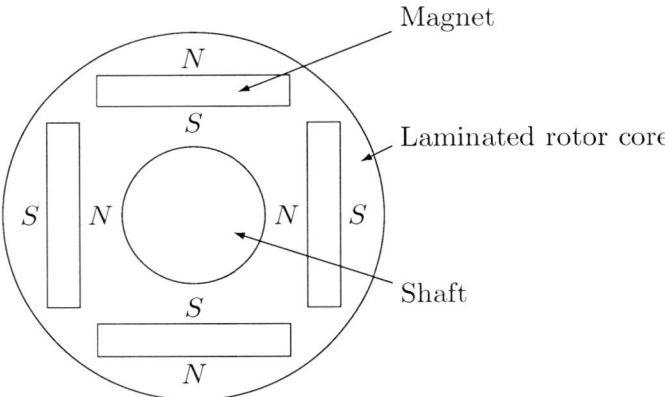

Figure 11.2: A permanent magnet machine rotor with the permanent magnets embedded into the rotor.

In Fig. 11.3 the magnets are also embedded into the rotor laminations. This construction is sometimes referred to as a spoke configuration since the magnets are oriented something like the spokes in a wheel. This construction allows for the magnet flux to be concentrated in the rotor poles since the field of two magnets combines to emerge from the rotor surface. In this construction, it is important for the laminations to neck down around the ends of the magnets so that the material is heavily saturated, thereby pushing the magnet flux out through the rotor surface rather than staying

11.2. A Physical Model

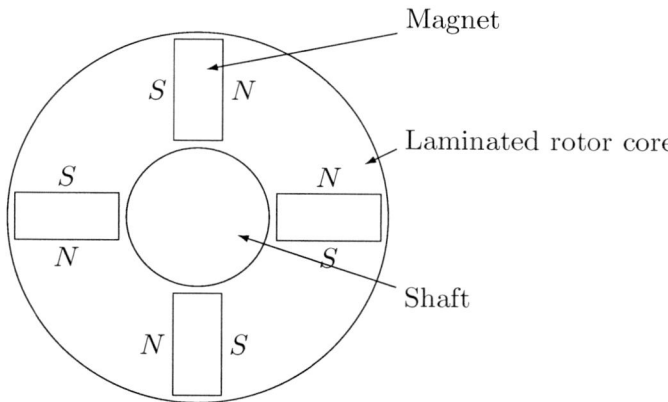

Figure 11.3: A permanent magnet machine rotor with the permanent magnets embedded into the rotor, like spokes in a wheel.

within the rotor.

To begin our construction of the physical model for the permanent magnet machine, we start with a fictitious set of windings to represent the presence of the permanent magnets. That is, we augment our model of physical windings with fictitious windings such that

$$\begin{bmatrix} \lambda(t) \\ \lambda_F(t) \end{bmatrix} = \begin{bmatrix} L(\theta_e) & L_F(\theta_e) \\ L_F^T(\theta_e) & L_{FF}(\theta_e) \end{bmatrix} \begin{bmatrix} i(t) \\ i_F(t) \end{bmatrix}, \qquad (11.1)$$

where the subscript F is used to indicate its association with the fictitious windings representing the permanent magnets such that L_F represents the coupling between the imaginary windings and the physical windings and L_{FF} represents the coupling among the imaginary windings. The torque expression is augmented to

$$T^e = \frac{1}{2} \begin{bmatrix} i^T & i_F^T \end{bmatrix} \frac{\partial}{\partial \theta} \begin{bmatrix} L(\theta_e) & L_F(\theta_e) \\ L_F^T(\theta_e) & L_{FF}(\theta_e) \end{bmatrix} \begin{bmatrix} i(t) \\ i_F(t) \end{bmatrix}. \qquad (11.2)$$

To get back to a model based on only physical windings, we define

$$k(\theta, t) = L_F(\theta) i_F(t) \qquad (11.3)$$

to be the magnet flux that links the stator windings, and

$$T_{FF}^e(\theta, t) = \frac{1}{2} i_F^T \frac{dL_{FF}(\theta)}{d\theta} i_F(t) \qquad (11.4)$$

to be the torque created by the interaction of the magnets among themselves. With these definitions we have

$$\lambda(t) = L(\theta)i(t) + k(\theta,t) \quad ; \tag{11.5}$$

$$\frac{d\lambda}{dt} = v - Ri(t) \quad ; \tag{11.6}$$

and

$$T^e = \frac{1}{2} i^T \frac{dL(\theta)}{d\theta} i + i^T \frac{\partial k(\theta,t)}{\partial \theta} + T^e_{FF}(\theta,t) \quad . \tag{11.7}$$

The first torque term is the reluctance torque created by the stator windings interacting with rotor saliency. The second term is the torque created by the interaction of the magnet and stator field. The torque created by the interaction among the permanent magnets, T^e_{FF}, is commonly referred to as the cogging torque. This torque would be zero if the stator were magnetically smooth since the magnets would not see any change in magnetic circuit as a function of rotor position. Since the stator windings are generally placed in slots with openings, there is a tendency of the magnets to put the rotor into a position where the field energy is maximized. The cogging torque is a reluctance torque that can be controlled by the magnetic design. Cogging torque is discussed in more detail in Sec. 11.4.

11.2.1 Sinusoidal Windings

For permanent magnet machines with sinusoidal back emfs, we must take into consideration all possible variations in rotor construction as suggested in Figs. 11.1 through 11.3. Accordingly, we need to deal with the single magnetic saliency created in the case where the magnets are embedded within the interior of the rotor. Because of the relatively low permeability of permanent magnets, the inductance seen by the stator phase windings is lowest when the magnet flux is aligned with the stator winding. In addition, the variation in stator winding inductance is going to vary at twice the rotational frequency of the back emf because the inductance is insensitive to the direction of the magnet flux; it only responds to the relative permeability.

From this discussion, physical reasoning suggests that the flux linking the stator windings is given by

$$\lambda_{abc} = L_{abc}(\theta) i_{abc} + k_{abc}(\theta) \quad , \tag{11.8}$$

where

$$L_{abc}(\theta_e) = \begin{bmatrix} L_{s_0} - L_{s_2}\cos(2\theta_e) & -L_{ss} - L_{s_2}\cos\left(2\theta_e - \frac{2\pi}{3}\right) \\ -L_{ss} - L_{s_2}\cos\left(2\theta_e - \frac{2\pi}{3}\right) & L_{s_0} - L_{s_2}\cos\left(2\theta_e + \frac{2\pi}{3}\right) \\ -L_{ss} - L_{s_2}\cos\left(2\theta_e + \frac{2\pi}{3}\right) & -L_{ss} - L_{s_2}\cos(2\theta_e) \end{bmatrix}$$

11.2. A Physical Model

$$\left.\begin{array}{c}-L_{ss}-L_{s_2}\cos\left(2\theta_e+\frac{2\pi}{3}\right)\\-L_{ss}-L_{s_2}\cos\left(2\theta_e\right)\\L_{s0}-L_{s_2}\cos\left(2\theta_e-\frac{2\pi}{3}\right)\end{array}\right], \tag{11.9}$$

where L_{s0} is the average value of stator inductance, L_{s_2} is the amplitude of the second harmonic in stator inductance, L_{ss} is the average value of the stator winding mutual inductance. For the back emf waveforms we have

$$k_{abc}(\theta) = K \begin{bmatrix} \cos\left(\theta_e\right) \\ \cos\left(\theta_e - \frac{2\pi}{3}\right) \\ \cos\left(\theta_e + \frac{2\pi}{3}\right) \end{bmatrix}, \tag{11.10}$$

where K is assumed to be the nominally constant amplitude. As usual, the electrical dynamics are given by

$$v_{abc} = \frac{d\lambda_{abc}}{dt} + R_{abc} i_{abc} \quad . \tag{11.11}$$

The torque is given by

$$T^e = \frac{1}{2} i_{abc}^T \frac{dL_{abc}(\theta)}{d\theta} i_{abc} + i_{abc}^T \frac{dk_{abc}(\theta)}{d\theta} + T_{FF}^e \quad , \tag{11.12}$$

and the mechanical dynamics by

$$H \frac{d\omega}{dt} = T^e - \tau_l \quad . \tag{11.13}$$

Equations 11.9 and 11.11 show that the nature of the reluctance and magnet torques are different. The reluctance torque varies with $2\theta_e$ because the torque is independent of the current direction. The magnet torque varies with θ_e because it is tied to coordinating magnet field with stator current direction.

As we did with the induction machine in Chapter 9, we would like to transform the three-phase coupled abc dynamics into the decoupled two-phase $\alpha\beta$ reference frame. Similarly, from there we will go to the dq reference frame.

11.2.2 Trapezoidal Windings

Permanent magnet machines that are designed to make use of trapezoidal windings typically have the permanent magnets mounted on the surface of the rotor, as in Fig. 11.1. Accordingly, the stator inductance has minimal (negligible) saliency. The stator flux linkage due to the permanent magnets is trapezoidal in shape, and so it is appropriate to represent it by a trapezoidal function.

Like the description of the permanent magnet machine with sinusoidal windings, the flux linking the stator windings is given by

$$\lambda_{abc} = L_{abc}(\theta)i_{abc} + k_{abc}(\theta) \quad , \tag{11.14}$$

where

$$L_{abc}(\theta_e) = \begin{bmatrix} L_s & -L_{ss} & -L_{ss} \\ -L_{ss} & L_s & -L_{ss} \\ -L_{ss} & -L_{ss} & L_s \end{bmatrix} \quad , \tag{11.15}$$

where L_s is the stator self inductance and L_{ss} is the stator winding mutual inductance. For the back emf waveforms we have

$$k_{abc}(\theta) = -\frac{12}{5\pi} K \begin{bmatrix} \int_{-\pi/2}^{\theta_e} \mathrm{trap}(\psi)d\psi \\ \int_{-\pi/2}^{\theta_e - \frac{2\pi}{3}} \mathrm{trap}(\psi)d\psi \\ \int_{-\pi/2}^{\theta_e + \frac{2\pi}{3}} \mathrm{trap}(\psi)d\psi \end{bmatrix} \quad , \tag{11.16}$$

where K is assumed to be the amplitude of the trapezoidal waveform.

To understand the origin of Eq. 11.16, first consider that the expected shape of the back emf is trapezoidal as shown in Fig. 11.4. The break points in the trapezoid are located at the ideal locations for operation with quasi-squarewaves of current. That is, the duration of the flat region of the back emf is 120° electrical. The underlying flux waveform is also shown in Fig. 11.4. Second, recall that the voltage is the time derivative of the flux linkage, so to determine the flux linkage we integrate the voltage. Third, $\psi = -\pi/2$ is the position where the flux linkage is zero. That is why the integration is chosen to start there. Fourth, the factor of $-12/5\pi$ forces the amplitude of the flux linkage to be K at $\psi = 0$.

For a two-phase permanent magnet machine, the back emf is ideally constant between $\psi = \pi/4$ and $\psi = 3\pi/4$, so the coefficient $12/5\pi$ in Eq. 11.16 would be replaced by $8/3\pi$. Through our modeling work we have seen the equivalence between two- and three-phase machines. The overwhelming majority of machines are three-phase, but there are some two-phase servo motors commercially available.

Consider the three-phase circuit model of the machine shown in Fig. 11.5. Each phase is modeled with a resistance, an inductance, and a back emf. The machine is modeled as a star (Y) connection. The back emf can be modeled with the $\mathrm{trap}(\theta_e)$ waveform of Fig. 11.4; numerical or experimental data could also be used if it is available. It will be appreciated that this is structurally the same model used to consider the current regulation problem

11.2. A Physical Model

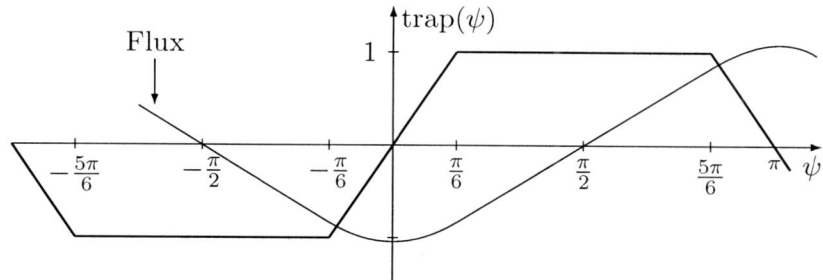

Figure 11.4: The assumed back emf waveform and the underlying flux linkage waveform for a permanent magnet machine with trapezoidal windings.

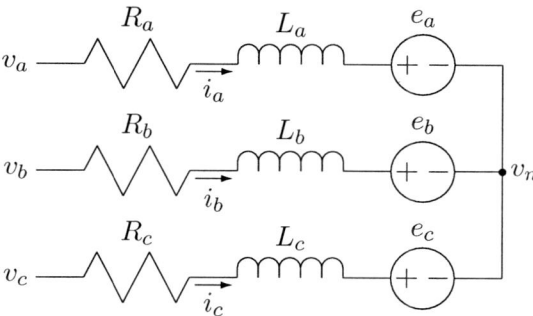

Figure 11.5: A circuit model for a three-phase permanent magnet machine with trapezoidal back emfs.

in Chapter 8. In a typical application the phase voltages applied to the machine are the voltages output by an inverter as shown in Fig. 8.3.

Consistent with Eq. 11.15,

$$L_a = L_b = L_c = L_s + L_{ss} \qquad (11.17)$$

since the three phase currents sum to zero. In addition, typically $R_a = R_b = R_c$ consistent with balanced windings. Applying Kirchhoff's voltage law to each phase gives

$$v_a - v_n = R_a i_a + L_a \frac{di_a}{dt} + e_a \quad , \qquad (11.18)$$

$$v_b - v_n = R_b i_b + L_b \frac{di_b}{dt} + e_b \quad , \qquad (11.19)$$

$$v_c - v_n = R_c i_c + L_c \frac{di_c}{dt} + e_c \quad . \tag{11.20}$$

Again, because the phase currents (and their time derivatives) sum to zero, if the phase voltages are balanced we conclude that

$$v_n = -\frac{e_a + e_b + e_c}{3} \quad . \tag{11.21}$$

Careful consideration of Eq. 11.21 reveals that v_n is comprised of triplen harmonics only. Harmonics that are not multiples of three will cancel due to the phase shift, whereas the triplen harmonics for each phase have the same phase relationship.

As discussed in Sec. 4.4, the shaft power is equal to the power delivered to the back emfs. It follows that the electromagnetic torque is

$$T^e = \frac{e_a i_a + e_b i_b + e_c i_c}{\omega} \quad . \tag{11.22}$$

It should be remembered that the back emfs are proportional to speed, keeping the torque finite at zero speed. For simulation purposes, it is numerically safer to use

$$T^e = i_a \frac{\partial \lambda_a}{\partial \theta_e} + i_b \frac{\partial \lambda_b}{\partial \theta_e} + i_c \frac{\partial \lambda_c}{\partial \theta_e} \tag{11.23}$$

so division by zero is never required. The ideal shape of $\partial \lambda_a / \partial \theta_e$ is trap(θ_e) with appropriate scaling, since the back emf is $e_a = \omega \partial \lambda_a / \partial \theta_e$, consistent with Eq. 4.45.

11.3 Two Phase Equivalent Models

In this section we build off of the physical models motivated in Sec. 11.2 for a permanent magnet machine with sinusoidal windings. First, we want to transform the three-phase coupled abc dynamics into the decoupled two-phase $\alpha\beta$ reference frame. Second, we want to express the machine model from the perspective of an observer that is sitting on the rotor. That is, we want to transform the model from the $\alpha\beta$ reference frame to the dq reference frame. Because of the methods of control to be applied to these models, we focus here on the sinusoidal permanent magnet machine.

11.3.1 The $\alpha\beta$ Model

To develop the $\alpha\beta$ model for the sinusoidal permanent magnet machine, we apply the transformations of Chapter 6 to the model summarized in Eqs. 11.8

11.3. Two Phase Equivalent Models

through 11.12. Applying our transformation T (Eq. 6.8) to Eq. 11.8 gives

$$\begin{aligned}\lambda_{\alpha\beta 0} &= TL_{abc}T^{-1}i_{\alpha\beta 0} + Tk_{abc} \\ &= L_{\alpha\beta 0}i_{\alpha\beta 0} + k_{\alpha\beta 0} \quad .\end{aligned} \quad (11.24)$$

Carrying out the indicated operations gives

$$L_{\alpha\beta 0}(\theta_e) = \begin{bmatrix} (L_{s_0} + L_{ss}) - \frac{3}{2}L_{s_2}\cos(2\theta_e) & -\frac{3}{2}L_{s_2}\sin(2\theta_e) \\ -\frac{3}{2}L_{s_2}\sin(2\theta_e) & (L_{s_0} + L_{ss}) + \frac{3}{2}L_{s_2}\cos(2\theta_e) \\ 0 & 0 \\ & \\ & 0 \\ & 0 \\ & (L_{s_0} - 2L_{ss}) \end{bmatrix}, \quad (11.25)$$

and

$$k_{\alpha\beta 0} = \sqrt{\frac{3}{2}}K \begin{bmatrix} \cos\theta_e \\ \sin\theta_e \\ 0 \end{bmatrix} \quad . \quad (11.26)$$

As we saw with the induction machine, the 0 (zero) phase does not contribute to torque production, so we do not want to excite this phase. Once we are finished transforming quantities from the abc frame to the $\alpha\beta$ frame we will drop the 0 phase.

Converting the abc quantities of Eq. 11.11 to $\alpha\beta 0$ quantities changes the voltage expression into

$$T^{-1}v_{\alpha\beta 0} = \frac{d\left(T^{-1}\lambda_{\alpha\beta 0}\right)}{dt} + R_{abc}T^{-1}i_{\alpha\beta 0} \quad . \quad (11.27)$$

Multiplying through by T gives

$$v_{\alpha\beta 0} = \frac{d\lambda_{\alpha\beta 0}}{dt} + TR_{abc}T^{-1}i_{\alpha\beta 0} \quad , \quad (11.28)$$

which reduces to

$$v_{\alpha\beta 0} = \frac{d\lambda_{\alpha\beta 0}}{dt} + R_{\alpha\beta 0}i_{\alpha\beta 0} \quad , \quad (11.29)$$

where $R_{\alpha\beta 0} = R_{abc}$.

Converting the abc quantities of Eq. 11.12 to $\alpha\beta 0$ quantities changes the torque expression into

$$T^e = \frac{1}{2}i_{\alpha\beta 0}^T T\frac{dL_{abc}(\theta)}{d\theta}T^{-1}i_{\alpha\beta 0} + i_{\alpha\beta 0}^T T\frac{dk_{abc}(\theta)}{d\theta} + T^e_{FF} \quad . \quad (11.30)$$

Because our transformation T does not depend on rotor position, we can pull the transformations on either side of the spatial derivatives inside the derivative. This gives

$$T^e = \frac{1}{2}i_{\alpha\beta 0}^T \frac{dL_{\alpha\beta 0}(\theta)}{d\theta}i_{\alpha\beta 0} + i_{\alpha\beta 0}^T \frac{dk_{\alpha\beta 0}(\theta)}{d\theta} + T^e_{FF} \quad . \quad (11.31)$$

Using Eq. 11.25 gives

$$T^e = \frac{3L_{s_2}N_p}{4} i_{\alpha\beta 0}^T \begin{bmatrix} \sin 2\theta_e & -\cos 2\theta_e & 0 \\ -\cos 2\theta_e & -\sin 2\theta_e & 0 \\ 0 & 0 & 0 \end{bmatrix} i_{\alpha\beta 0} +$$

$$\sqrt{\frac{3}{2}} \frac{KN_p}{2} i_{\alpha\beta 0}^T \begin{bmatrix} -\sin\theta_e \\ \cos\theta_e \\ 0 \end{bmatrix} + T_{FF}^e \quad . \tag{11.32}$$

Equation 11.32 can be reduced further. At this point it is straightforward to see that the torque is comprised of a term that involves the interaction among the currents on the α and β axes, a term tied to the interaction of the permanent magnets with the stator currents, and the cogging torque.

While it may not be immediately clear why we would want to do this, expanding Eq. 11.32 gives

$$T^e = \frac{3L_{s_2}N_p}{4} \left[\sin 2\theta_e \left(i_\alpha^2 - i_\beta^2\right) - 2\cos 2\theta_e i_\alpha i_\beta\right] +$$

$$\sqrt{\frac{3}{2}} \frac{KN_p}{2} \left[\cos\theta_e i_\beta - \sin\theta_e i_\alpha\right] + T_{FF}^e \quad . \tag{11.33}$$

From our flux linkage expressions in Eq. 11.24 we have

$$\lambda_\alpha = \left[(L_{s_0} + L_{ss}) - \frac{3}{2}\cos 2\theta_e\right] i_\alpha - \frac{3}{2}\sin 2\theta_e i_\beta + \sqrt{\frac{3}{2}} K \cos\theta_e \quad , \tag{11.34}$$

which can be solved for i_β to give

$$i_\beta = -\frac{1}{\frac{3}{2}L_{s_2}\sin 2\theta_e} \times$$

$$\left[\lambda_\alpha - \left[(L_{s_0} + L_{ss}) - \frac{3}{2}\cos 2\theta_e\right] i_\alpha - \sqrt{\frac{3}{2}} K \cos\theta_e\right] \quad . \tag{11.35}$$

Similarly, on the β axis we have

$$\lambda_\beta = -\frac{3}{2}\sin 2\theta_e i_\alpha + \left[(L_{s_0} + L_{ss}) + \frac{3}{2}\cos 2\theta_e\right] i_\beta + \sqrt{\frac{3}{2}} K \sin\theta_e \quad , \tag{11.36}$$

which can be solved for i_α to give

$$i_\alpha = -\frac{1}{\frac{3}{2}L_{s_2}\sin 2\theta_e} \times$$

$$\left[\lambda_\beta - \left[(L_{s_0} + L_{ss}) + \frac{3}{2}\cos 2\theta_e\right] i_\beta - \sqrt{\frac{3}{2}} K \sin\theta_e\right] \quad . \tag{11.37}$$

11.3. Two Phase Equivalent Models

First, substituting Eq. 11.35 for one i_β in the i_β^2 term in Eq. 11.33, then substituting Eq. 11.35 for one i_α in the i_α^2 term in Eq. 11.33, and finally reducing gives

$$T^e = \frac{N_p}{2}[\lambda_\alpha i_\beta - \lambda_\beta i_\alpha] + T_{FF} \quad . \tag{11.38}$$

This shows that the torque can be viewed as the cross product between the stator flux and current vectors. This torque expression will be useful when we look at controlling the permanent magnet machine in Chapter 12.

Since the zero phase does not contribute to torque production, we do not want to excite this phase, and we will now drop it from our model.

To formulate equivalent circuit models for the α and β axes, we need to revisit the time rate of change in the phase flux linkages:

$$\frac{d\lambda_{\alpha\beta}}{dt} = \frac{d}{dt}(L_{\alpha\beta}i_{\alpha\beta} + k_{\alpha\beta}) \quad . \tag{11.39}$$

Because the inductance depends on rotor position, we must apply the chain rule when taking the time derivative of the first term on the right. It follows that

$$\frac{d\lambda_{\alpha\beta}}{dt} = L_{\alpha\beta}\frac{di_{\alpha\beta}}{dt} + \frac{dL_{\alpha\beta}}{d\theta_e}i_{\alpha\beta}\frac{d\theta_e}{d\theta}\frac{d\theta}{dt} + \frac{dk_{\alpha\beta}}{d\theta_e}\frac{d\theta_e}{d\theta}\frac{d\theta}{dt} \quad . \tag{11.40}$$

Recognizing that $d\theta_e/d\theta = N_p/2$ is the number of pole pairs and $d\theta/dt = \omega_m$ is the rotor speed, and extracting the position dependent terms from Eq. 11.25, Eq. 11.40 becomes

$$\frac{d\lambda_{\alpha\beta}}{dt} = L_{\alpha\beta}\frac{di_{\alpha\beta}}{dt} - \frac{3}{2}N_p L_{s2}\begin{bmatrix} -\sin 2\theta_e & \cos 2\theta_e \\ \cos 2\theta_e & \sin 2\theta_e \end{bmatrix}i_{\alpha\beta}\,\omega_m +$$
$$\sqrt{\frac{3}{2}}\frac{KN_p}{2}\begin{bmatrix} -\sin\theta_e \\ \cos\theta_e \end{bmatrix}\omega_m \quad . \tag{11.41}$$

Using the voltage expression Eq. 11.29 in conjunction with the time rate of change in flux linkage given by Eq. 11.41, we have the basis for equivalent circuit models for the α and β axes. Figures 11.6 and 11.7 give the circuits for the α and β axes, respectively. It is worth noting that there is still coupling between these two axes. In addition, there is significant coupling between the electrical dynamics and the mechanical dynamics.

11.3.2 The dq Model

The strong coupling between the α and β axes makes the $\alpha\beta$ model of limited usefulness. However, the construction of the $\alpha\beta$ model is a useful first step

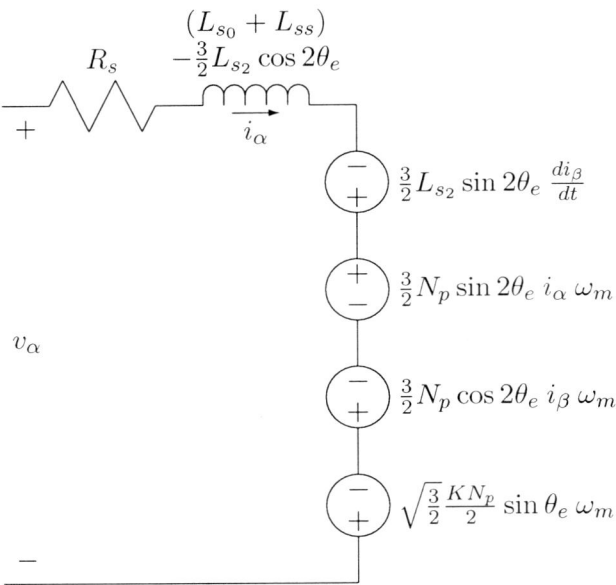

Figure 11.6: The equivalent circuit model for the α axis of the permanent magnet machine.

in developing the dq model of the machine, where the dynamics are viewed from the perspective of an observer riding through the air gap at the same speed as the excitation. In the permanent magnet machine, it is standard practice to align the direct axis with the field of the permanent magnet. The relationship between the $\alpha\beta$ and dq reference frames for the permanent magnet machine is shown in Fig. 11.8.

Our transformation between the $\alpha\beta$ reference frame and the dq reference frame is given by Eq. 6.14 such that

$$x_{\alpha\beta} = e^{-J\theta_e} x_{dq} \quad . \tag{11.42}$$

Accordingly, our flux linkage expression given in Eq. 11.24 becomes

$$e^{-J\theta_e} \lambda_{dq} = L_{\alpha\beta} e^{-J\theta_e} i_{dq} + k_{\alpha\beta} \quad . \tag{11.43}$$

Rearranging by premultiplying both sides by $e^{J\theta_e}$ gives

$$\lambda_{dq} = e^{J\theta_e} L_{\alpha\beta} e^{-J\theta_e} i_{dq} + e^{J\theta_e} k_{\alpha\beta} \quad , \tag{11.44}$$

or

$$\lambda_{dq} = L_{dq} i_{dq} + k_{dq} \quad , \tag{11.45}$$

11.3. Two Phase Equivalent Models

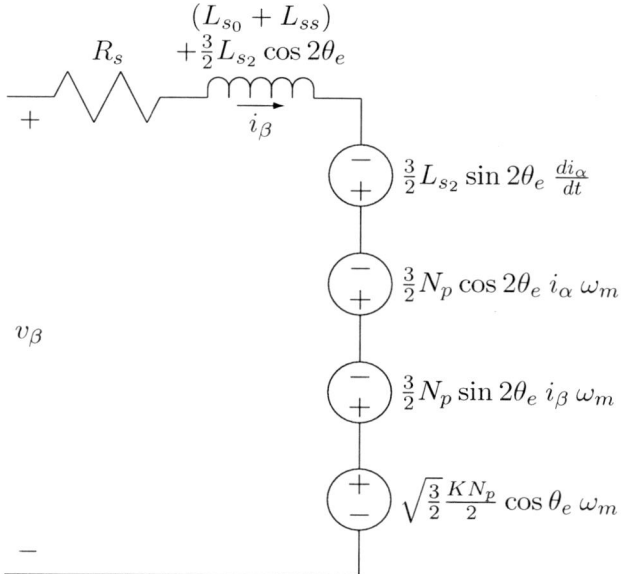

Figure 11.7: The equivalent circuit model for the β axis of the permanent magnet machine.

where

$$\begin{aligned}
L_{dq} &= e^{J\theta_e} L_{\alpha\beta} e^{-J\theta_e} \\
&= \begin{bmatrix} (L_{s0} + L_{ss}) - \frac{3}{2}L_{s2} & 0 \\ 0 & (L_{s0} + L_{ss}) + \frac{3}{2}L_{s2} \end{bmatrix} \\
&= \begin{bmatrix} L_d & 0 \\ 0 & L_q \end{bmatrix},
\end{aligned} \qquad (11.46)$$

and

$$k_{dq} = e^{J\theta_e} k_{\alpha\beta} = \sqrt{\frac{3}{2}} K \begin{bmatrix} 1 \\ 0 \end{bmatrix}. \qquad (11.47)$$

The voltage equation Eq. 11.29 becomes

$$e^{-J\theta_e} v_{dq} = \frac{d\lambda_{\alpha\beta}}{dt} + R_{\alpha\beta} e^{-J\theta_e} i_{dq}. \qquad (11.48)$$

The time derivative of flux linkage was left as it is, because we need to be careful not to lose the dependence on rotor position. To do so would mean losing motional voltages that are important to the description of the machine. Rearranging Eq. 11.48 by premultiplying both sides by $e^{J\theta_e}$ gives

$$v_{dq} = e^{J\theta_e} \frac{d\lambda_{\alpha\beta}}{dt} + e^{J\theta_e} R_{\alpha\beta} e^{-J\theta_e} i_{dq}. \qquad (11.49)$$

280 Chapter 11. Permanent Magnet Machine Models

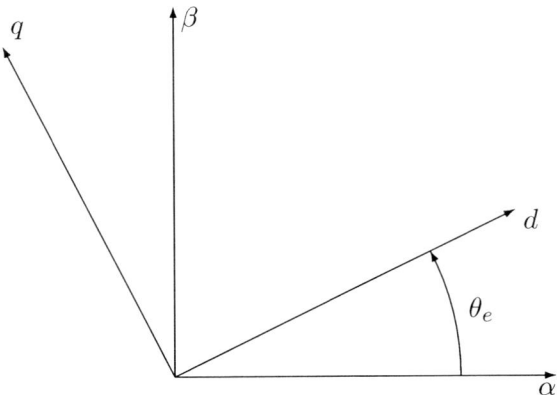

Figure 11.8: The relationship among the $\alpha\beta$ axes and the dq reference frame in the permanent magnet machine. The direct axis is usually aligned with the rotor flux.

Using Eq. 11.41, the flux linkage term becomes

$$e^{J\theta_e}\frac{d\lambda_{\alpha\beta}}{dt} = e^{J\theta_e} L_{\alpha\beta} \frac{d\left(e^{-J\theta_e} i_{dq}\right)}{dt} - $$
$$\frac{3}{2} N_p L_{s2}\, e^{J\theta_e} \begin{bmatrix} -\sin 2\theta_e & \cos 2\theta_e \\ \cos 2\theta_e & \sin 2\theta_e \end{bmatrix} e^{-J\theta_e} i_{dq}\, \omega_m + $$
$$\sqrt{\frac{3}{2}}\frac{K N_p}{2}\, e^{J\theta_e} \begin{bmatrix} -\sin\theta_e \\ \cos\theta_e \end{bmatrix} \omega_m \quad, \tag{11.50}$$

which expands into

$$e^{J\theta_e}\frac{d\lambda_{\alpha\beta}}{dt} = e^{J\theta_e} L_{\alpha\beta} \frac{de^{-J\theta_e}}{dt} i_{dq} + e^{J\theta_e} L_{\alpha\beta} e^{-J\theta_e} \frac{di_{dq}}{dt} - $$
$$\frac{3}{2} N_p L_{s2}\, e^{J\theta_e} \begin{bmatrix} -\sin 2\theta_e & \cos 2\theta_e \\ \cos 2\theta_e & \sin 2\theta_e \end{bmatrix} e^{-J\theta_e} i_{dq}\, \omega_m + $$
$$\sqrt{\frac{3}{2}}\frac{K N_p}{2}\, e^{J\theta_e} \begin{bmatrix} -\sin\theta_e \\ \cos\theta_e \end{bmatrix} \omega_m \quad, \tag{11.51}$$

which can be simplified somewhat by substituting the definition of L_{dq} such that

$$e^{J\theta_e}\frac{d\lambda_{\alpha\beta}}{dt} = -\frac{N_p}{2} L_{dq} J\, i_{dq}\, \omega_m + L_{dq} \frac{di_{dq}}{dt} - $$
$$\frac{3}{2} N_p L_{s2}\, e^{J\theta_e} \begin{bmatrix} -\sin 2\theta_e & \cos 2\theta_e \\ \cos 2\theta_e & \sin 2\theta_e \end{bmatrix} e^{-J\theta_e} i_{dq}\, \omega_m + $$

11.3. Two Phase Equivalent Models

$$\sqrt{\frac{3}{2}}\frac{KN_p}{2}e^{J\theta_e}\begin{bmatrix} -\sin\theta_e \\ \cos\theta_e \end{bmatrix}\omega_m \quad . \tag{11.52}$$

Carrying out the indicated operations gives

$$e^{J\theta_e}\frac{d\lambda_{\alpha\beta}}{dt} = -\frac{N_p}{2}L_{dq}J\,i_{dq}\,\omega_m + L_{dq}\frac{di_{dq}}{dt} -$$
$$\frac{3}{2}N_pL_{s2}\begin{bmatrix} 0 & 1 \\ 1 & 0 \end{bmatrix}i_{dq}\,\omega_m +$$
$$\sqrt{\frac{3}{2}}\frac{KN_p}{2}\begin{bmatrix} 0 \\ 1 \end{bmatrix}\omega_m \quad . \tag{11.53}$$

Dropping this into the voltage equation gives

$$v_{dq} = R_{dq}i_{dq} - \frac{N_p}{2}L_{dq}J\,i_{dq}\,\omega_m + L_{dq}\frac{di_{dq}}{dt} -$$
$$\frac{3}{2}N_pL_{s2}\begin{bmatrix} 0 & 1 \\ 1 & 0 \end{bmatrix}i_{dq}\,\omega_m +$$
$$\sqrt{\frac{3}{2}}\frac{KN_p}{2}\begin{bmatrix} 0 \\ 1 \end{bmatrix}\omega_m \quad . \tag{11.54}$$

It may not be obvious that there is additional simplification that can be accomplished. However,

$$-\frac{N_p}{2}L_{dq}J\,i_{dq}\,\omega_m - \frac{3}{2}N_pL_{s2}\begin{bmatrix} 0 & 1 \\ 1 & 0 \end{bmatrix}i_{dq}\,\omega_m \tag{11.55}$$

can be reduced because $L_q - L_d = 3L_{s2}$ so Eq. 11.55 can be written as

$$-\frac{N_p}{2}\begin{bmatrix} 0 & L_d \\ -L_q & 0 \end{bmatrix}i_{dq}\,\omega_m - \frac{N_p}{2}\begin{bmatrix} 0 & L_q - L_d \\ L_q - L_d & 0 \end{bmatrix}i_{dq}\,\omega_m \quad , \tag{11.56}$$

which reduces to

$$\frac{N_p}{2}\begin{bmatrix} 0 & -L_q \\ L_d & 0 \end{bmatrix}i_{dq}\,\omega_m \quad , \tag{11.57}$$

so the voltage equation becomes

$$v_{dq} = R_{dq}i_{dq} + L_{dq}\frac{di_{dq}}{dt} + \frac{N_p}{2}\begin{bmatrix} 0 & -L_q \\ L_d & 0 \end{bmatrix}i_{dq}\,\omega_m +$$
$$\sqrt{\frac{3}{2}}\frac{KN_p}{2}\begin{bmatrix} 0 \\ 1 \end{bmatrix}\omega_m \quad . \tag{11.58}$$

We see several important characteristics in this expression. First, by projecting the dynamics onto the dq reference frame we have eliminated explicit

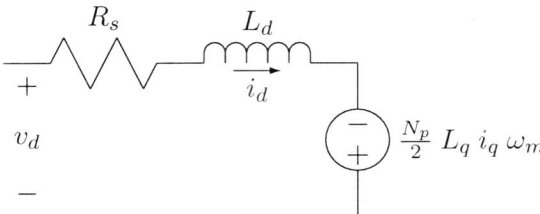

Figure 11.9: The equivalent circuit model for the d axis of the permanent magnet machine.

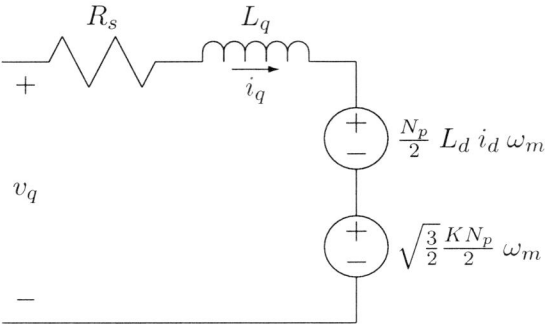

Figure 11.10: The equivalent circuit model for the q axis of the permanent magnet machine.

dependence on rotor position. Second, there is still coupling between the direct and quadrature axes. This coupling shows up in the term that contains J, and also in the term that represents the induced voltages created by the saliency. Third, aligning the magnet flux with the direct axis has put all of the magnet induced voltage into the quadrature axis, since voltage leads flux linkage by 90° electrical. Finally, there is substantially less coupling between the two phases than we saw in the $\alpha\beta$ model.

The first row of Eq. 11.58 gives rise to the circuit model of Fig. 11.9, and the bottom row gives rise to the circuit model of Fig. 11.10. These circuit models still reflect coupling between the two axes, but are substantially simpler than their $\alpha\beta$ counterparts.

Our final piece of business in developing the dq model of the permanent magnet machine is the torque expression. Starting with Eq. 11.32 and transforming the $\alpha\beta$ quantities into the dq reference frame gives

$$T^e = \frac{3L_{s2}N_p}{4} i_{dq}^T e^{J\theta_e} \begin{bmatrix} \sin 2\theta_e & -\cos 2\theta_e \\ -\cos 2\theta_e & -\sin 2\theta_e \end{bmatrix} e^{-J\theta_e} i_{dq} + $$
$$\sqrt{\frac{3}{2}} \frac{KN_p}{2} i_{dq}^T e^{J\theta_e} \begin{bmatrix} -\sin\theta_e \\ \cos\theta_e \end{bmatrix} + T_{FF}^e \quad . \tag{11.59}$$

Carrying out the indicated operations gives

$$T^e = \frac{N_p}{2}(L_d - L_q) i_d i_q + \sqrt{\frac{3}{2}} \frac{KN_p}{2} i_q + T_{FF}^e \quad . \tag{11.60}$$

We see that quadrature axis current dictates the sign and magnitude of the torque produced by the permanent magnet. The reluctance term depends on the magnitude of the saliency and the product of the direct and quadrature currents. For efficient operation of the machine, it is important to coordinate the reluctance and magnet torques. We will examine how this is accomplished in Chapter 12.

11.4 Cogging Torque

11.4.1 Overview

As discussed in the last section, cogging torque is taken as the torque created by the interaction between the magnets on the rotor and the stator teeth. Cogging torque gives rise to tangential vibration; it is usually not sufficiently strong to create substantial radial vibration unless there is appreciable rotor eccentricity. The cogging torque model developed in this section strives to take into consideration both skew and rotor eccentricity. The approach is based on flux tube analysis. A description based on nonlinear magnetics is beyond our scope and would typically need numerical tools, such as finite element analysis. Other approaches are based on a field description within the air gap of the machine, with corrections for slotting[1].

---

[1] See, for example, Z. Q. Zhu, D. Howe, E. Bolte, B. Ackermann, "Instantaneous magnetic field distribution in brushless permanent magnet DC motors. I. Open-circuit field," *IEEE Trans. on Magnetics*, Vol. 29, pp. 124-135, 1993, Z. Q. Zhu and D. Howe, "Instantaneous magnetic field distribution in brushless permanent magnet DC motors. II. Armature-reaction field," *IEEE Trans. on Magnetics*, Vol. 29, pp. 136-142, 1993, Z. Q. Zhu and D. Howe, "Instantaneous magnetic field distribution in brushless permanent magnet DC motors. III. Effect of stator slotting," *IEEE Trans. on Magnetics*, Vol. 29, pp. 143-151, 1993, and Z. Q. Zhu and D. Howe, "Instantaneous magnetic field distribution in permanent magnet brushless DC motors. IV. Magnetic field on load," *IEEE Trans. on Magnetics*, Vol. 29, pp. 152-158, 1993.

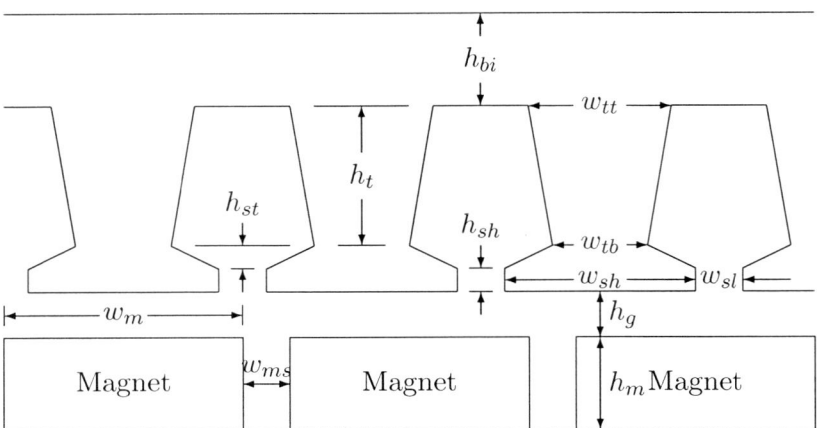

Figure 11.11: The assumed geometry for analysis of cogging torque.

Figure 11.11 gives the machine geometry used as the basis of the cogging torque model. The cogging torque is determined using a flux tube analysis, similar to what was used to model slotting at the air gap in Sec. 4.2. The magnets are the source of field within each flux tube, and simple models for the magnet, air gap, and the space between the stator teeth are compiled to determine the energy stored in the machine. Once we have the energy, the cogging torque can be determined.

The approach used to develop the cogging torque uses:

1. A rectangular geometry so that flux tubes have parallel sides. This is a reasonable approximation since the combined thickness of the magnets and air gap is typically small compared to the active radius.

2. A shape function that describes the length of flux paths between the stator teeth.

3. A shape function that describes the length of the air gap. This shape function can easily accommodate rotor eccentricity. In the absence of eccentricity this function is just a constant length.

4. A shape function that describes the shape of the permanent magnets. This model deals specifically with magnets that are mounted on the surface of the rotor. This model assumes the magnets have rectangular cross-section, consistent with Fig. 11.11. Other magnet shapes, such as breadloaf magnets, can be handled in a similar manner.

5. A function that describes the magnetic flux density of the permanent magnets. This function assumes that the flux density is symmetrical

11.4. Cogging Torque

with respect to each pole face, but it does allow for lower flux density at the edges of the magnet relative to the center of the pole. The model assumes that the flux density is directed straight across the air gap.

6. Proper combination of the shape and magnet functions to determine the magnetic flux within each flux tube.

7. Using the flux and shape functions within each flux tube to determine the magnetic energy stored within the volume of the flux tube.

8. Summing the energy within all flux tubes to determine the total magnetic energy within the machine.

9. Taking the spatial derivative of the energy to determine the cogging torque.

10. Applying the filtering effect created by skewed stator teeth.

The following subsections provide the details of the cogging torque model.

11.4.2 Flux Tube Preliminaries

At any rotor position, the incremental energy stored in the field is

$$w = \frac{B^2}{2\mu} dV \quad , \tag{11.61}$$

where B is the magnetic flux density, μ is the appropriate permeability, and dV is the incremental volume. The incremental volume is equal to the tangential width of the flux tube times its axial length[2].

The magnetic flux density and the magnetic field intensity are related through appropriate constituitive relations. In air,

$$B_a = \mu_0 H_a \quad . \tag{11.62}$$

In the magnet,

$$B_m = \mu_R H_m + B_r \quad , \tag{11.63}$$

---

[2]There are different formulations of the incremental energy in a magnetic field. A formulation commonly used is $w = \left[\int H dB\right] dV$, but this formulation has problems with hard magnetic materials described by Eq. 11.63, giving a term that will actually cause the energy to be negative because the H and B fields are oppositely directed in the magnet. The formulation used here is designed to give the same energy in a flux tube, regardless of whether the energy is created by a permanent magnet or a fictitious winding that produces the same magnetic flux density. This approach gives confidence that the energy within the magnet is being determined properly, motivated by the discussion in Sec. 1.4.

where μ_R is the recoil permeability of the magnet material and B_r is the residual flux density of the magnet material.

Given the material models, the incremental energy in air is given by

$$w_{\text{air}} = \frac{B_a^2}{2\mu_0} dV \quad . \tag{11.64}$$

In the magnet material the incremental energy is

$$w_{\text{magnet}} = \frac{B_m^2}{2\mu_R} dV \quad . \tag{11.65}$$

Ampere's law along a flux tube from the rotor to the stator requires

$$H_m \ell_m + H_a \ell_a = 0 \quad , \tag{11.66}$$

where H_m is the magnetic field intensity in the magnet, H_a is the magnetic field intensity in the air, ℓ_m is the magnetic path length through the magnet, and ℓ_a is the magnetic path length through the air.

Gauss's law requires that the net flux leaving the rotor equal zero. In addition, the magnetic flux density is constant along a flux tube. Using the material models of Eqs. 11.62 and 11.63 in Eq. 11.66 gives

$$\left(\frac{B_m - B_r}{\mu_R}\right) \ell_m + \frac{B_a}{\mu_0} \ell_a = 0 \quad . \tag{11.67}$$

Conservation of flux along the flux tube requires $B_m = B_a = B$. Solving for B gives

$$B\left(\frac{\ell_m}{\mu_R} + \frac{\ell_a}{\mu_0}\right) = B_r \frac{\ell_m}{\mu_R} \quad , \tag{11.68}$$

which reduces to

$$B = B_r \frac{1}{1 + \mu_R \ell_a / \mu_0 \ell_m} \quad . \tag{11.69}$$

Using Eq. 11.69, the energy density in the air portion of each flux tube is

$$w_{\text{air}} = \frac{1}{2\mu_0} \left[\frac{B_r}{1 + \mu_R \ell_a / \mu_0 \ell_m}\right]^2 dV_{\text{air}} \quad . \tag{11.70}$$

The energy density in the magnet portion of each flux tube is

$$w_{\text{magnet}} = \frac{1}{2\mu_R} \left[\frac{B_r}{1 + \mu_R \ell_a / \mu_0 \ell_m}\right]^2 dV_{\text{magnet}} \quad . \tag{11.71}$$

The following subsections develop more detailed descriptions of the air and magnet regions of the problem, thereby enabling a determination of the magnetic energy in the system due to the permanent magnets. Once we have this, the cogging torque can be determined.

11.4.3 The Stator Slot Shape Function

The stator slot region is modeled using flux tubes that are circular arcs starting in the plane at the air gap side of the stator teeth, and terminating on the stator tooth. Examination of Fig. 11.11 shows that there are two cases to consider. If the height of the stator pole shoe is taller than one half of the slot opening, then all of the flux tubes in the slot will terminate on the side of the pole shoe. Otherwise, the flux tubes will terminate on the back side of the pole shoe.

Without loss of generality, our model of the stator slots assumes:

1. There are N_s slots and teeth.

2. The position $x = 0$ aligns with the middle of slot 1.

3. Flux lines follow circular arcs.

Regardless of the relationship between h_{sh} and w_{sl}, we should expect periodicity such that

$$w_{sh} + w_{sl} = \frac{2\pi R}{N_s} , \qquad (11.72)$$

where R is the effective radius used to convert angular measure to rectangular measure.

Case 1: $w_{sl} \leq 2h_{sh}$ For this case the flux lines will end on the sides of the pole shoes, with each flux tube completing one quarter of a circular arc. The air gap function describing the length of the flux tubes in the slots is shown in Fig. 11.12.

By inspection, this function has a dc component and even symmetry. An appropriate representation is then

$$f_{sl}(x) = a_0 + \sum_{n=1}^{\infty} a_n \cos\left(\frac{nN_s x}{R}\right) . \qquad (11.73)$$

Solving for a_0 and a_n gives

$$a_0 = \frac{N_s w_{sl}^2}{16R} , \qquad (11.74)$$

and

$$a_n = \frac{R}{n^2 N_s} \left[1 - \cos\left(\frac{nN_s w_{sl}}{2R}\right)\right] . \qquad (11.75)$$

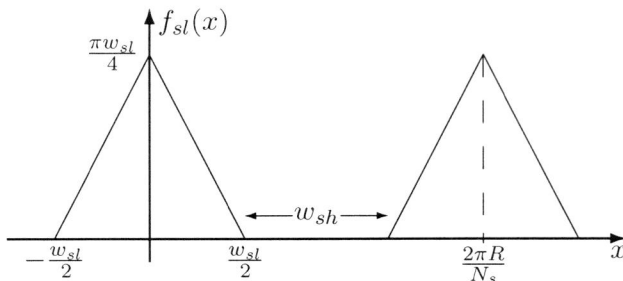

Figure 11.12: The shape function describing the length of flux tubes in the stator slots when the flux lines end on the side of the pole shoe.

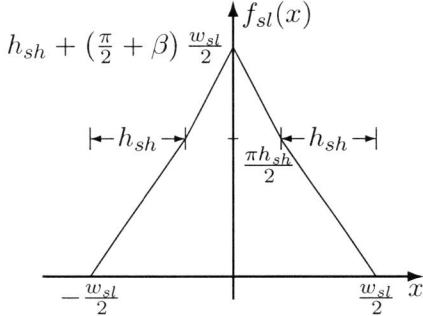

Figure 11.13: The shape function describing the length of flux tubes in the stator slots when the flux tubes end on the back of the stator shoes.

Case 2: $w_{sl} > 2h_{sh}$ For this case the flux lines will end on the back of the pole shoes, with some flux tubes completing one quarter of a circular arc while others complete more than one quarter of a circular arc. The air gap function describing the length of the flux tubes in the slots is shown in Fig. 11.13.

The form of this function is the same as the previous case:

$$f_{sl}(x) = a_0 + \sum_{n=1}^{\infty} a_n \cos\left(\frac{nN_s x}{R}\right) \quad . \tag{11.76}$$

Solving for a_0 and a_n gives

$$a_0 = \frac{N_s w_{sl}^2}{16R} + \frac{N_s \left(\frac{w_{sl}}{2} - h_{sh}\right)\left(h_{sh} + \beta \frac{w_{sl}}{2}\right)}{2\pi R} \quad , \tag{11.77}$$

and

$$a_n = \frac{R}{n^2 N_s}\left[1 - \cos\left(\frac{nN_s w_{sl}}{2R}\right)\right] + \frac{h_{sh} + \beta\frac{w_{sl}}{2}}{\frac{w_{sl}}{2} - h_{sh}}\frac{2R}{\pi n^2 N_s}\left[1 - \cos\left(\frac{nN_s\left(\frac{w_{sl}}{2} - h_{sh}\right)}{R}\right)\right] \quad . \tag{11.78}$$

The angle β is defined as

$$\beta = \frac{\pi}{2} - \tan^{-1}\left(\frac{2h_{st}}{w_{sh} - w_{tb}}\right) \quad . \tag{11.79}$$

The structure of Eqs. 11.77 and 11.78 can be recognized as the superposition of two triangular functions, the first being the same as for the case where $h_{sh} > w_{sl}/2$ and the second covering the additional length added by the flux tubes curving around the back of the pole shoe.

11.4.4 The Air Gap Shape Function

Normally the air gap region is of thickness h_g. However, rotor eccentricity is probable. This eccentricity will tend to increase the air gap in some areas and decrease the air gap in other areas. The eccentricity will go through one cycle per revolution. The eccentricity will not necessarily line up with any of the rotor poles or stator teeth. It follows that a reasonable model for the air gap height is

$$f_g(x) = h_g + \varepsilon_g \sin\left(\frac{2\pi x}{R} + \gamma_g\right) \quad , \tag{11.80}$$

where h_g is the nominal air gap height, ε_g is the magnitude of the eccentricity, and γ_g is the phase of the eccentricity.

11.4.5 The Magnet Shape Function

The magnet shape function captures the rectangular shape of the magnets shown in Fig. 11.11. Unlike the stator, the rotor structure is dependent on the rotor position θ. Figure 11.14 shows the assumed physical geometry of the magnets. While the model given here assumes regular spacing of the magnet poles, it can be extended to include situations where the placement of the magnet poles is varied to reduce cogging torque[3]. Slight shifting of the

---

[3] See, for example, N. Bianchi and S. Bolognani, "Design techniques for reducing the cogging torque in surface-mounted PM motors," *IEEE Trans. on Industry Applications*, Vol. 38, pp. 1259-1265, 2002. The paper R. Lateb, N. Takorabet, and F. Meibody-Tabar, "Effect of magnet segmentation on the cogging torque in surface-mounted permanent-magnet motors," *IEEE Trans. on Magnetics*, Vol. 42, pp. 442-445, 2006 also discusses similar issues.

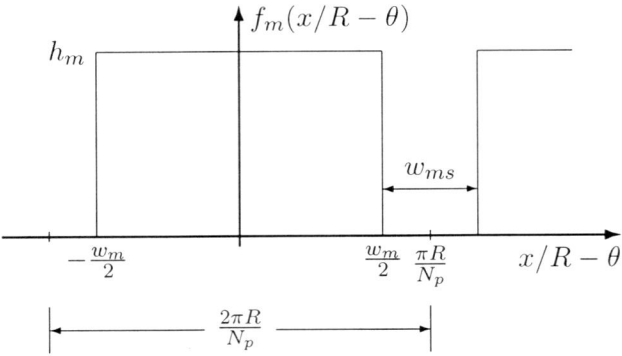

Figure 11.14: The shape function describing the space taken up by the permanent magnets.

magnet poles off of their normal regular spacing interferes with the tendency of the multiple magnets to grab multiple stator poles in exactly the same way, thereby reducing the peak cogging torque.

Similar to the models for the stator slot regions and the air gap, the physical shape of the magnets is described by a Fourier series:

$$f_m(x/R - \theta) = \frac{h_m w_m N_p}{2\pi R} + \sum_{n=1}^{\infty} \frac{2h_m}{n\pi} \sin\left(\frac{n w_m N_p}{2R}\right) \cos\left(n N_p (x/R - \theta)\right) \quad (11.81)$$

This model only captures the actual space taken up by the magnets. We do not need to worry about the space between the magnets because there is no magnetization within these regions to produce field.

11.4.6 The Magnet Magnetization Function

Describing the magnetization of the magnets is a little more involved than modeling the space taken up by the magnets for two reasons. First, the magnetization alternates. While this does not really impact the cogging torque because the energy always depends on the square of the field, it is useful to be able to generate a picture of the air gap field due to the magnets. Second, the model allows for the magnetization strength to be weaker at the edges of the magnets than at their center. Figure 11.15 shows the assumed magnetization profile.

11.4. Cogging Torque

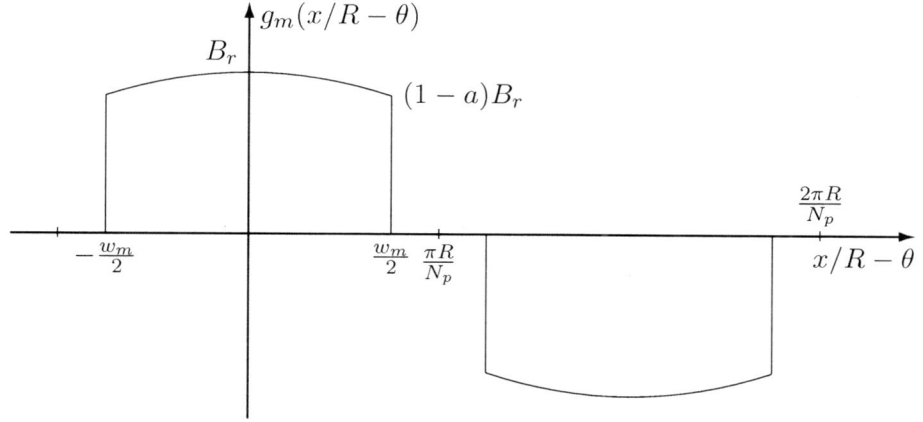

Figure 11.15: The function describing the magnetization provided by the permanent magnets.

Consistent with Fig. 11.15, the magnetization function is

$$g_m(x/R - \theta) = \sum_{\substack{n=1 \\ n \text{ odd}}}^{\infty} \left\{ \frac{4(1-a)B_r}{n\pi} \sin\left(\frac{nw_m N_p}{2R}\right) - \frac{32 a B_r R}{n^2 \pi w_m N_p} \sin\left(\frac{nw_m N_p}{4R}\right) + \frac{128 a B_r R}{n^3 \pi w_m^2 N_p^2} \cos\left(\frac{nw_m N_p}{4R}\right) \right\} \cos\left(nN_p(x/R - \theta)\right) \quad . \tag{11.82}$$

By virtue of the assumed regular spacing of the magnets, the magnetization function possesses both even and half-wave symmetry. In the limit as $a \to 0$, the Fourier coefficient collapses to that expected for an alternating quasi-square waveform.

11.4.7 Energy and Cogging Torque

Along a flux tube, Ampere's law requires

$$H_m f_m + H_g (f_g + f_{sl}) = 0 \quad , \tag{11.83}$$

where f_m, f_g, and f_{sl} correspond to the flux path lengths through the magnet, air gap, and slot, respectively. Conservation of flux along the tube requires

$$H_m = \frac{B}{\mu_R} - \frac{g_m}{\mu_R} \qquad (11.84)$$

in the magnet, where g_m is the magnet flux density, and

$$H_g = \frac{B}{\mu_0} \qquad (11.85)$$

in the air portion of the flux tube. Substituting Eqs. 11.84 and 11.85 into Eq. 11.83 gives

$$B = \frac{\mu_0 g_m f_m}{\mu_0 f_m + \mu_R (f_g + f_{sl})} \qquad . \qquad (11.86)$$

Gauss' law also requires that the net flux leaving the rotor surface be zero. It follows that if $B(x, \theta, \gamma_g)$ as given by Eq. 11.86 has some offset due to eccentricity, a constant needs to be added to force conversation of flux. Formally, we require

$$\int_0^{2\pi R} [B(x, \theta, \gamma_g) - K] \, dx = 0 \quad , \qquad (11.87)$$

so

$$K = \frac{1}{2\pi R} \int_0^{2\pi R} B(x, \theta, \gamma_g) dx \qquad (11.88)$$

and

$$B(x, \theta, \gamma_g) = \frac{\mu_0 g_m f_m}{\mu_0 f_m + \mu_R (f_g + f_{sl})} - \frac{1}{2\pi R} \int_0^{2\pi R} \left[\frac{\mu_0 g_m f_m}{\mu_0 f_m + \mu_R (f_g + f_{sl})} \right] dx \quad . \qquad (11.89)$$

The incremental energy stored in each flux tube is

$$w = \frac{B^2}{2\mu} dV \quad . \qquad (11.90)$$

It follows that the incremental energy in the air portion of the flux tube is given by

$$w_{\text{air}} = \frac{B^2(x, \theta, \gamma_g)}{2\mu_0} (f_g + f_{sl}) L_s dx \quad . \qquad (11.91)$$

In the magnet portion of the flux tube the incremental energy is

$$w_{\text{magnet}} = \frac{B^2(x, \theta, \gamma_g)}{2\mu_R} f_m L_s dx \quad . \qquad (11.92)$$

11.4. Cogging Torque

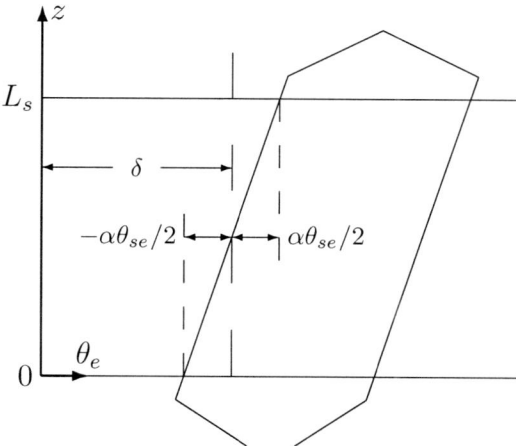

Figure 11.16: One coil showing the introduction of skew angle α.

The total energy in the machine due to the magnets is

$$W = \int_0^{2\pi R} B^2(x,\theta,\gamma_g) \left\{ \frac{(f_g + f_{sl})}{2\mu_0} + \frac{f_m}{2\mu_R} \right\} L_s dx \quad . \tag{11.93}$$

From conservation of energy, the cogging torque is determined as

$$T_{\text{cogging}} = -\frac{\partial W}{\partial \theta} \quad . \tag{11.94}$$

We note that the integration indicated in Eq. 11.93 is carried out over x, resulting in $W = W(\theta, \gamma_g)$.

11.4.8 Skew

If the stator laminations of the machine are skewed, the energy in the machine will be different from that predicted by Eq. 11.93. Skewing offsets each lamination by a small angular displacement relative to its neighbors. The effect is one of twisting the lamination stack over a total angle $\alpha\theta_{se} = \alpha\pi R N_p/N_s$ in electrical measure, where α is the amount of skew normalized to one slot pitch. Figure 11.16 depicts the introduction of skew into one of the stator coils, as was considered in Sec. 3.6.

If the angular position of a stator tooth is taken as $\alpha\theta_{se}/2$ at one end of the stack and $-\alpha\theta_{se}/2$ at the other end of the stack, the incremental energy in the machine as a function of axial position along the stack is given by

$$w(z,\theta,\gamma_g) = \Re\left\{\frac{W(\theta,\gamma_g)}{L_s}\exp\left[-\jmath\frac{\alpha\theta_{se}}{2}\left(1-\frac{2z}{L_s}\right)\right]dz\right\} \quad , \qquad (11.95)$$

where $W(\theta,\gamma_g)$ is given by Eq. 11.93, L_s is the axial length of the stack, and the exponential term $\exp\left[-\jmath\alpha\theta_{se}/2\left(1-2z/L_s\right)\right]$ represents the spatial shifting in the θ direction created by the skew at axial position z. The total influence of the skew is found by summing the energy distribution over the length of the machine:

$$W(\theta,\gamma_g) = \int_0^{L_s} w(z,\theta,\gamma_g) \quad . \qquad (11.96)$$

Substituting Eq. 11.95 gives

$$W(\theta,\gamma_g) = \frac{W(\theta,\gamma_g)}{L_s}\Re\left\{\int_0^{L_s}\exp\left[-\jmath\frac{\alpha\theta_{se}}{2}\left(1-\frac{2z}{L_s}\right)\right]dz\right\} \quad . \qquad (11.97)$$

Carrying out the indicated integration and making use of the relationship between complex exponentials and the circular sine function[4] gives

$$W(\theta,\gamma_g) = W\frac{\sin(\alpha\theta_{se}/2)}{(\alpha\theta_{se}/2)} \quad , \qquad (11.98)$$

where W is given by Eq. 11.93. The effect of skew manifests itself as a low pass filter and as a modulation function; it is shown in Fig. 11.17. The filtering effect on the energy in the machine passes right through to the cogging torque. The reduction of cogging torque is a significant motivation for using skew. This same effect was found in Sec. 3.6 when considering the effect of skew on the air gap mmf produced by windings.

11.4.9 Model Results

The elements of the model can be programmed to determine the cogging torque. Table 11.1 lists the parameters used to exercise the cogging torque model. Figure 11.18 shows the air gap flux density as a function of rotor position. The pitch of the magnets and the influence of the slots are apparent. Figure 11.19 shows the air gap energy density for rotor position $\theta = 0$. The energy density within each flux tube is always positive, regardless of the direction of flux density. The air gap eccentricity causes the modulation in the energy density. This modulation is also in the underlying flux density, but it is more difficult to see.

---
[4]$\sin\alpha\theta_{se} = [\exp(\jmath\alpha\theta_{se}) - \exp(-\jmath\alpha\theta_{se})]/(\jmath 2)$.

11.4. Cogging Torque

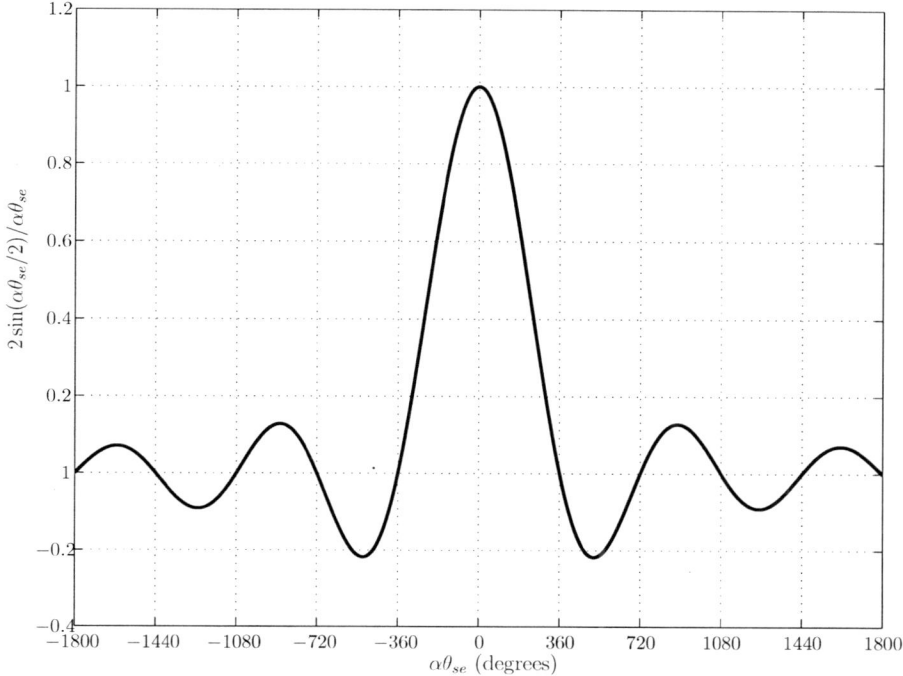

Figure 11.17: The filter function produced by skewing the stator slots.

Figure 11.20 shows the cogging torque for the permanent magnet machine described by the parameters in Table 11.1. The cogging torque waveform is comprised of positive and negative pulses that form doublets, created as magnets approach and leave each stator slot. Because the machine under consideration is an integral slot machine, all magnets are approaching and leaving slots in synchronism, thereby increasing the cogging torque.

The amplitude of the cogging torque is due in part to the rapid transitions in air gap flux density and the corresponding energy density. Figure 11.18 shows the air gap flux density underlying the cogging torque calculation. In reality one would expect smoother edges on the flux density waveform, thereby substantially reducing the amplitude of the cogging torque. To show this effect, Fig. 11.21 shows the air gap flux density when only one harmonic is carried in each of the Fourier series representations. That is, N_h in Table 11.1 is reduced from 100 to 1. Figure 11.22 shows that the amplitude of the cogging torque drops from nearly 1500 N to only about 40 N, a more reasonable value for this size machine.

In addition to softened edges on the radial magnetic flux density waveform, which is suggested in going from Fig. 11.18 to Fig. 11.21, cogging

Table 11.1: The machine parameters used to exercise the cogging torque model.

Parameter	Symbol	Value
Number of poles	N_p	8
Number of stator slots	N_s	48
Number of phases	N_ϕ	3
Number of Fourier harmonics	N_h	100
Effective rotor radius	R	0.3048 m
Width of stator pole shoe	w_{sh}	0.03391 m
Width of stator slot opening	w_{sl}	0.00598 m
Height of stator pole shoe	h_{sh}	0.01 m
Height of stator pole shoe taper	h_{st}	0.01 m
Width of stator pole at shoe	w_{tb}	0.01696 m
Nominal air gap height	h_g	0.64 mm
Magnitude of air gap eccentricity	ε_g	0.128 mm
Location of rotor eccentricity	γ_g	45°
Radial height of magnets	h_m	3.81 mm
Width of magnets	w_m	0.2063 m
Magnet residual flux density	B_r	1.35 T
Reduction in flux density at magnet edges	a	10%
Recoil permeability of magnets	μ_R	1.05 μ_0 H/m
Active stack length	L_s	0.2 m

torque in a real machine is influenced by the nature of the tangential fields within the air gap. The cogging torque model developed here includes only radial fields. One would expect that the alternating magnetic poles on the rotor would create a tangential component of the air gap magnetic field. This additional field component would have some impact on the cogging torque profile. Further, local saturation in the corners of the stator pole shoes would tend to lessen the cogging forces.

A fractional slot machine would exhibit less cogging torque because not all magnets are approaching the slots in synchronism. This is often used in smaller machines where it is difficult to include a large number of slots and there is an interest in reducing the cost of the machine by using fewer windings. While cogging torque benefits from a fractional slot winding, a fractional slot winding tends to introduce some asymmetries that have negative performance consequences, as discussed in Sec. 4.5.

11.5. Summary

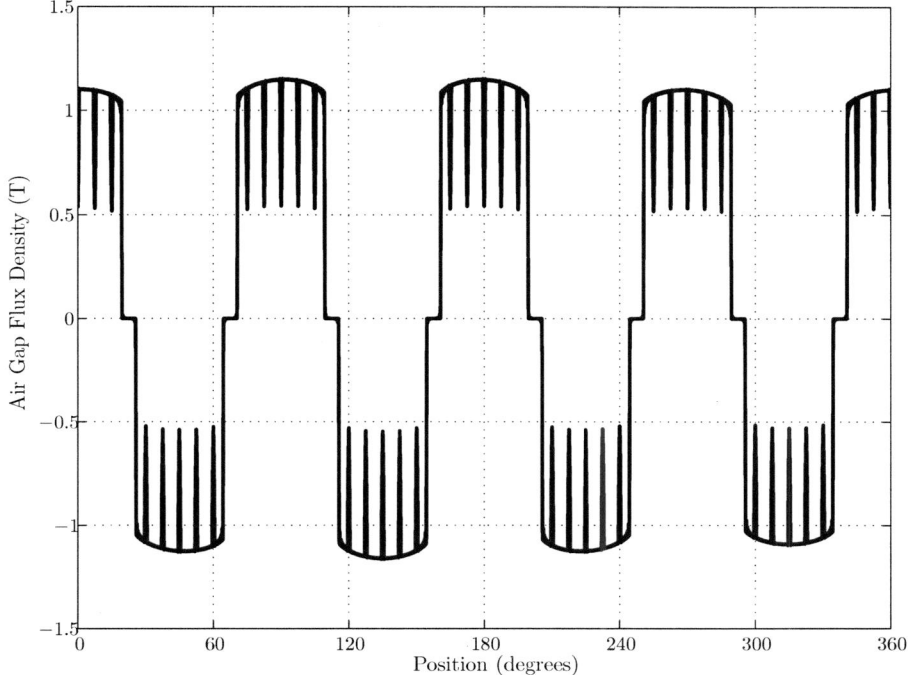

Figure 11.18: The air gap flux density as a function of rotor position. The model uses 100 harmonics in representing the machine features.

11.5 Summary

This chapter has focused on the permanent magnet machine and the development of models that not only support a physical understanding of the machine, but can be used for understanding how to control the machine. The physical model of the permanent magnet machine was motivated by capturing the interaction between the stator circuits and the permanent magnets that is embodied in the flux linkage expression.

To move toward a representation of the machine useful for control, we employed the transformations developed in Chapter 6 to move currents and flux linkages from one reference frame to another. In the $\alpha\beta$ model, stator quantities were taken from a coupled abc representation to a decoupled $\alpha\beta$ representation. The 0 phase components were suppressed because they do not contribute to torque production.

The dq model was developed by projecting stator quantities of the $\alpha\beta$ model into a common reference frame that is allowed to rotate. It is quite

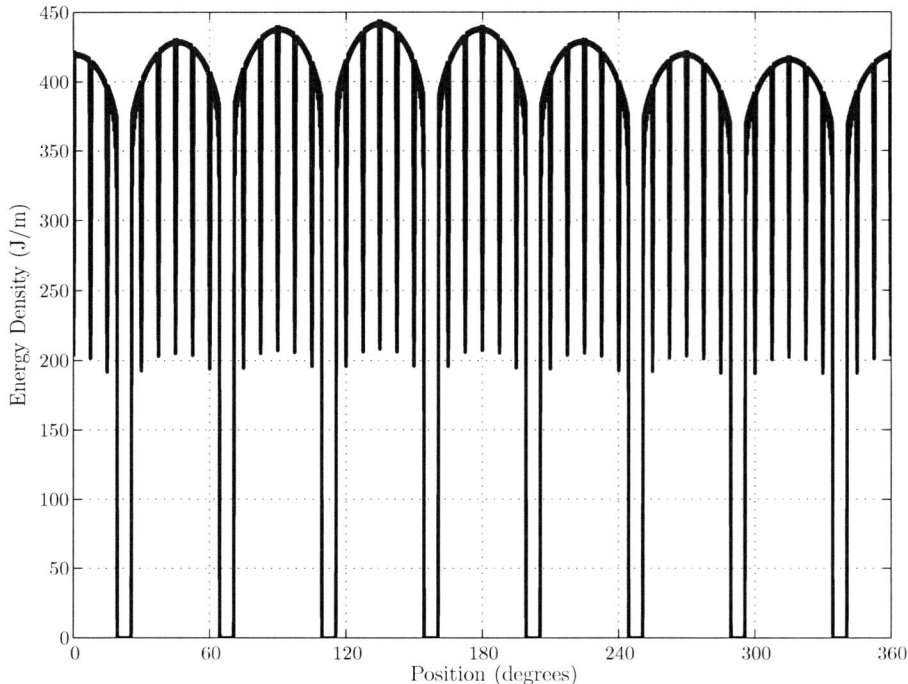

Figure 11.19: The air gap energy density for $\theta = 0$.

natural to tie this dq reference frame to the permanent magnet flux. As we will see in Chapter 12, the saliency included in our models is important since saliency is often used to gain greater control over the torque production under a wider range of operating conditions.

11.5. Summary

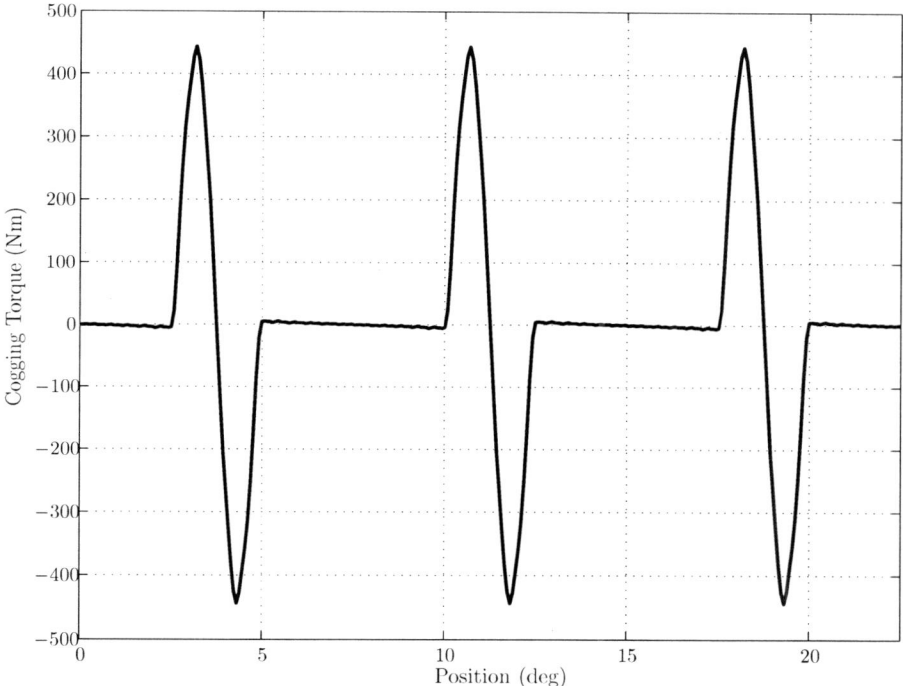

Figure 11.20: A detailed view of the cogging torque for the machine described by the parameters in Table 11.1, showing the doublet in cogging torque created by each tooth.

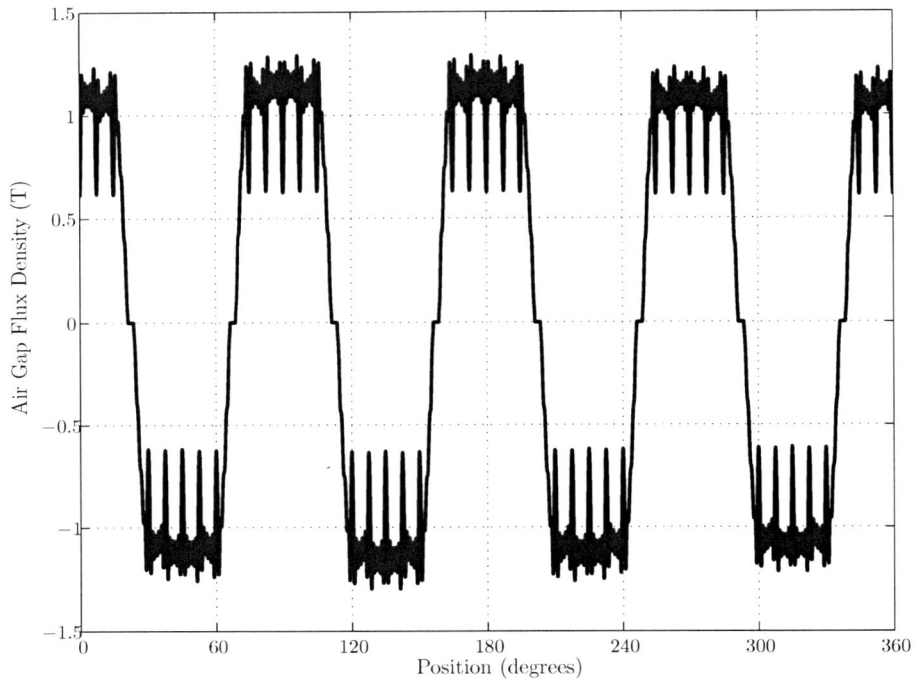

Figure 11.21: The air gap flux density for $\theta = 0$ when N_h is reduced from 100 to 5.

11.5. Summary

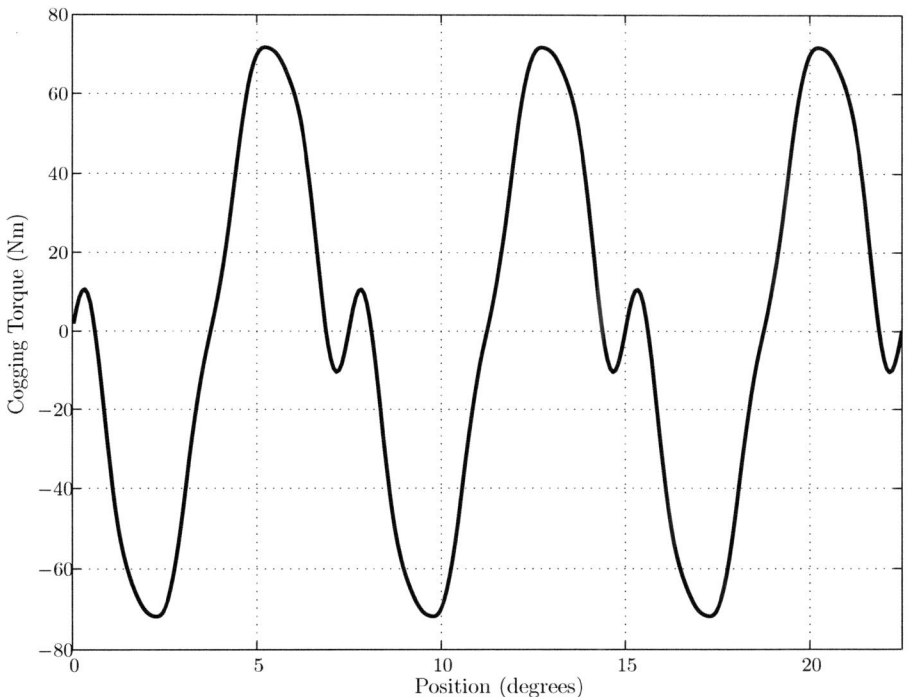

Figure 11.22: The cogging torque for the machine described by the parameters in Table 11.1 when N_h is reduced from 100 to 5.

Chapter 12

Control of Permanent Magnet Machines

12.1 Introduction

This chapter examines the control of permanent magnet machines, building off of the models developed in Chapter 11. Our development will separately consider machines with sinusoidal and trapezoidal back emf profiles.

Machines with sinusoidal back emf profiles are used in two distinct circumstances. In the first, the sinusoidal back emf profile is used, perhaps in conjunction with saliency, to facilitate operation over a wide speed range. The sinusoidal back emf is particularly amenable to field weakening, where a portion of the stator current is used to effectively reduce the strength of the magnet field. This is necessary to extend the operating speed range in the face of a limited supply voltage.

Machines with sinusoidal back emfs are also used in instances where precise control over torque and speed is required. Using field orientation principles in a manner similar to those presented for induction machines in Chapter 10, precise control over the three phase currents allows suppression of torque fluctuations that plague simpler control structures, particularly those applied to machines with trapezoidal back emfs.

Permanent magnet machines with trapezoidal back emfs are sometimes called "brushless dc machines." This terminology is unfortunate because it is not sufficiently descriptive to even indicate the fundamental behavior of the machine! Many types of machines could be considered to be brushless dc machines when operated through an inverter sourced by a dc supply, including: induction, permanent magnet (with either back emf profile), synchronous reluctance, or variable-reluctance.

The intent behind brushless dc machines is to emulate a brushed dc machine. This is accomplished by viewing the field to be set up by the permanent magnets on the rotor. The stator behaves as the armature, with phase currents commutated electronically rather than mechanically. Operation in this mode is characterized by phase currents that ideally have quasi-square profiles that are synchronized to the back emfs. Ideally two and only two phases conduct at a time and produce a torque that is constant. Unfortunately, reality can be well removed from the ideal, and operation at high speed only exacerbates the situation.

12.2 Sinusoidal Permanent Magnet Machines

As we consider permanent magnet machines with sinusoidal back emfs, we use the general case of a machine with saliency consistent with machines that contain magnets embedded within the rotor as shown in Figs. 11.2 and 11.3. Of course, the case of a machine without saliency is achieved by setting $L_d = L_q$ consistent with the rotor of Fig. 11.1.

We begin our discussion by examining the torque expression, Eq. 11.60, and seek to maximize torque production per Ampere of current supplied to the machine. Subsequently we consider the issues associated with the interaction among the voltage supply and the operation of the machine. This will motivate the need for field weakening operation. In our discussions, we ignore the cogging torque term T_{FF}^e since we cannot control it directly.

It was stated explicitly during our development of the dq model for the permanent magnet machine that the direct axis is coincident with the magnet flux. Accordingly, the creation of the dq model forms an implicit basis for field oriented control of the permanent magnet machine.

12.2.1 Maximizing Torque Per Ampere

In any electric machine system, loss is associated with the flow of current. It follows that minimizing current flow subject to providing the required torque is a desirable thing. This suggests we want to maximize the torque produced for every Ampere of current supplied to the machine.

The torque produced by our permanent magnet machine is given by Eq. 11.60, which is repeated here for convenience:

$$T^e = \frac{N_p}{2}(L_d - L_q)i_d i_q + \sqrt{\frac{3}{2}}\frac{KN_p}{2}i_q + T_{FF}^e \quad . \quad (12.1)$$

12.2. Sinusoidal Permanent Magnet Machines

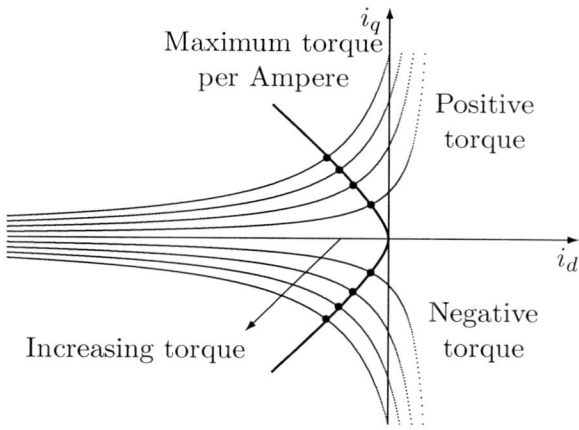

Figure 12.1: Constant torque loci for a permanent magnet machine with saliency. Heavy points denote maximum torque per Ampere.

As observed in Chapter 11, in order for the first two terms to produce torque of the same sign, we need the sign of i_d to be the same as the sign of the term $(L_d - L_q)$, which is typically negative. Note that if i_d is negative, the magnet field is being weakened and we conclude that an interior permanent magnet machine is always most efficient with some degree of field weakening.

Figure 12.1 shows loci of constant torque as functions of i_d and i_q based on Eq. 12.1. In the absence of saliency these loci become horizontal lines. Note that i_q is positive for positive torque (motor operation) and i_q is negative for negative torque (generator operation). Given the shape of the constant torque loci, a legitimate question is by what basis should we select the operating point?

The magnitude of the current supplied to the machine is

$$|I_s| = \sqrt{i_d^2 + i_q^2} \quad . \tag{12.2}$$

This is the vectorial distance from the origin to a point on the desired constant torque locus. It follows that maximizing the torque per Ampere is tantamount to minimizing the length of $|I_s|$ for the desired torque. The enlarged points in Fig. 12.1 represent the points for which $|I_s|$ is minimized. Unfortunately, there is not a simple relationship between the torque and the appropriate values of i_d and i_q. In fact, as shown in Fig. 12.2 the relationship among the currents and torque are nonlinear. The enlarged points in Fig. 12.2 correspond to the enlarged points in Fig. 12.1. It is seen that i_d is not simply the negative of i_q. For negative torque, both currents would be negative.

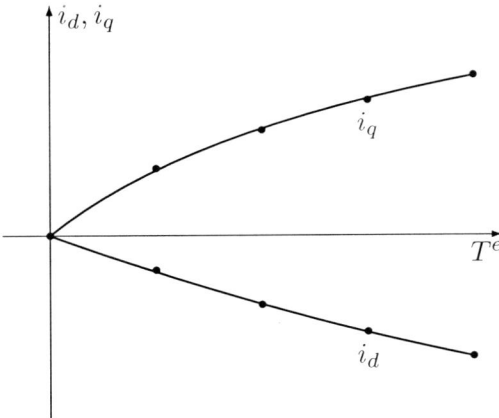

Figure 12.2: The direct and quadrature current commands as a function of torque for an interior permanent magnet machine.

Figure 12.3 shows how the information contained in Figs. 12.1 and 12.2 would be integrated into the speed control of a permanent magnet machine in order to maximize the torque per Ampere. The speed error is used to establish a torque command. Using the dependence of torque on current as shown in Fig. 12.2, commands for i_d and i_q are determined. These currents are then transformed into commands for i_a, i_b, and i_c first using rotation from the dq frame to the $\alpha\beta$ frame before transformation to the abc frame. As discussed in Chapter 8, it is also possible to perform the current regulation in the $\alpha\beta$ or the dq reference frames.

The rotation from the dq frame to the $\alpha\beta$ frame is facilitated by knowing the electrical rotor angle between the d and α axes. This angle information can be obtained by direct measurement through an encoder or resolver. It can also be obtained by observing the back emf waveforms or the α and β inductances.

12.2.2 Direct Torque Control

An alternative to field oriented control is direct torque control. As discussed in Chapter 10, the objective of direct torque control is to manipulate the stator flux vector in a hysteretic manner to regulate the torque produced by the machine. For the permanent magnet machine, the vector diagram is shown in Fig. 12.4 where the magnet flux is along the direct axis such that

$$\vec{\lambda}_m = \sqrt{\frac{3}{2}} K \hat{d} \quad , \tag{12.3}$$

12.2. Sinusoidal Permanent Magnet Machines

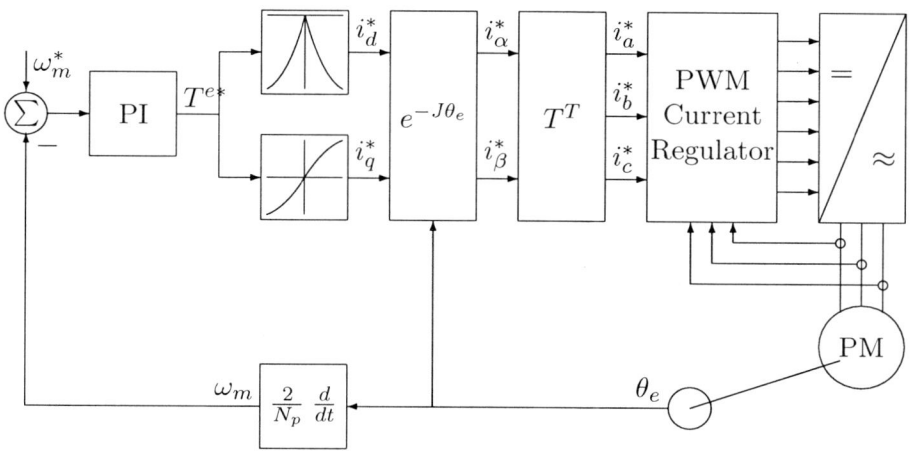

Figure 12.3: A block diagram indicating how field oriented control is implemented for a permanent magnet machine.

and
$$\vec{\lambda}_s = \lambda_{ds}\,\hat{d} + \lambda_{qs}\,\hat{q} \quad, \tag{12.4}$$
where
$$\lambda_{ds} = \left|\vec{\lambda}_s\right|\cos\delta \quad, \tag{12.5}$$
and
$$\lambda_{qs} = \left|\vec{\lambda}_s\right|\sin\delta \quad. \tag{12.6}$$

Substituting Eqs. 12.3 through 12.6 into Eq. 12.1 gives the torque to be

$$T^e = \frac{N_p}{2}\left\{\frac{(L_d - L_q)}{2L_d L_q}\left|\vec{\lambda}_s\right|\sin 2\delta + \frac{1}{L_d}\lambda_m\left|\vec{\lambda}_s\right|\sin\delta\right\} \quad. \tag{12.7}$$

This torque expression suggests the torque can be viewed as the vector cross product between fluxes on the direct and quadrature axes.

The implementation of direct torque control for the permanent magnet machine is the same as that shown in Figs. 10.12, 10.13, and 10.14 for the induction machine. The torque calculation given in Fig. 10.13 is consistent with Eq. 11.38 in which we showed that the torque can be viewed as the vector cross product between flux and current on the α and β axes.

12.2.3 Field Weakening

The discussion in the previous subsections ignored the voltage limitation imposed by the supply. Figure 12.1 does not reflect the interaction between

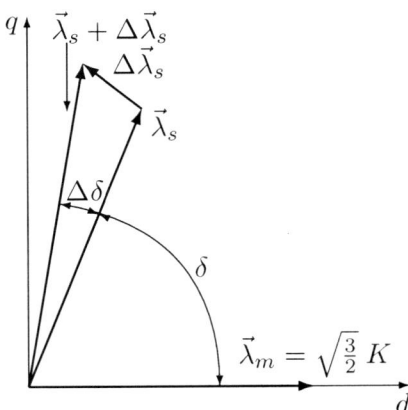

Figure 12.4: The flux vectors showing how changes in torque and stator flux are generated in the permanent magnet machine.

machine speed and the ability to control the currents fed to the machine. We see from the equivalent circuit models of Figs. 11.9 and 11.10 that there are a number of voltage sources that depend on machine speed. Accordingly, the voltage difference available for changing the currents drops as the speed increases.

For a voltage source inverter supplied with voltage V_{dc}, the maximum stator voltage that can be applied to the machine is $\sqrt{2/3}V_{dc}$ in six-step mode. (Recall that the magnitude of each voltage space vector for a two level inverter is $\sqrt{2/3}V_{dc}$ as we saw in Chapters 7 and 8.) The fundamental magnitude is smaller, namely $\left|\vec{V}_s\right| = 2V_{dc}/\pi$. Ignoring stator resistance for constant stator currents, at the maximum operating speed we have the motional voltage equal to the applied voltage such that

$$v_d = -\frac{N_p}{2} L_q i_q \omega_m \quad , \tag{12.8}$$

and

$$v_q = \frac{N_p}{2}\left(L_d i_d + \sqrt{\frac{3}{2}}K\right)\omega_m \quad . \tag{12.9}$$

We also have

$$\left|\vec{V}_s\right|^2 = v_d^2 + v_q^2 \quad , \tag{12.10}$$

or

$$\left[\frac{2V_{dc}}{\pi}\right]^2 = \left[\frac{N_p L_q \omega_m}{2}\right]^2 i_q^2 + \left[\frac{N_p L_d \omega_m}{2}\right]^2 \left(i_d + \sqrt{\frac{3}{2}}\frac{K}{L_d}\right)^2 \quad . \tag{12.11}$$

12.2. Sinusoidal Permanent Magnet Machines

Rearranging this gives

$$\frac{i_q^2}{\left[\frac{4V_{\text{dc}}}{\pi N_p \omega_m L_q}\right]^2} + \frac{\left(i_d + \sqrt{\frac{3}{2}}\frac{K}{L_d}\right)^2}{\left[\frac{4V_{\text{dc}}}{\pi N_p \omega_m L_d}\right]^2} = 1 \quad . \tag{12.12}$$

The structure of Eq. 12.12 can be recognized as describing an ellipse where

$$\frac{4V_{\text{dc}}}{\pi N_p \omega_m L_q} \tag{12.13}$$

is the length of the semi-minor axis,

$$\frac{4V_{\text{dc}}}{\pi N_p \omega_m L_d} \tag{12.14}$$

is the length of the semi-major axis, and

$$-\sqrt{\frac{3}{2}}\frac{K}{L_d} \tag{12.15}$$

is the offset of the ellipse center on the direct axis. Note that in the absence of saliency our ellipse collapses to a circle with $L_d = L_q$.

From Eqs. 12.13 and 12.14 we see that the lengths of the semi-minor and semi-major axes are inversely dependent on speed. Because Eq. 12.15 does not depend on speed, the location of the ellipse is independent of speed. Accordingly, we end up with a series of concentric ellipses that get progressively smaller with increasing speed. These ellipses describe the operating limitation imposed by a limited supply voltage. In addition, there is a thermal limitation on the magnitude of the currents fed to the machine.

The voltage and current operating limits are superimposed on the current plane in Fig. 12.5. Also shown is a constant torque locus and the maximum torque per Ampere relationship. The implication of the voltage limit is that as speed increases we must add additional (negative) direct axis current to maintain the same torque. This gives us additional speed range at the expense of no longer producing maximum torque per Ampere. The addition of (negative) direct axis current opposes the magnet field. This is known as field weakening.

For example, consider the middle ellipse and its intersection with the locus of constant torque. The limitation imposed by voltage forces operation to move from the line of maximum torque per Ampere to the point where the torque locus and the voltage ellipse intersect. In this example both operating points are within the current limit of the machine. However, it is possible for the current limit to become more restrictive than the voltage

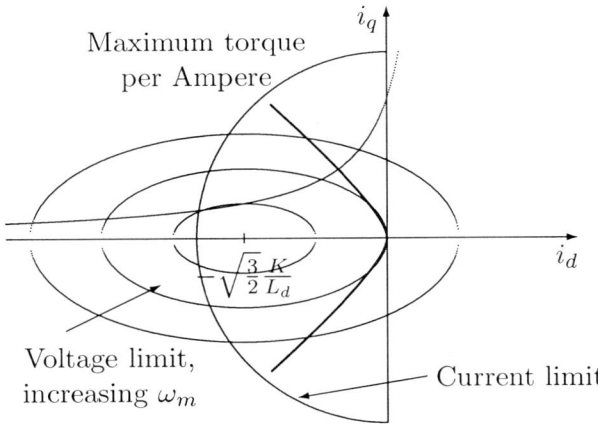

Figure 12.5: The interaction of voltage and current limits on torque production.

limit. Under this situation, the voltage limit is pushing for larger direct axis current, while the current will force the quadrature current to decrease to compensate for the increase in i_d. This will typically force the available torque to roll off even more quickly with speed, thereby defining the upper extent of the constant power operating region.

12.3 Trapezoidal Permanent Magnet Machines

The operation of permanent magnet machines with trapezoidal back emf profiles is generally quite a bit different from the control of machines with sinusoidal back emfs. The intent is to have only two phases conducting at a time, with each phase supporting piecewise constant current that is synchronized to the phase back emfs.

With constant current in the phase windings while the corresponding phase back emfs are constant, the machine theoretically produces constant torque. Figure 12.6 shows the theoretical waveforms associated with machine operation. The idea is to electronically shift dc currents among the phases in a manner analogous to what a mechanical commutator does in a conventional dc machine. As previously stated, this type of machine is often referred to as a brushless dc machine. Unfortunately, this terminology is often confusing because many ac machines are brushless. including induction, sinusoidal permanent magnet, trapezoidal permanent magnet, synchronous reluctance, and variable reluctance.

12.3. Trapezoidal Permanent Magnet Machines

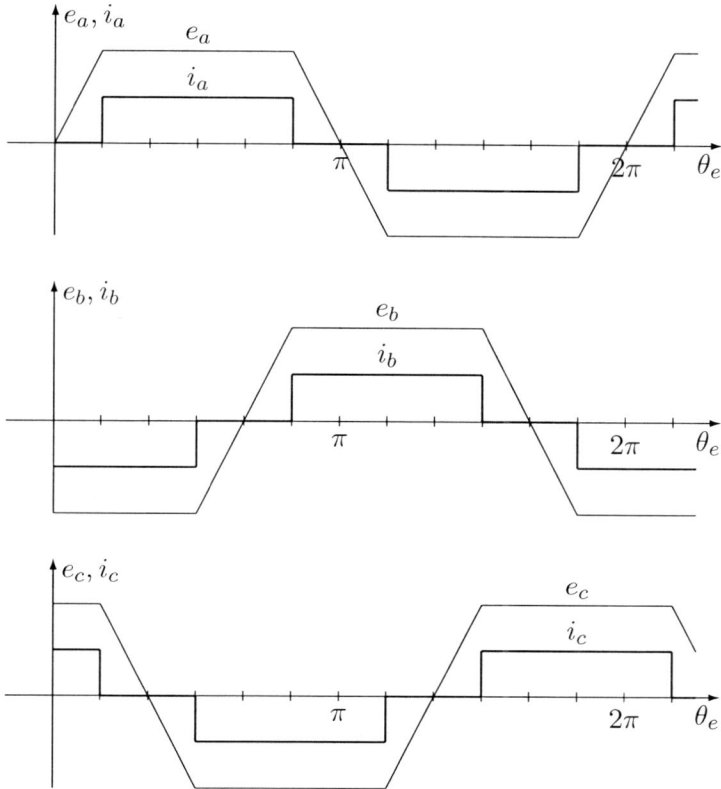

Figure 12.6: The ideal back emf and phase current waveforms for a permanent magnet motor with trapezoidal back emfs.

Figure 12.7 shows the structure of a speed controller for a brushless dc machine. Speed error is used to generate a current command. An electronic commutator converts the current command into reference current waveforms for each phase. These current waveforms are imposed on the machine through a switching inverter, in which the switching function for each switch is determined by a current regulator.

To illustrate the operation of a brushless dc machine and some of the issues that arise, a brushless dc machine drive has been simulated. The parameters associated with the machine and the controller are given in Table 12.1. The details of the simulation are provided in Appendix C. To provide something of an upper limit on the performance, the back emf profile of the machine is taken to be ideally trapezoidal consistent with Figs. 11.4 and 11.5. In practice, the back emf profile will not be perfectly trapezoidal, showing some curvature in transition regions. The result is that torque dips

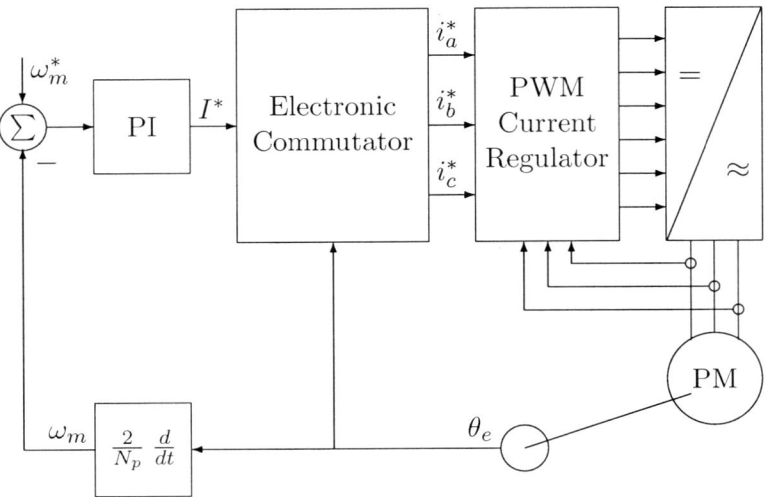

Figure 12.7: A block diagram showing how speed is typically controlled for a trapezoidal permanent magnet machine.

will occur during the commutation regions. These effects have been intentionally left out of the simulation.

The simulation considers the machine as it drives a passive load. The load torque is proportional to the square of the shaft speed, consistent with a fan type load. The commanded speed of the machine is 100 rad/s. Consistent with sound engineering practice, the peak currents applied to the machine phases are constrained to remain below an upper limit. In this example the limit is 70 A. This limit is usually selected as the lower of the machine phase current limit based on machine overheating, or the upper limit of the phase currents that can be supported by the inverter.

There are integral controllers involved in the speed controller that generates the desired phase current amplitude, and in the current regulator that generates the voltage applied to the phase windings. These integral controllers make use of the same anti-windup approach shown in Fig. 8.20. The absence of anti-windup provisions would hurt dynamic performance of the system.

Figure 12.8 shows the machine phase currents while the rotor is still coming up to speed. Also shown in the figure is a curve representing the commanded current amplitude. For this machine drive, this limit is set at 70 A to avoid thermal overload of the windings. In an electronic machine drive, current limitation through the inverter is important for protecting both the

12.3. Trapezoidal Permanent Magnet Machines

Table 12.1: Parameters for the brushless dc machine system simulation.

Parameter	Symbol	Value	Units
Speed Control Proportional Gain	$K_{p\omega}$	314.1	As/rad
Speed Control Integral Gain	$K_{i\omega}$	19,739	A/rad
Current Regulator Proportional Gain	K_{pi}	56.3	Ω
Current Regulator Integral Gain	K_{ii}	56,297	Ω/s
Inverter Switching Frequency	f_s	16	kHz
Dc Bus Voltage	V_{dc}	625	V
Phase Resistance	R	0.7	Ω
Phase Inductance	L	5.6	mH
Back Emf Constant	K_e	2.2	Vs/rad
Number of Poles	N_p	6	
Rotor Inertia	H	0.5	kg m^2
Load Torque Constant	k_T	0.008	Nm s^2/rad^2
Speed Regulator Bandwidth	$\omega_{c\omega}$	160	Hz
Current Regulator Bandwidth	ω_{ci}	1.6	kHz
Motor Electrical Frequency	$R/2\pi L$	19.9	Hz

inverter and the machine. It will be observed that the phase currents do not precisely follow the commanded current value, particularly during the commutation intervals. When the currents are below the commanded value, there is no apparent increase in the command because the command is already saturated.

Figure 12.9 focuses in on the time interval where the current command is coming out of saturation. As the commanded speed is reached, the current command is coming out of saturation. With the speed controller out of saturation, it is now possible to see increases in the commanded phase current at each commutation interval. The drop in torque production created by the commutation process is causing the speed controller to respond with an increase in current command. Such a response in the current command implies that the speed controller is able to detect the change in rotor speed as the torque drops during the commutation interval. A larger load inertia might reduce the ability of the controller to detect a drop in torque. The rotor inertia effectively acts as a low pass filter that tries to reject torque ripple.

Figure 12.10 shows the phase a current and back emf as the rotor is nearing the commanded speed. The back emf in the machine model is nearly

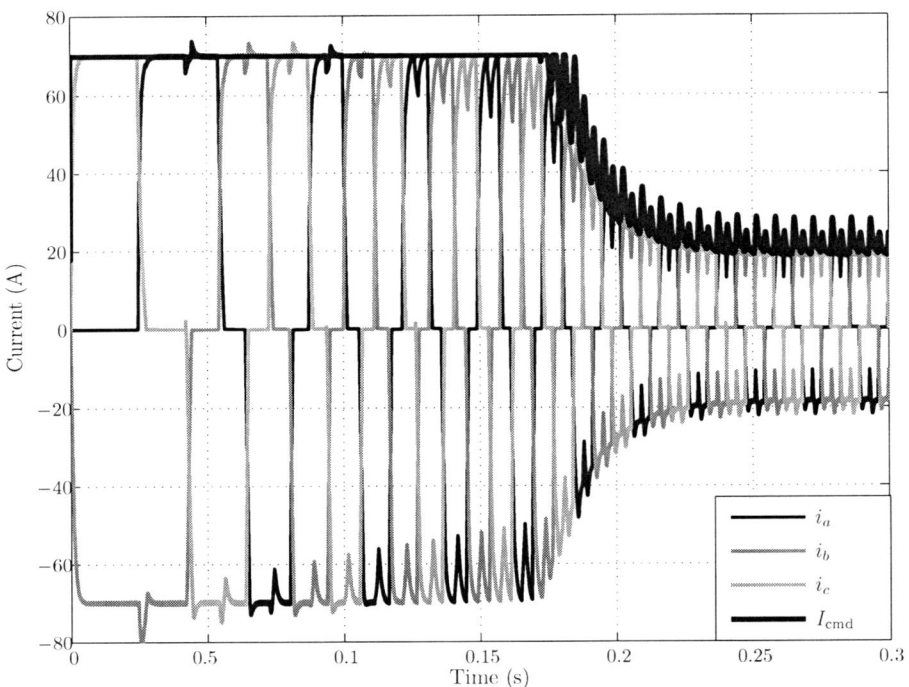

Figure 12.8: The phase currents and the current command for the simulation of a brushless dc machine drive during the time interval 0 s to 0.3 s.

ideal, in that it is represented as a trapezoidal waveform that is nominally constant for 120° electrical at constant speed. Since the phase power being converted to mechanical form is equal to the product of the phase current and the back emf, the waveforms in Fig. 12.10 suggest fluctuation in the phase power during commutation intervals, even when those intervals are intended to affect only the other phases.

Figure 12.11 shows the electromagnetic torque, the load torque, and the rotor speed during the starting transient. The rate of change in speed is equal to the difference between the electromagnetic and load torques. As the rotor speed approaches its commanded value, the electromagnetic torque is reduced through the current command. Significant ripple in the electromagnetic torque is apparent. This should not be surprising given the shape of the phase current waveforms shown in Figs. 12.9 and 12.8. Figure 12.12 shows the three phase currents and the resulting electromagnetic torque. This figure drives home the tight correlation between current ripple and torque ripple. After all, the electric machine is a converter of current into torque.

12.3. Trapezoidal Permanent Magnet Machines

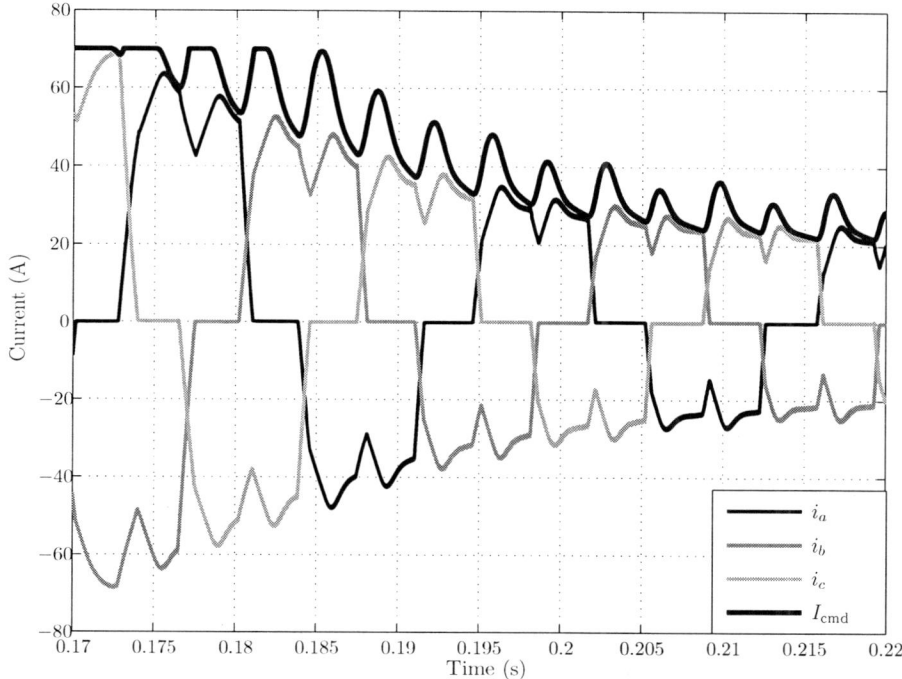

Figure 12.9: The phase currents and the current command for the simulation of a brushless dc machine drive during the time interval 0.17 s to 0.22 s.

Figure 12.13 shows the phase voltage applied to phase a in comparison to the phase a back emf. The phase voltage shows the switching actions within the inverter, in which the phase voltages are being manipulated to regulate the phase currents. It will be observed that there is something of a triangular waveform underlying the phase voltage waveform. This triangular waveform appears in Fig. 12.14, showing the voltage at the common connection among the three machine phases. The shape of this waveform is due to the superposition of the phase back emfs as discussed in Sec. 11.2.2. The voltage at the common connection would be zero if not for the triplen harmonics in the back emf waveforms.

It will be observed in Fig. 12.13 that there are intervals when the phase voltages are tightly correlated to the phase back emf. During these intervals the inverter switches for phase a are not conducting, so the phase a voltage is the back emf plus the voltage at the common connection of the phase windings. It follows that it is possible to measure the back emf of each phase while the respective current is zero. Tracking the zero crossings of each back emf provides valuable rotor information. Indeed, this is sufficient

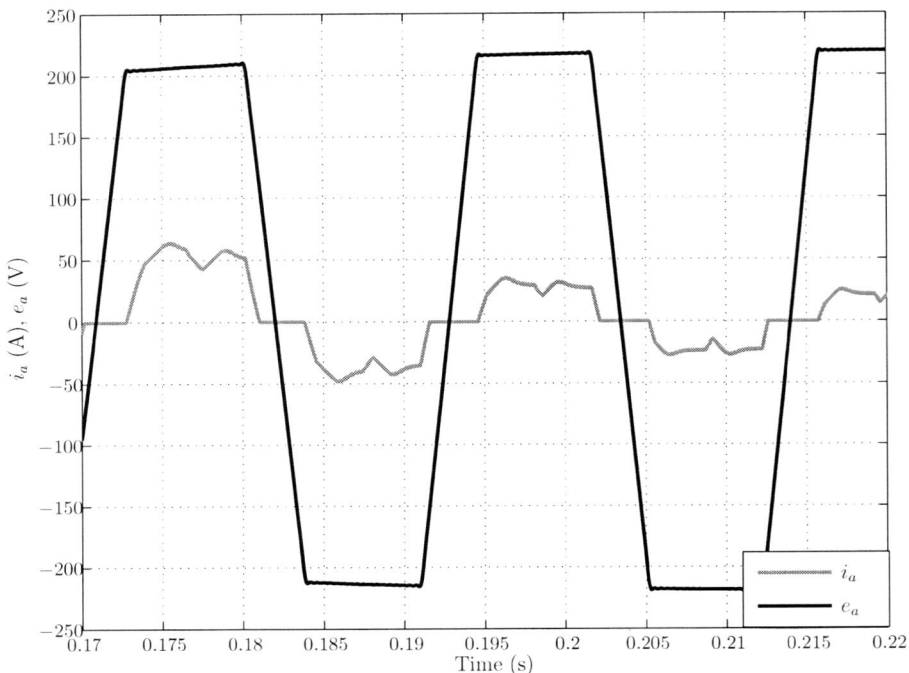

Figure 12.10: The phase a current and back emf for the simulation of a brushless dc machine drive during the time interval 0.17 s to 0.22 s.

information to allow elimination of the mechanical rotor position sensor.

Figure 12.15 shows a close-up of the commutation process as current is being transferred from phase b to phase c while phase a is intended to carry constant current. First, it will be observed that the magnitude of the phase a current is reduced because the sum of the currents in phases b and c is reduced during the commutation process. Since the machine is a three-phase three-wire device, the sum of the three phase currents must always sum to zero. Second, it will be observed that the current in phase b drops faster than the current in phase c increases. This is what causes the magnitude of the phase a current to fall. Third, once the current in phase b has reached zero, phases a and c are essentially in series and the two phase currents recover to their commanded values together. This slight overshoot is partly due to the current regulator coming out of saturation, and partly due to the recovery in the commanded phase current magnitude coming out of the speed controller. Fourth, it will be observed that the commutation intervals between positive currents are slightly different from the commutation intervals between negative currents. These differences are

12.3. Trapezoidal Permanent Magnet Machines

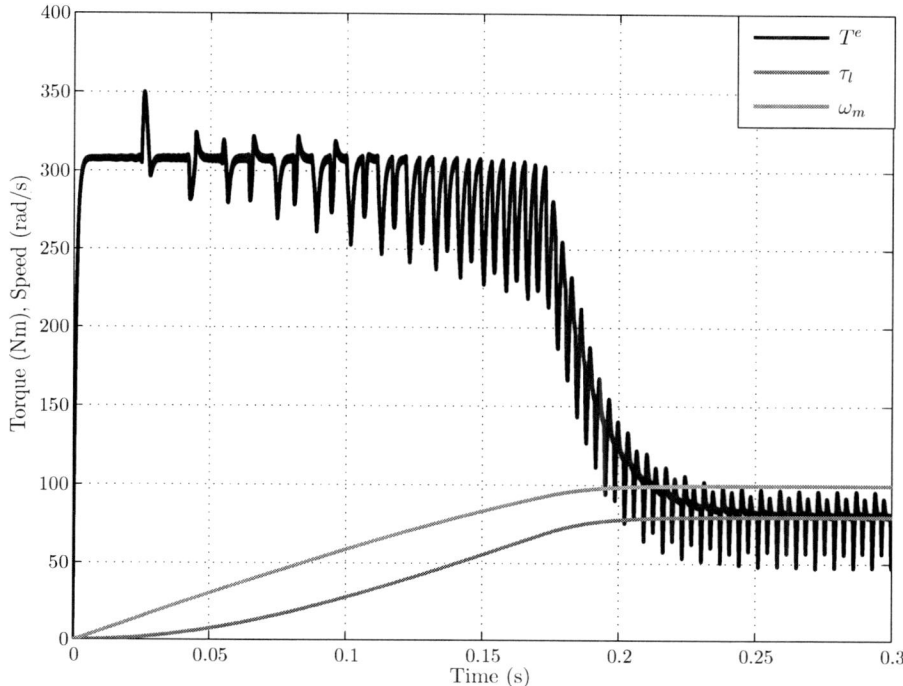

Figure 12.11: The electromagnetic torque, the load torque, and the rotor speed for the simulation of a brushless dc machine drive during the time interval 0 s to 0.3 s.

reflected in the current magnitude command.

The simulation example discussed here is intended to drive home the issue of torque ripple and its fundamental link to the phase currents. It also illustrates the folly of trying to change the phase currents instantaneously in the face of the phase inductance. If one is interested in much more precise control of the torque produced by the machine, it is necessary to use a more sophisticated reference current waveform generator in conjunction with a sufficiently fast speed controller.

The more complex reference current generator can address the issue illustrated in Fig. 12.15 where the rise and fall of the on-coming and off-going phases is not equal. If the phase current reference waveform were based on applying equal ramps to the currents, it would create a symmetric drop in the shaft torque. At the same time, the speed controller can increase the amplitude of the reference currents during the commutation interval. The combination of the speed controller working in conjunction with the reference

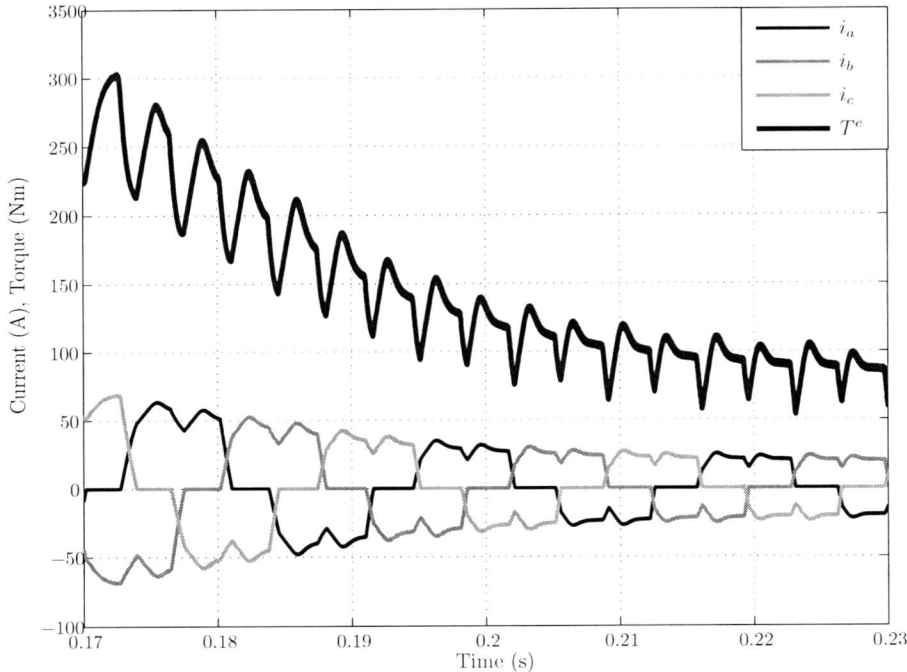

Figure 12.12: The phase currents and the electromagnetic torque for the simulation of a brushless dc machine drive during the time interval 0.17 s to 0.23 s.

current generator can reduce the torque ripple substantially[1].

12.4 Summary

This chapter has examined the control of permanent magnet machines. Field-oriented and direct torque control for a sinusoidal permanent magnet machine bear close resemblance to the same techniques when applied to induction machines. However, the alignment of the direct axis with the magnet flux by virtue of the synchronous nature of the permanent magnet machine simplifies the discussion significantly.

The control of permanent magnet machines with trapezoidal back emfs can be structurally much simpler than for machines with sinusoidal back emfs. However, the simplicity in control structure is commensurate with the

---

[1] See, for example, Y. Sozer and D. A. Torrey, "Adaptive torque ripple control of permanent magnet brushless dc motors," *Proc. of the IEEE Applied Power Electronics Conference*, Anaheim, CA, February 1998, pp. 86-92.

12.4. Summary

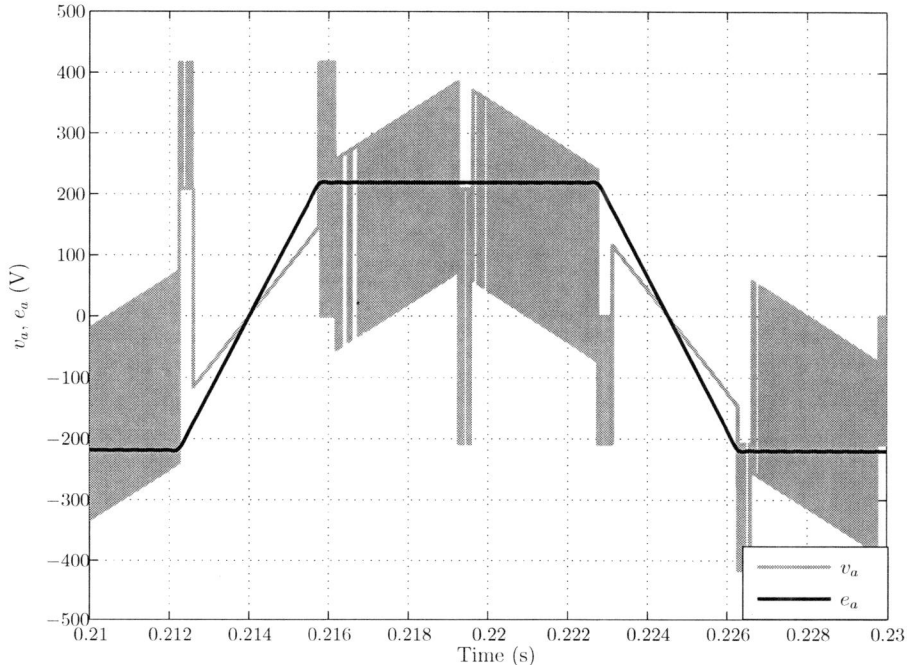

Figure 12.13: The phase a voltage and the phase a back emf for the simulation of a brushless dc machine drive during the time interval 0.21 s to 0.23 s.

quality of the torque produced. Indeed, many have found that applying sinusoidal currents to machines with trapezoidal back emfs results in smoother torque than when a simple control strategy is applied. Despite this, there are a large number of applications for which control of average speed is sufficient to weigh in favor of the simpler control method.

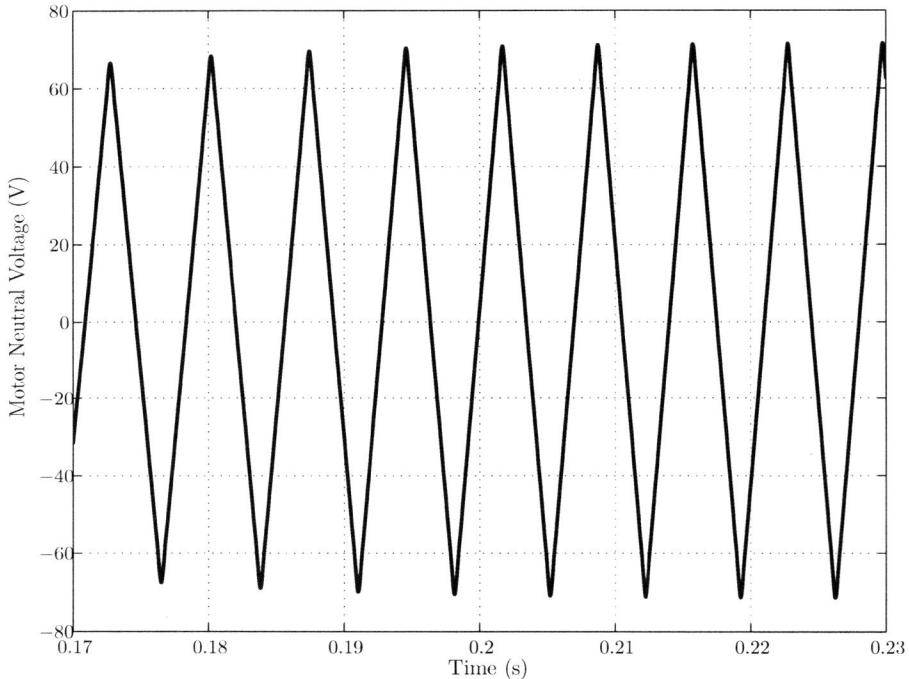

Figure 12.14: The voltage at the common connection among the three machine phases for the simulation of a brushless dc machine drive during the time interval 0.16 s to 0.23 s.

12.4. Summary

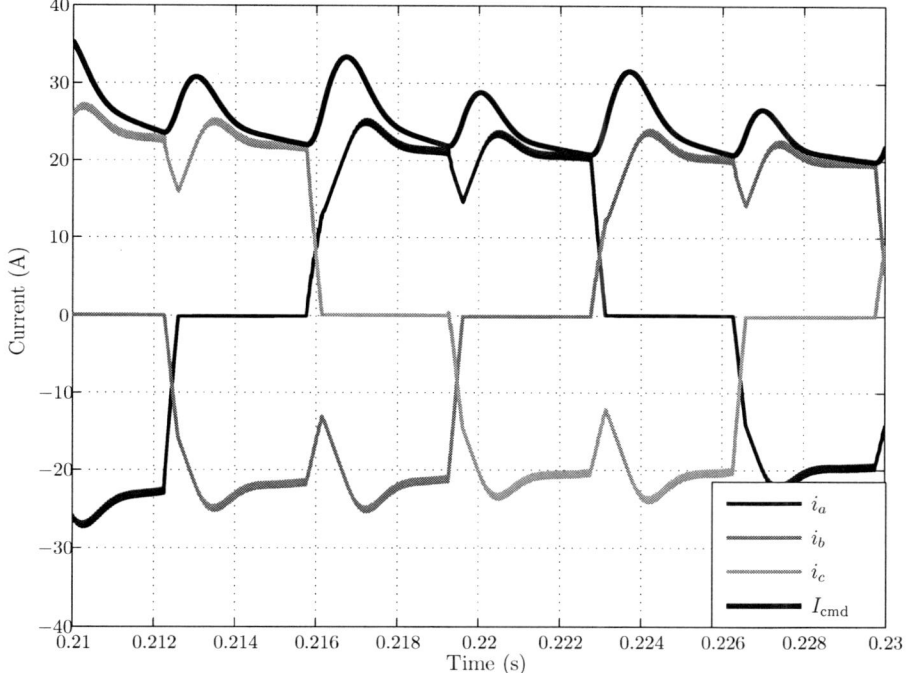

Figure 12.15: The phase currents for the simulation of a brushless dc machine drive during the time interval 0.21 s to 0.23 s.

Chapter 13

Variable-reluctance Machines

13.1 Introduction

This chapter examines variable-reluctance machines (VRMs). This type of machine is commonly referred to as a switched-reluctance machine (SRM) because the machine is always operated through a switched inverter. In this machine, torque is produced entirely by reluctance torque as a result of the rotor poles trying to align with the nearest excited stator poles.

The operation of a VRM is relatively simple to understand, though its design is deceptively challenging because of the strong interplay between the magnetic and structural design. This is generally because the magnetic design of the VRM suggests insufficient back iron for adequate structural stiffness. VRMs often generate two impulsive responses from people that have some familiarity with them: they are noisy and they have large torque ripple. Proper magnetic and control design can, however, produce superior performance in some applications. It should be pointed out that any poorly designed machine can be noisy or have large torque ripple; remember the example of the brushless dc machine in Sec. 12.3.

The VRM is compatible with demanding applications. The absence of windings and permanent magnets on the rotor supports both high rotational speeds and high temperature operation. Further, the absence of windings on the rotor helps to keep the majority of the losses within the stator, making the VRM relatively easy to cool. The switched nature of the VRM makes it well suited to any application that requires variable speed operation. In the case of aerospace and automotive applications, variable speed operation is needed for compatibility with the source or load. In other applications, variable speed operation is needed to extract additional energy, and to lessen the mechanical stresses within the system.

The VRM is under development for variable speed applications where its inherent characteristics make commercial sense. To date, these applications include sourcing aerospace power systems[1], starter/alternators for hybrid vehicles[2], and wind turbine applications[3]. The aerospace and automotive applications are generally characterized by high speed operation. The wind energy application is characterized by low speed, high torque operation.

Unlike the induction and permanent magnet machines considered earlier, the nature of the variable-reluctance machine is such that while there is duality between motoring and generating operation, the details of control implementation are not duals of one another. Accordingly, we will discuss motor and generator control separately.

13.2 Modeling VRM Systems

In this section we focus on the electromechanics of the VRM. The intent is to provide an understanding of the energy conversion process so that we can better understand the control of the VRM. For illustrative purposes, we will use the VRM of Fig. 13.1 as a running example. The VRM of Fig. 13.1 is a three-phase design having 24 stator poles and 16 rotor poles, a 24/16 VRM that has been developed for automotive applications. Its performance has been documented in the technical literature[4]. From experience, it is known

---

[1] See, for example, S. R. MacMinn and J. W. Sember, "Control of a switched-reluctance aircraft starter-generator over a very wide speed range," *Proc. of the Intersociety Energy Conversion Engineering Conf.*, pp. 631-638, 1989, A. Radun, "Generating with the switched-reluctance motor," *Proc. of the IEEE Applied Power Electronics Conf.*, pp. 41-47, 1994, and D. E. Cameron and J. H. Lang, "The control of high-speed variable-reluctance generators in electric power systems," *IEEE Trans. on Industry Applications*, Vol. 29, pp. 1106-1109, 1993.

[2] See, for example, J. M. Kokernak, D. A. Torrey, and M. Kaplan, "A switched reluctance starter/alternator for hybrid electric vehicles," *Power Electronics Proc. (PCIM Conference)*, pp. 74-80, 1999, E. Mese, Y. Sozer, J. M. Kokernak, and D. A. Torrey, "Optimal excitation of a high speed switched reluctance generator," *Proc. of the IEEE Applied Power Electronics Conf.*, pp. 362-368, 2000, and M. Besbes, M. Gabsi, E. Hoang, M. Lecrivain, B. Grioni, and C. Plasse, "SRM design for starter-alternator system," *Proc. of the International Conference on Electric Machines*, pp. 1931-1935, 2000.

[3] See, for example, D. A. Torrey, "Variable-reluctance generators in wind-energy systems," *Proc. of the IEEE Power Electronics Specialists Conf.*, pp. 561-567, 1993, and R. Cardenas, W. F. Ray and G. M. Asher, "Switched reluctance generators for wind energy applications," *Proc. of the IEEE Power Electronics Specialists Conf.*, pp. 559-564, 1995.

[4] See J. M. Kokernak, D. A. Torrey, and M. Kaplan, "A switched reluctance starter/alternator for hybrid electric vehicles," *Power Electronics Proc. (PCIM Confer-*

13.2. Modeling VRM Systems

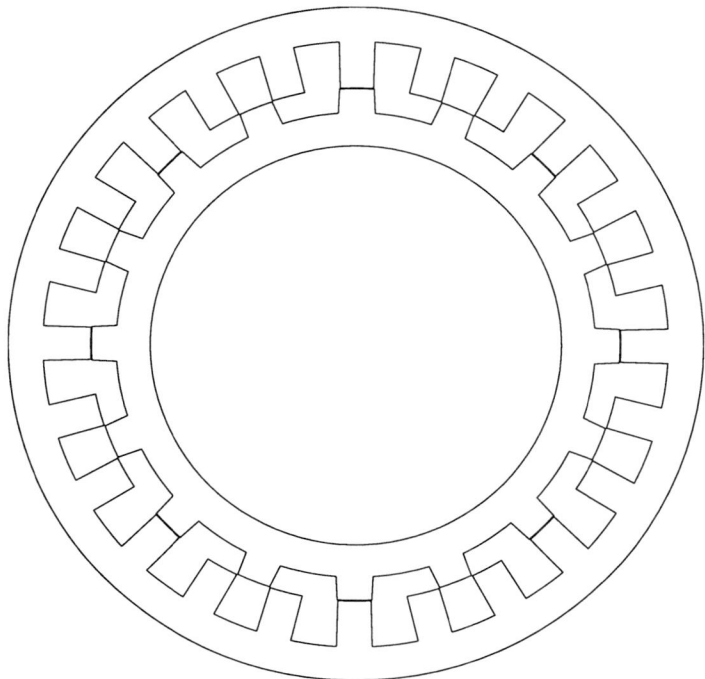

Figure 13.1: A three-phase 24/16 VRM. Phase windings are not shown, but would be concentrated around each stator pole. The windings within a phase could be connected in series or parallel, or some combination thereof.

that this VRM is representative of typical VRM behavior.

This VRM has steel laminations on the rotor and stator. There are concentrated windings placed around each salient pole on the stator. The coils around the individual poles are connected to form the phase windings. There are no windings or permanent magnets on the rotor, making the structural integrity of the rotor compatible with operation at very high speeds. The design of the laminations and phase windings is beyond the scope of this discussion. The interested reader is referred to standard references for a discussion of the issues related to magnetic design of the VRM[5].

---

ence), pp. 74-80, 1999, and E. Mese, Y. Sozer, J. M. Kokernak and D. A. Torrey, "Optimal excitation of a high speed switched reluctance generator," *Proc. of the IEEE Applied Power Electronics Conf.*, pp. 362-368, 2000

[5]See, for example, T. J. E. Miller, *Switched Reluctance Motors and Their Control*, Oxford University Press, 1993, and P. J. Lawrenson, et al., "Variable-speed switched reluctance motors," *IEE Proc.*, Vol. 127, pt. B, no. 4, pp. 253-265, 1980.

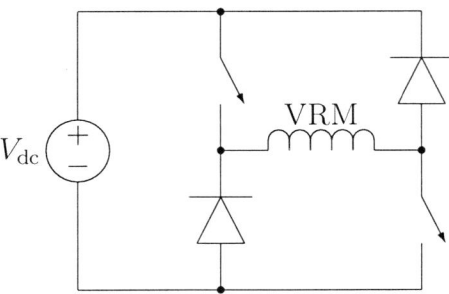

Figure 13.2: One phase of the inverter used to excite the VRM of Fig. 13.1.

The design of the VRM of Fig. 13.1 was the compromise of two competing functions. The first function is that of starting an internal combustion engine by providing high torques at low speeds. The second function is providing electrical energy through generation at speeds ranging from engine idle to maximum speed. The compromise of these two competing functions resulted in a magnetic design that supported generation from 500 rpm through 5000 rpm at a maximum power of 6 kW. Because of the influence of the high starting torque requirement, generation between 500 rpm and 1000 rpm required current regulation. Generation above 1000 rpm did not require current regulation.

The VRM of Fig. 13.1 is excited through a common asymmetrical bridge. Figure 13.2 shows one phase of this inverter that uses the same dc source for exciting each VRM phase through two controllable switches, and demagnetizing the same phase through the diodes. There are other inverter topologies that may be useful for motor and generator systems, and may involve segregating the excitation and demagnetization functions[6]. The phase excitation is based on a turn-on angle (θ_{on}) and a turn-off angle (θ_{off}). The conduction angle (θ_{cond}) is the interval over which the phase is excited: $\theta_{\text{cond}} = \theta_{\text{off}} - \theta_{\text{on}}$. Some authors refer to the conduction angle as the dwell angle.

The switches and diodes within the inverter of Fig. 13.2 must support the maximum dc bus voltage and the maximum phase current. For the VRM of Fig. 13.1, the maximum phase currents occurred during motoring operation at low speeds where the phase currents are regulated. The maximum voltage occurs when the VRM is operating as a generator.

---

[6]See, for example, A. Radun, "Generating with the switched-reluctance motor," *Proc. of the IEEE Applied Power Electronics Conf.*, pp. 41-47, 1994 and T. J. E. Miller, *Switched Reluctance Motors and Their Control*, Oxford University Press, 1993, and the references therein for a description of these other topologies

13.2.1 Energy Conversion

Energy conversion within the VRM was already discussed in Sec. 2.6 in a simplistic way. For a VRM, mechanical energy is converted between electrical and mechanical form by virtue of proper synchronization of phase currents with rotor position. During motoring, the VRM produces torque in the direction of rotation. During generation, the variable reluctance generator (VRG) produces negative torque that is trying to oppose rotation, thereby extracting energy from the prime mover. It is the responsibility of a commutator to excite the phases in proper order to support continuous energy conversion.

Torque in the VRM is created by the natural tendency of the excited stator poles to attract the nearest rotor poles. If the phase is excited before the rotor poles come into alignment with the stator poles, the rotor experiences torque in the direction of rotation consistent with operation as a motor. If the phase is excited as the rotor poles move through the aligned position, the rotor experiences torque opposing rotation consistent with generator operation. Figure 13.3 shows the relationship between the idealized inductance profile and phase currents for motoring and generating above base speed. For both the VRG and the VRM, base speed is the speed at which phase currents are nominally constant without the need of current regulation. Base speed can be taken to be the speed where the phase back emf balances the source voltage and resistive drop. Base speed is slightly different for the VRM and VRG due to the different signs on the resistive drop. For the VRG, the nominally constant phase currents are supported by the diodes of Fig. 13.2 for reasons that will become clear below.

More generally, the VRM has both spatial and magnetic nonlinearities that must be considered when designing both the magnetic structure and the control. The flux linking a phase of the VRM of Fig. 13.1 is shown in Fig. 13.4 where phase flux linkage (λ) is shown as a function of phase current (i) for a number of different rotor positions. The rotor positions are given in mechanical measure, with $0°$ corresponding to alignment between stator and rotor for phase a. Rotor positions in electrical measure can be found by multiplying the mechanical position by the number of rotor poles (N_r), 16 in this case[7]. The magnetic nonlinearity is evident particularly near the aligned position; there is no perceptible nonlinearity in the unaligned position.

The data for Fig. 13.4 were obtained through finite element analysis and

---

[7]Unlike induction and permanent magnet machines that complete $N_p/2$ electrical cycles in each mechanical revolution, the VRM completes one electrical cycle for each rotor pole.

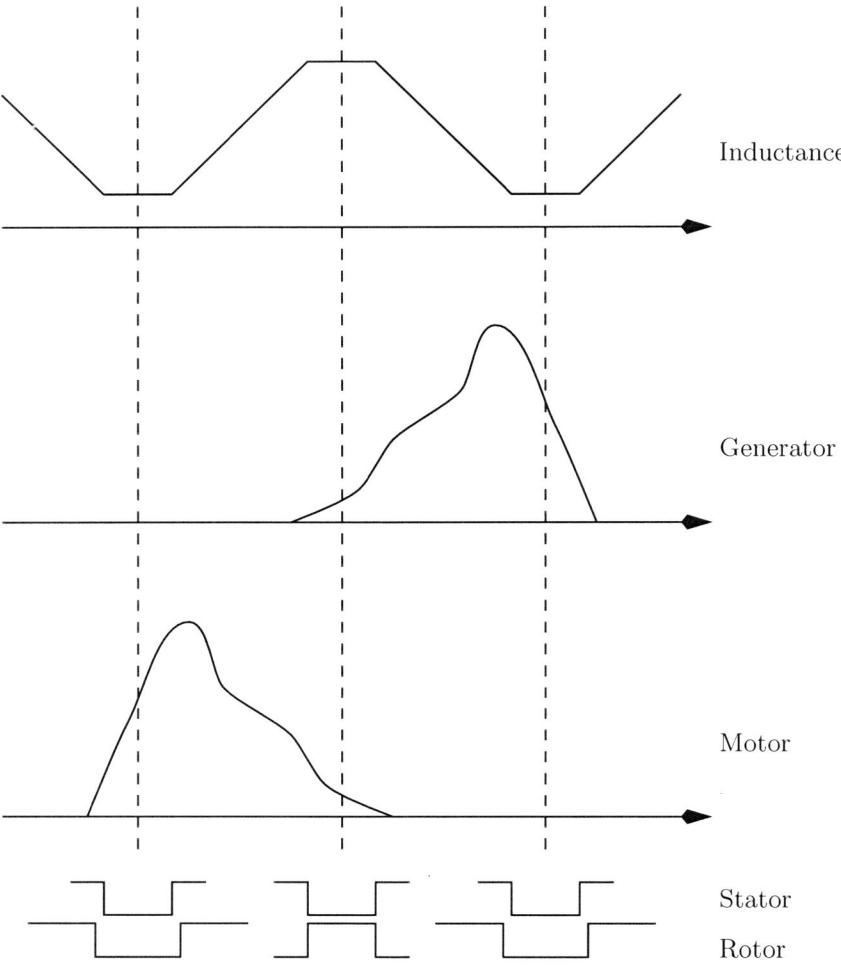

Figure 13.3: Idealized inductance variation and the currents used to support motor and generator operation.

subsequently modeled using the relationships[8]

$$\lambda(i,\theta) = a_1(\theta)\left(1 - e^{a_2(\theta)i}\right) + a_3(\theta)i \quad ; \tag{13.1}$$

$$a_j(\theta) = \sum_{k=0}^{N} A_{jk} \cos(kN_r\theta) \quad . \tag{13.2}$$

Another insightful way of representing the data of Fig. 13.4 is shown in Fig. 13.5 where the flux linkage is shown as a function of rotor position for a

---

[8]See D. A. Torrey and J. H. Lang, "Modelling a nonlinear variable-reluctance motor drive," *IEE Proc.*, Vol. 137, pt. B, pp. 315-326, 1990.

13.3. Control of VR Motors

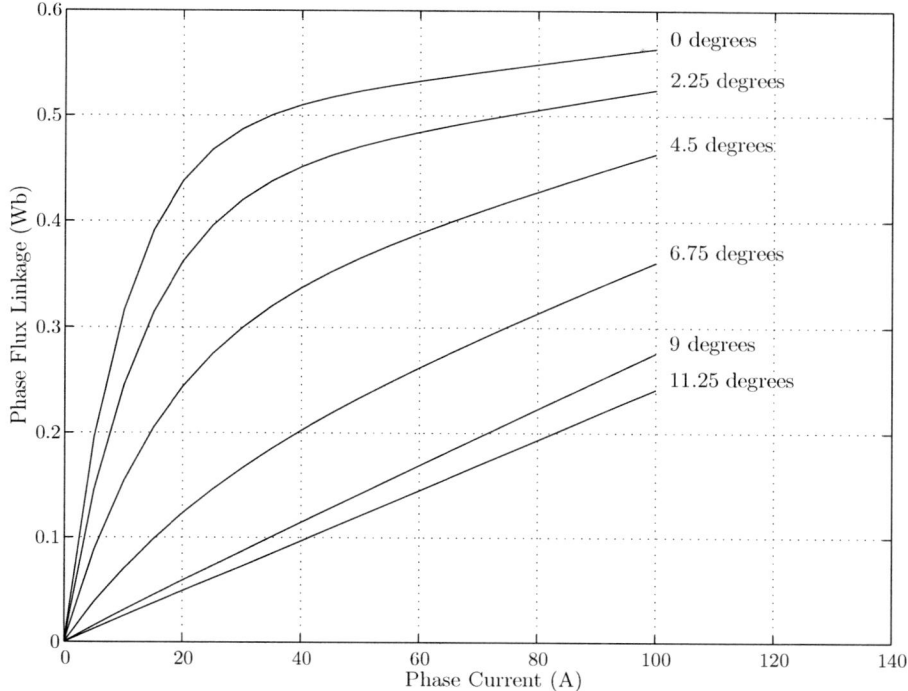

Figure 13.4: The phase flux linkage as a function of current and rotor position for the VRM of Fig. 13.1.

number of different phase currents[9]. Figure 13.5 shows both the spatial and magnetic nonlinearities found in a practical VRG.

13.3 Control of VR Motors

The VRM produces torque through excitation that is synchronized to rotor position. The excitation is generally described by three excitation parameters: the turn-on angle θ_{on}, the turn-off angle θ_{off}, and the reference current I_{ref}. A control algorithm would typically use the same excitation parameters for each phase, implemented with the spatial shift consistent with the symmetrically displaced phase structure. Control of the excitation angles results in either positive net torque for motoring, or negative net torque for generating.

Efficient operation of the VRM, or any motor drive, is always impor-

---

[9]See T. J. E. Miller and M. I. McGilp, "Nonlinear theory of the switched reluctance motor for rapid computer-aided design," *IEE Proc.*, Vol. 137, pt. B, pp. 337-347, 1990.

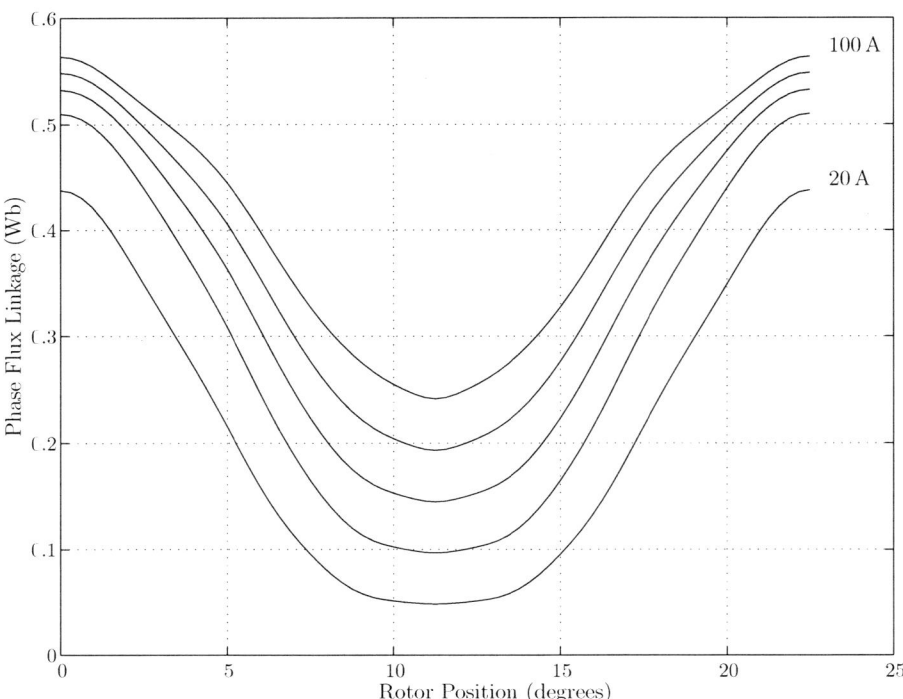

Figure 13.5: The phase flux linkage as a function of rotor position and current for the VRG of Fig. 13.1.

tant Inefficiency leads to larger size, increased weight, and increased energy consumption. In order to maximize motor efficiency, we seek to maximize the average torque to RMS phase current, T_{avg}/I_{phrms}. This ratio captures our intended goal of providing the required electromechanical output with the minimum electrical input. Reducing the RMS phase current reduces the losses within the VRM, as well as driving down the conduction and switching losses within the VRM inverter.

This section presents a closed loop controller for the VRM turn-on and turn-off excitation angles, in order to get the most efficient operation over the entire operating region. This is a distinct alternative to the self-tuning optimization of excitation parameters that is characterized by a relatively slow approach to the optimal excitation[10]. The turn-on angle is driven by

---

[10] See, for example, B. Fahimi, G. Suresh, J. P. Johnson, M. Ehsani, M. Arefeen, and I. Panahi, "Self-tuning control of switched reluctance motors for optimized torque per Ampere at all operating points," *Proc. of the IEEE Applied Power Electronics Conf.*, pp. 778-783, 1998, and K. Russa, I. Husain, and M. Elbuluk, "A self-tuning controller for switched reluctance motors," *IEEE Trans. on Power Electronics*, Vol. 15, pp. 545-552,

13.3. Control of VR Motors

Table 13.1: The specifications for the experimental VRM.

Quantity	Value	Units
Rated Power	1000	W
Base Speed	1500	rpm
Maximum Speed	3,000	rpm
Dc Voltage	12	V
Number of Rotor Poles	12	
Number of Stator Poles	16	
Number of Phases	4	
Aligned Phase Inductance	0.228	mH
Unaligned Phase Inductance	0.0226	mH
θ_g	142	° (electrical)
θ_m	218	° (electrical)

closed loop control. It automatically adjusts the turn-on angle by using the first peak of the phase current and its magnitude. The algorithm produces the turn-off angle based on experimental characterization at only four operating points, representing all combinations of low speed, high speed, low phase current and high phase current. The method does not need machine modeling or extensive simulations. Because these operating points can be characterized experimentally, it is not necessary to characterize them analytically. The algorithm is easy to implement and does not need look up tables for excitation parameters. Implementation of the algorithm is demonstrated in simulation for a four phase VRM[11]. The VRM used as a running example in this section is a 16/12 four-phase VRM designed for a 1 kW 12 V automotive application; Table 13.1 gives the parameters of the VRM.

13.3.1 The Algorithm

The objectives of the algorithm are best explained through consideration of the linear inductance profile for the VRM shown in Fig. 13.6. The minimum inductance region is defined by the angular interval over which the rotor poles do not overlap the stator poles. The maximum inductance region is defined by the angular interval over which there is complete overlap between the stator and rotor poles. The regions of increasing and decreasing inductance

---

2000.

[11] For experimental documentation of the algorithm, see Y. Sozer and D. A. Torrey, "Optimal turn-off angle control in the face of automatic turn-on angle control for switched-reluctance motors," *IET Electric Power Applications*, Vol. 1, pp. 395-401, 2007.

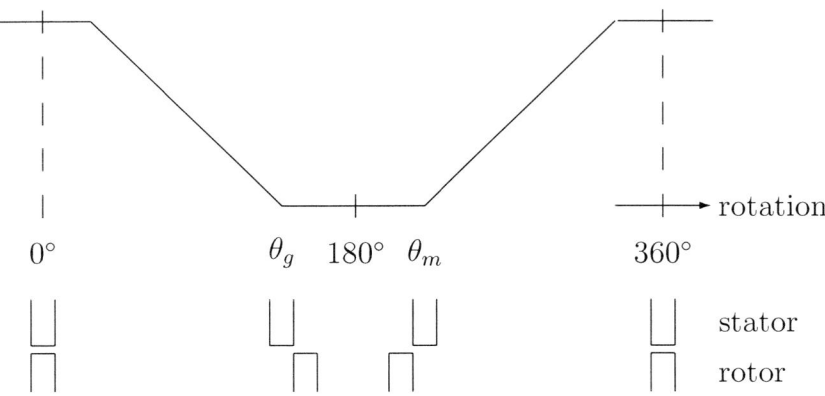

Figure 13.6: The linear inductance profile of the VRM showing θ_g and θ_m.

correspond to varying overlap between the stator and rotor poles.

For operation of the VRM as a motor, phase current must be present in the phase winding as the inductance is increasing in the direction of rotation. For operation of the VRM as a generator, phase current must be present in the phase winding as the inductance is decreasing in the direction of rotation. The polarity of current is immaterial, so we assume that the phase current is always positive. The turn-on angle is the electrical position where we start exciting the motor phase. On the other hand the turn-off angle is the electrical angle where we completely shut off the phase for that cycle, though current will continue to flow beyond turn-off. The placement of the excitation angles is extremely important in producing torque efficiently at any operating point. As long as continuous conduction is prevented during one electrical cycle, turn-on and turn-off angles can be controlled separately.

Turn-on Control

If one were to examine the static torque curve for a typical VRM, it would be observed that the maximum torque for a given amount of current occurs as the rotor begins to move out of the minimum inductance position. This observation suggests that maximum torque per Ampere is produced upon leaving the minimum inductance position. Iron permeance causes torque production to fall off as overlap between the stator and rotor poles increases. In applications where average torque is of primary importance, it is important to make the most of the region near the unaligned position. Because it takes time to build the phase currents, we must anticipate the arrival of the torque production region. We must, therefore, turn on the phase windings

13.3. Control of VR Motors

before the angle marked θ_m in Fig. 13.6 so that the current is at I_{ref} when the rotor reaches θ_m.

The conventional approach to determining θ_{on} is to work backward from θ_m:

$$\theta_{on} = \theta_m - \frac{L_{min} I_{ref} \omega}{V_{dc}}, \qquad (13.3)$$

where L_{min} is the minimum inductance, V_{dc} is the dc bus voltage, ω is the rotor speed, and I_{ref} is the chopping level. Equation 13.3 assumes the inductance is constant during the region $[\theta_g, \theta_m]$. The inductance, however, is a function of the phase current and rotor position. At low speed this method can give reasonable performance. For operation over a wide speed range Eq. 13.3 starts to break down as the turn-on angle is pushed to occur before the start of the minimum inductance position (θ_g in Fig. 13.6). It is desired to have closed loop control that adjusts the turn-on angle to force the first peak of the phase current to occur at θ_m without the need of accurate motor parameters and measurement of the dc bus voltage.

The proposed closed loop control algorithm continuously monitors the position of the first peak of the phase current (θ_p). The turn-on is advanced or retarded automatically according to the error between θ_p and θ_m. This piece of the controller successfully places θ_p at θ_m. Above base speed the peak current naturally tends to occur near θ_m. At these speeds θ_{on} has little impact on θ_p but significant impact on the magnitude of the current at θ_p. This phenomenon can be observed from Fig. 13.7. The VRM is simulated at 2500 rpm with two different turn-on angles. For each of the turn-on angles θ_p occurs approximately at the same place with different current magnitudes. To reflect this, the algorithm forces the peak phase current to match the commanded phase current. Feed-forward control of θ_{on} using Eq. 13.3 is used to speed convergence to the correct value of θ_{on}.

If the controller is in current regulation mode, I_p occurs close to I_{ref} so the error between I_p and I_{ref} does not have any effect on the command for θ_{on}. Below base speed, the piece of the controller responsible for keeping θ_p at θ_m effectively works to achieve the control objective. At higher speeds where the controller is in voltage control mode, θ_p naturally occurs at θ_m. The piece of the controller responsible for forcing I_p to track I_{ref} effectively works to advance the turn-on angle to keep I_p close to I_{ref}. If the reference current or the motor speed is reduced the drive enters into current regulation mode and θ_p occurs before θ_m. The piece of the controller responsible for forcing $\theta_p = \theta_m$ becomes active and brings θ_p to θ_m by retarding θ_{on}.

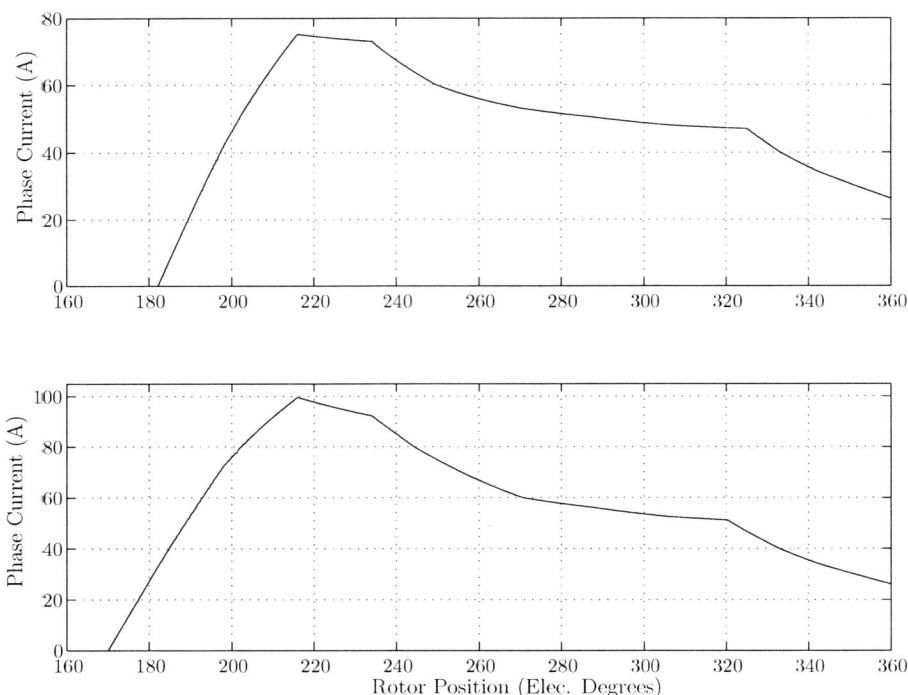

Figure 13.7: The phase currents at 2500 rpm with different turn-on angles.

Turn-off Control

Once we have established the phase current after turning on the phase excitation we need to turn off the excitation at the optimum place to produce the maximum amount of torque with a minimum electrical input for a given peak phase current. From the turn-on control we have the optimum turn-on angle for the operating point. For a given speed, peak phase current, and turn-on angle from the turn-on angle controller, it is possible to simulate the drive with every possible turn-off angle. Simulations are performed at every 500 rpm between 500 rpm and 3,000 rpm and every 25 A between 25 A and 150 A. Figure 13.8 shows the motor torque, RMS phase current, and $T_{\text{avg}}/I_{ph\text{rms}}$ versus turn-off angle for 1000 rpm and 75 A. Torque per Ampere peaks at a turn-off angle of 333.6°. Extending the turn-off angle farther enables us to produce more torque at the expense of increased RMS phase current. At this operating point using 333.6° we can produce 3.67 Nm which requires 44.97 A of RMS phase current. Extending turn-off angle to 342° enables us to produce 3.76 Nm which requires 47.3 A of RMS phase current. By keeping the turn off angle at 333.6° we needed to increase the reference current level to 76.1 A to get 3.76 Nm which requires 45.7 A RMS phase cur-

13.3. Control of VR Motors

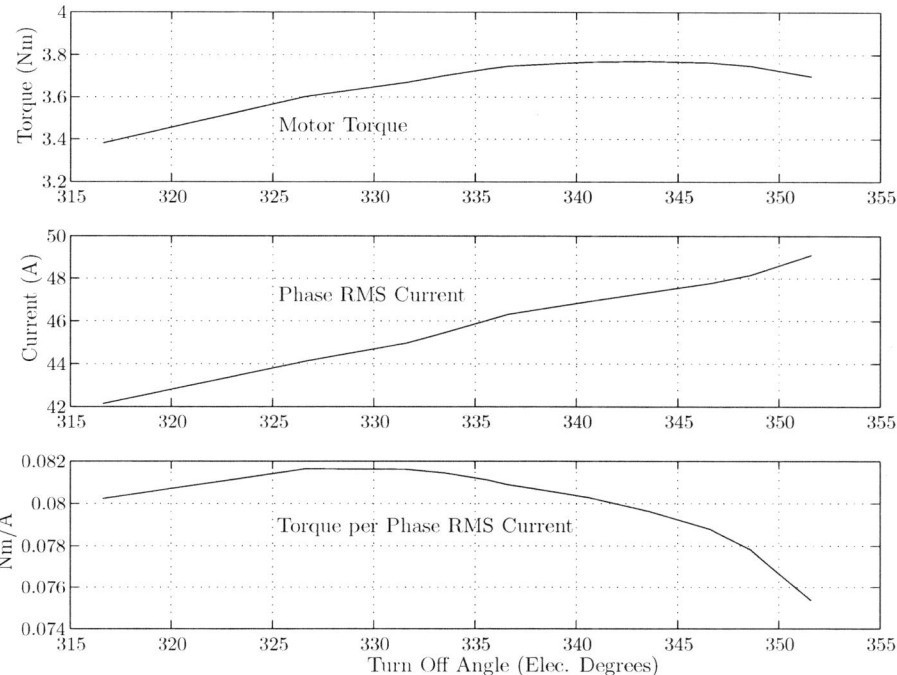

Figure 13.8: Motor torque, RMS phase current, and $T_{\text{avg}}/I_{ph\text{rms}}$ versus turn-off angle at 1000 rpm, 75 A.

rent. We can conclude from this example that it is more efficient to increase reference current rather than extending the turn-off angle beyond most efficient turn-off angle. This is because a higher current command pushes the torque production toward θ_m where the torque per Ampere is higher.

Among all simulated data, the optimum turn-off angles are selected for a given speed and peak phase current. The criteria is to find the turn-off angle that maximizes $T_{\text{avg}}/I_{ph\text{rms}}$ for a given speed and peak phase current. Figure 13.9 shows optimal turn-off angles as a function of speed and peak phase current.

As we have seen, the optimum turn-off angles are a function of operating speed and peak phase current level. The optimum turn-off angles can be represented as a function of the form

$$\theta_{\text{off}} = k_1 \omega I_{\text{ref}} + k_2 \omega + k_3 \sqrt{I_{\text{ref}}} + k_4 \quad , \tag{13.4}$$

where ω is rotor speed, I_{ref} is the reference peak phase current and k_1, k_2, k_3 and k_4 are curve fit parameters. For our simulation work, the curve fit parameters are based on a least-squares fit to the collection of optimal turn-

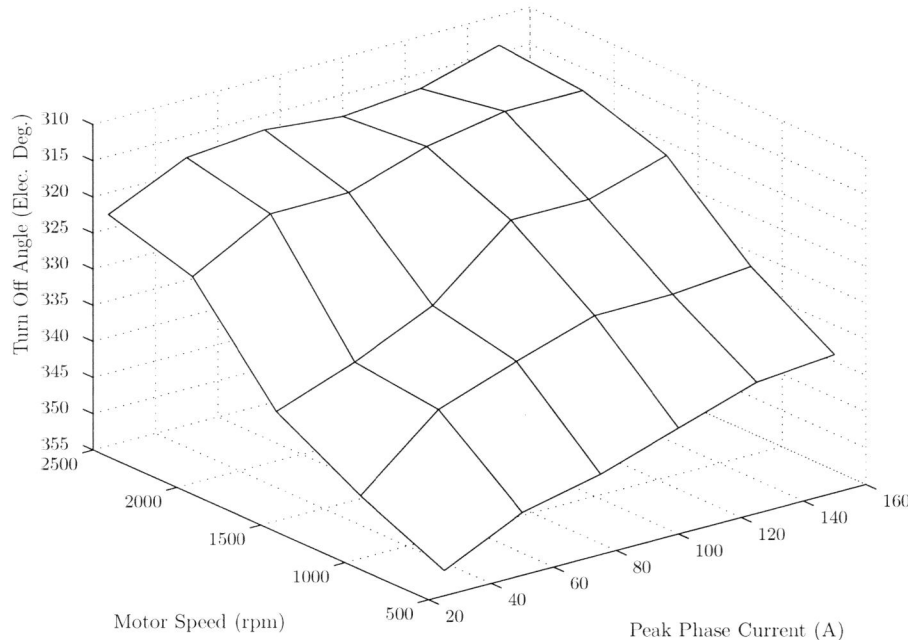

Figure 13.9: Optimal-efficiency turn-off angles as a function of speed and peak phase current.

off angles over all operating points. For our experimental work, the curve fit parameters are based on the optimized data for four operating points representing all combinations of low speed, high speed, low current and high current. Figure 13.10 shows the curve fit of the turn-off angles using all data and data from only four operating points. As seen from these figures, the curve fit using only four data points gives results that are acceptably close to the actual optimized turn-off angles. This is quite significant because it dramatically reduces the amount of experimental data that are required to optimize the turn-off angle. The integrated control of θ_{on} and θ_{off} is summarized in Fig. 13.11.

Algorithm Simulation

The algorithms motivated in Sec. 13.3.1 were implemented in simulation to confirm proper operation before being experimentally implemented on the physical system. The VRM to which the simulation is applied is a 16/12 four-phase VRM designed for a 1 kW 12 V automotive application. Table 13.1 gives the parameters of the VRM used in this work. The VRM magnetics are modeled analytically based on data collected through finite element

13.3. Control of VR Motors

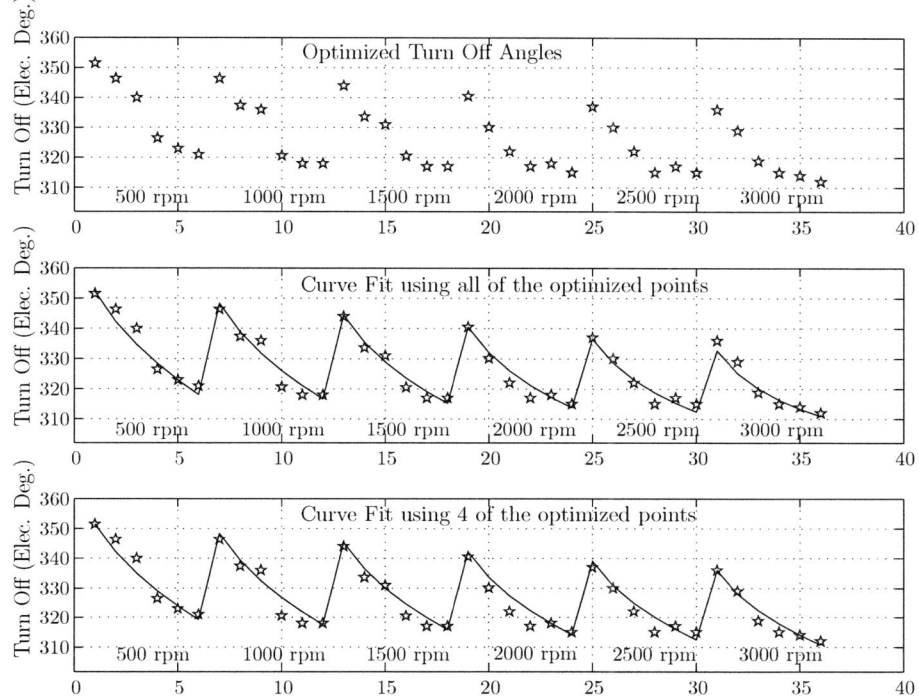

Figure 13.10: Optimal efficiency turn-off angles for different speeds and phase current levels along with the comparison of the curve fits using all of the optimized turn-off angles and using only four points from the optimized data.

analysis. Figure 13.12 shows the result of the angle control technique for 125 A reference current and a varying speed profile. Figure 13.13 shows the simulation results of the implemented control technique at 2500 rpm where the reference current is changed from 25 A to 140 A at 0.3 s. These figures show the turn-on controller properly placing the peak current at θ_m and producing the required peak current. The turn-off controller is also properly adjusting the turn-off angle for the operating speed and peak phase current level. To the extent that θ_{on} and θ_{off} correspond to that required for peak efficiency based on the control algorithm, the required torque is produced with the highest efficiency possible. The parameters for the controller are given in Table 13.2.

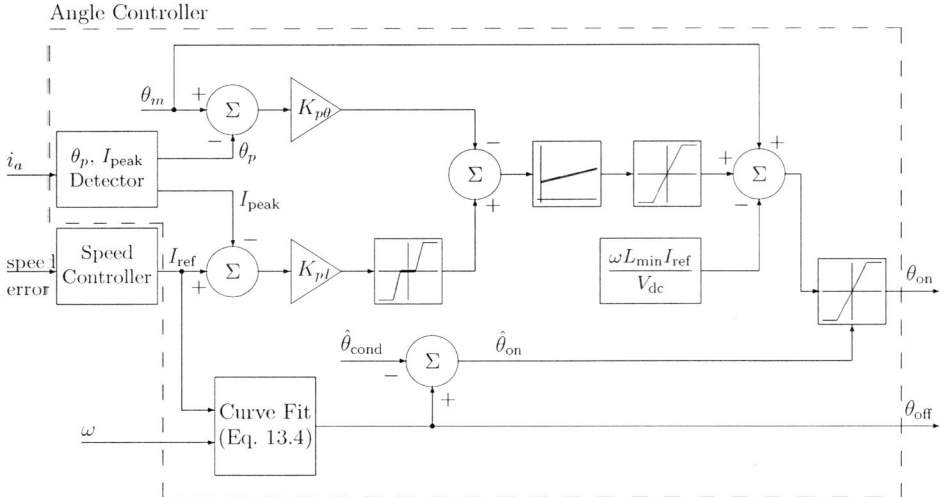

Figure 13.11: The algorithm used to automatically adjust the excitation angles.

Table 13.2: The parameters for the controller of Fig. 13.11.

Quantity	Value	Units
$K_{p\theta}$	0.033	
K_{pI}	1.899	°/A (electrical)
$\hat{\theta}_{\text{cond}}$	178	° (electrical)
k_1	-8.106e-5	°s/A (electrical)
k_2	-1.397e-2	°s (electrical)
k_3	-9.278e-1	°/√A (electrical)
k_4	3.556e2	° (electrical)
Proportional Gain	0	
Integral Gain	78	s^{-1}

13.4 Control of VR Generators

The variable-reluctance generator (VRG) can be considered the dual of the variable-reluctance motor (VRM), though there are some important differences in control objectives and control implementation. The goal of this section is to elucidate the control issues for the VRG and summarize typical VRG behavior. This is accomplished by drawing on the previous discussion

13.4. Control of VR Generators

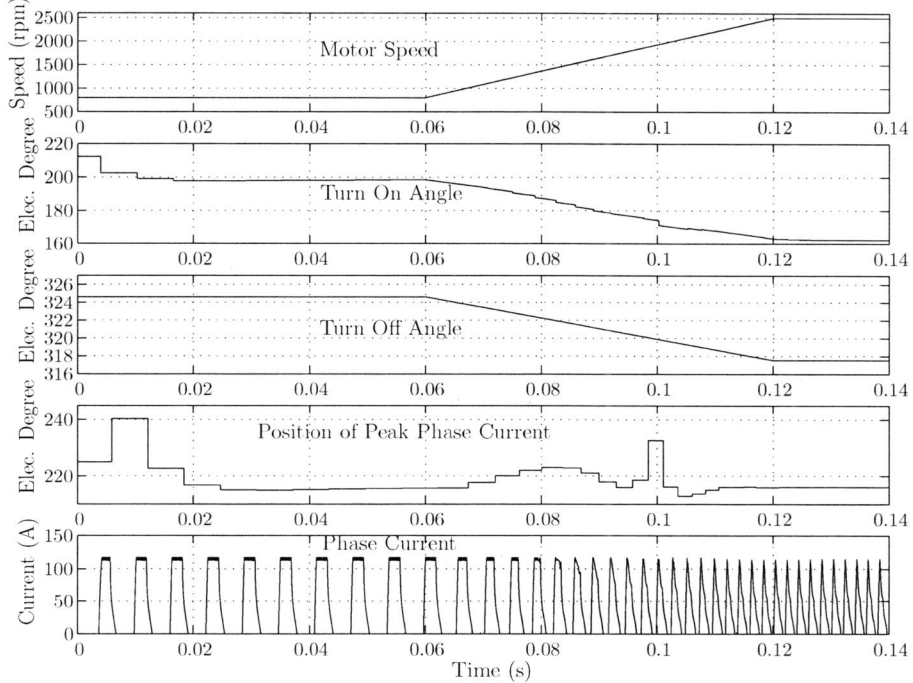

Figure 13.12: Simulation results of the implemented control technique with 125 A reference current with a varying speed profile.

of energy conversion in the VRM, identifying the important features of the VRG that must be considered in designing its control.

As with motor systems, the proper application of a generator to a system requires an understanding of the characteristics of the prime mover. In the aerospace and automotive applications, the prime mover is able to provide essentially constant power over a wide speed range. In the wind energy application, the shaft power is proportional to the cube of speed, implying a substantial increase in torque and power as the speed increases. The prime mover torque-speed characteristic must be considered carefully when determining the electromechanical specifications of the generator. Equally important are the characteristics of the electric power system into which the VRG provides energy.

13.4.1 Excitation and Generation

The VRG requires a source of excitation in order to generate electrical energy. This excitation is derived from a switching inverter, such as that shown

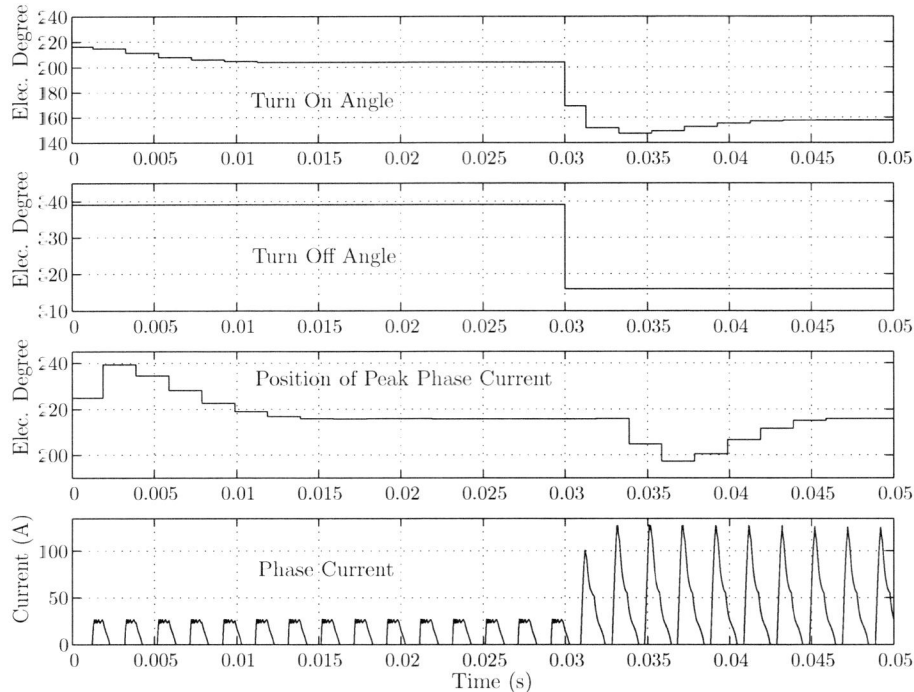

Figure 13.13: Simulation results of the implemented control technique with 2500 rpm motor speed and current reference changed from 25 A to 140 A at 0.3 s.

in Fig. 13.2. When the controllable switches are closed, current builds in the VRG phase winding. For generator operation, excitation generally begins near the aligned position for relatively low speed operation. The excitation is often advanced with increasing speed so that excitation begins before the aligned position. This is analogous to the advance introduced in the control of the VRM. After the controllable switches are turned off, more energy is returned to the source than was provided for excitation.

Figure 13.14 shows two energy conversion cycles for the VRG of Fig. 13.1 operating at two different speeds. Each loop is traversed in a clockwise manner. The areas enclosed by the loops correspond to the energy converted from mechanical to electrical form for the two cases. There are $N_\phi N_r$ energy conversion cycles in each revolution of the rotor; N_ϕ is the number of phases. The circles in Fig. 13.14 indicate the point where the controllable switches are turned off and phase current is supported by the diodes of Fig. 13.2.

To better understand the issues in exciting a VRG and extracting energy through the phase winding, it is instructive to look at the back emf coefficient

13.4. Control of VR Generators

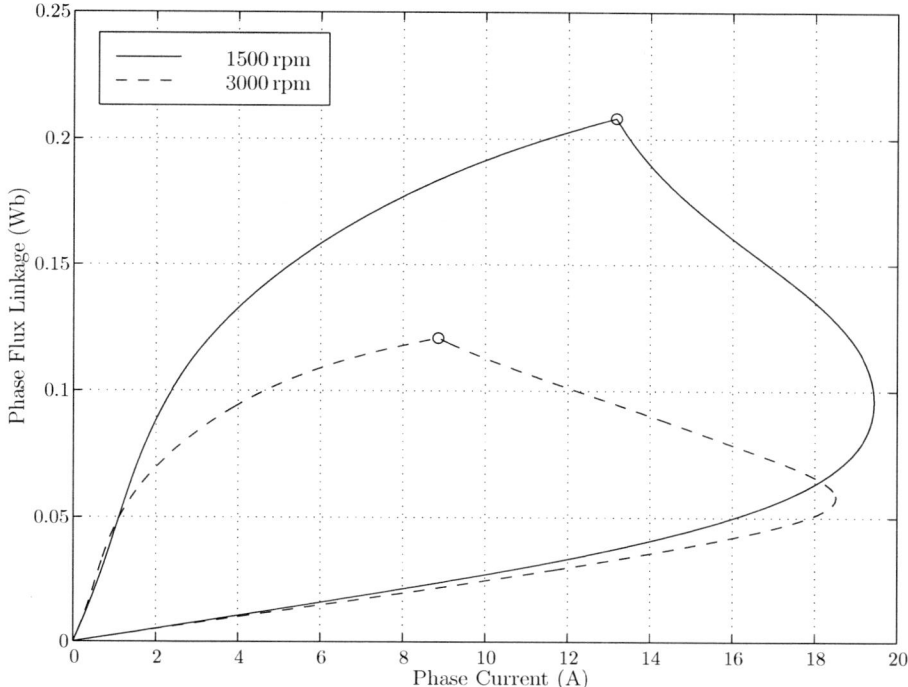

Figure 13.14: Energy conversion cycles for the VRG of Fig. 13.1 at 1500 rpm and 3000 rpm. The circles indicate where the controllable switches are turned off.

of the VRG. To begin, the electrical dynamics of a VRG phase are

$$\frac{d\lambda}{dt} = v - Ri \quad . \tag{13.5}$$

Because phase flux linkage is a function of current and position, Eq. 13.5 can be expanded and rewritten as

$$v = \frac{\partial \lambda}{\partial \theta}\frac{d\theta}{dt} + \frac{\partial \lambda}{\partial i}\frac{di}{dt} + Ri \quad , \tag{13.6}$$

or

$$v = \omega \frac{\partial \lambda}{\partial \theta} + \frac{\partial \lambda}{\partial i}\frac{di}{dt} + Ri \quad . \tag{13.7}$$

The first term on the right side of Eq. 13.7 represents the back emf presented by the phase winding. The second term on the right side of Eq. 13.7 represents the voltage dropped across the phase inductance. The back emf coefficient for the VRG is $\partial \lambda / \partial \theta$ where it is implied that the partial derivative is taken with current held constant. It follows that the back emf coefficient of

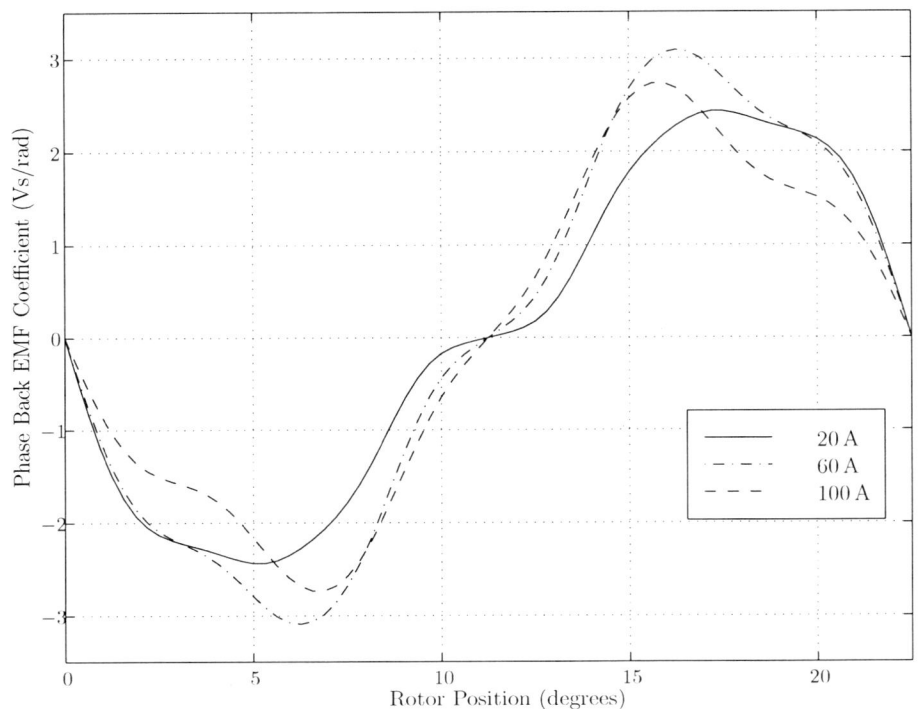

Figure 13.15: The back emf coefficient for the VRG of Fig. 13.1 for three levels of phase current.

the VRG is the slope of the flux linkage versus position information given in Fig. 13.5. Figure 13.15 gives the back emf coefficient for the VRG for three values of phase current. In other machines, such as a brushless dc motor, this $\partial\lambda/\partial\theta$ term is sometimes called the back emf constant; this terminology is not used here because it is hard to consider this term constant in view of Fig. 13.15.

There are several features of Fig. 13.15 that are worthy of note. First, the back emf coefficient is negative during the region of decreasing phase inductance and positive during the region of increasing phase inductance. Second, the back emf coefficient retains the spatial and magnetic nonlinearities that are present in the flux linkage data of Figs. 13.4 and 13.5. Third, the peak back emf coefficient increases with current up to a certain point; additional increases in phase current actually serve to reduce the back emf coefficient.

Figures 13.16 and 13.17 show the inverter circuit topologies for the excitation interval and demagnetization interval, respectively. The dynamics

13.4. Control of VR Generators

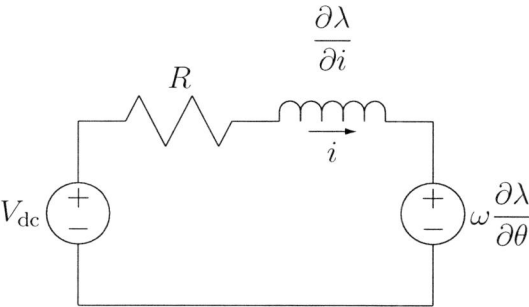

Figure 13.16: The inverter topology during phase excitation.

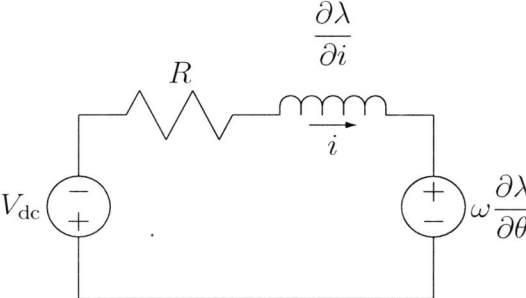

Figure 13.17: The inverter topology during phase demagnetization.

of the phase current can be inferred from the relative sign and magnitude of the back emf compared to the applied voltage. For example, during excitation prior to the aligned position, phase current is building up in the face of the back emf reducing the effectiveness of the source voltage. This drives significant advancement in θ_{on} to have adequate phase current as the rotor enters the region of decreasing phase inductance.

Behavior during demagnetization can be assessed by comparing the relative magnitude of the back emf and the source voltage. If the source voltage has larger magnitude than the back emf, the phase current will decrease. This may necessitate multiple periods of excitation when generating at low speed, leading to a current waveform that is regulated to maintain adequate excitation as the rotor moves from the aligned position to the unaligned position. At high speeds the back emf can serve to drive increases in phase current in the face of negative source voltage and decreasing flux linkage after the controllable switches have been turned off. This occurs

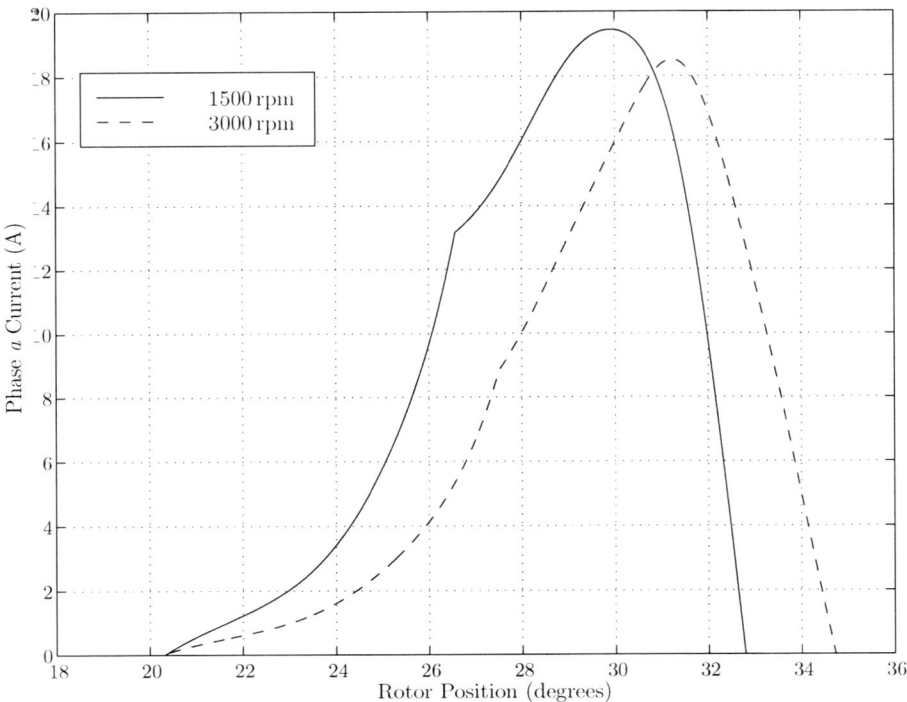

Figure 13.18: The phase current waveforms as a function of rotor position for the energy conversion cycles of Fig. 13.14.

when $\omega (\partial \lambda / \partial \theta) < -V_{\rm dc}$.

Phase current waveforms tend toward exaggerated peaks at increasing speeds. Figure 13.18 shows the phase current waveforms for the energy conversion cycles given in Fig. 13.14. The waveform for operation at 3000 rpm has a higher crest factor than the waveform at 1500 rpm due to the back emf being substantially higher; the crest factor for operation at 1500 rpm is 2.22 versus 2.36 for operation at 3000 rpm. The crest factor is defined as the ratio of the peak phase current to the RMS phase current. Note that the rotor position of the peak current varies relatively little despite the large difference in speed, and corresponds to a rotor position just before the idealized phase inductance decreases into the minimum inductance region, position θ_g in Fig. 13.6. This characteristic is typical in the VRG, just as the peak phase current for motoring operation above base speed is always located near the position where the phase inductance begins increasing from its minimum value. This phenomenon has its origins in the rapid change in back emf coefficient on either side of the aligned and unaligned positions as shown in Fig. 13.15.

13.4. Control of VR Generators

In addition to increased crest factor at higher speeds, the VRG has a tendency toward open loop instability when the back emf magnitude is sufficiently larger than the source voltage. The essence of the instability is that increased source voltage tends to increase the excitation current, thereby increasing the energy extracted from the prime mover. The increased energy conversion typically outpaces the increase in load associated with the increase in excitation voltage. Depending on the nature of the load, this extra energy conversion may go into further increasing the excitation voltage[12]. The VRG of Fig. 13.1 has been found unstable even with a resistive load where the power consumed is proportional to the square of the excitation voltage.

Further, the tendency toward instability contributes to the self-excitation of the VRG. Residual magnetism is often sufficient to start the self-excitation process. With the controllable switches disabled, the VRG of Fig. 13.1 was able to charge the dc bus to a small voltage, typically less than 10 V for a VRG intended to feed power into a 300 V dc bus. Enabling the controllable switches applies the small bus voltage to developing excitation current which, in conjunction with the open loop instability, quickly charged the dc bus to its intended operating voltage. There are abundant anecdotal data that confirm self-excitation of the VRG, though there has not been careful analysis of this phenomenon. More work is necessary to thoroughly characterize the conditions under which the VRG can self-excite.

13.4.2 Control Implications

The previous subsection alluded to several excitation issues that have implications for the control implementation. This subsection summarizes these points for clarity as we prepare to discuss how the VRG is controlled. These points have duals with regard to the control of the VRM.

1. We have no control over where the peak phase current occurs once the VRG has entered single pulse operation above base speed. This leaves us with many combinations of turn-on angle and conduction angle that will yield the same output power. An important issue is how to best choose the excitation parameters. This is more challenging for the VRG than the VRM because the peak phase current for the VRG occurs while both controllable switches are off. For the VRM, the

---

[12]This phenomenon is discussed in A. Radun, "Generating with the switched-reluctance motor," *Proc. of the IEEE Applied Power Electronics Conf.*, pp. 41-47, 1994, and D. E. Cameron and J. H. Lang, "The control of high-speed variable-reluctance generators in electric power systems," *IEEE Trans. on Industry Applications*, Vol. 29, pp. 1106-1109, 1993.

turn-on angle can be used to directly control the peak current thereby partitioning the responsibilities of the turn-on angle and conduction angle. For the VRG, the turn-on angle and conduction angle control the peak phase current jointly and severally.

2. For operation below base speed, it is necessary to regulate the phase currents. It is usual to use unison operation of the phase switches to regulate phase currents during generation. There is little value in having the VRG generate into its phase windings.

3. The turn-on and turn-off angles must be advanced to support constant power operation above base speed. The rate of advance is generally different for turn-on and turn-off until the maximum conduction angle is reached.

13.4.3 VRG Control

Building upon the electromechanics of the VRG and the energy conversion principles discussed above, we can now examine the control of the VRG. We begin with a discussion of how a typical VRG controller is structured. We then turn our attention to methods for choosing "optimal" excitation parameters. Finally, we discuss some specific control implementations.

Controller Structures

The structure of the VRG controller is similar in nature to that of the VRM. For operation from below base speed to above base speed, there must be a commutator that determines the appropriate turn-on and turn-off angles. Below base speed current regulation is required; this is generally accomplished by logically combining the commutation signals with a signal that reflects the relative relationship between the desired current and the actual current. The commutator and the current regulator operate on a very short time scale, suggesting these loops constitute the innermost elements of the control.

The outer loop is usually concerned with either regulating the speed of the VRG, or in regulating the average power supplied by the VRG. An application such as wind energy would focus on regulating the speed of the VRG relative to that of the wind stream in order to force peak aerodynamic efficiency[13]. An application such as an aircraft power system would require

---

[13]See, for example, D. A. Torrey, "Variable-reluctance generators in wind-energy systems," *Proc. of the IEEE Power Electronics Specialists Conf.*, pp. 561-567, 1993, and R. Cardenas, W. F. Ray, and G. M. Asher, "Switched reluctance generators for wind energy

13.4. Control of VR Generators

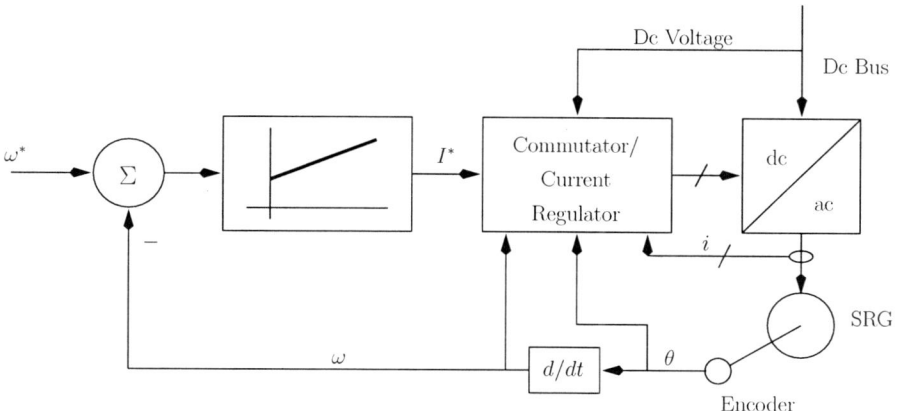

Figure 13.19: The structure of the controller for regulating VRG speed.

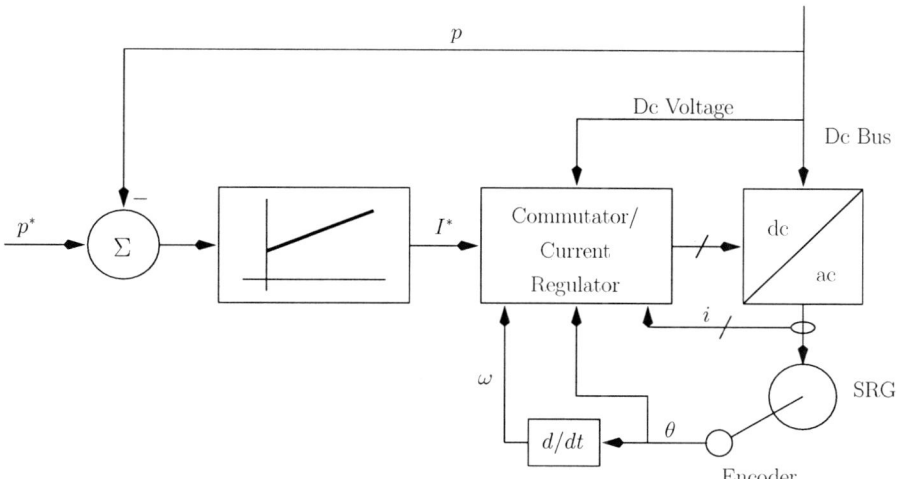

Figure 13.20: The structure of the controller for regulating VRG power.

regulation of average power (or current) output. Figures 13.19 and 13.20 summarize the structure of controllers for regulating VRG speed and power, respectively.

The design of the controller for the VRG is challenging for three reasons. First, the output of the VRG tends to be open loop unstable. As discussed above, as the speed of the VRG increases beyond the base speed, the output dc current increases with increasing voltage. Such a situation demands stabilization through application of closed-loop control. Second,

---

applications," *Proc. of the IEEE Power Electronics Specialists Conf.*, pp. 559-564, 1995.

the VRG system is highly nonlinear making the selection of suitable control gains problematic. Third, the excitation parameters needed to support a particular output power are not unique.

If the VRG is supplying energy to a sufficiently stiff voltage source, it may not be necessary to close the outer control loop that regulates the average power delivered to the load. For example, the battery in an automotive application is sufficiently stiff to prevent the system voltage from running away. Proper charge control of the battery, however, would dictate that some effort be taken to control the average power generated by the VRG.

Optimal Selection of Excitation Parameters

The most challenging piece within the controllers of Figs. 13.19 and 13.20 is the commutator that is responsible for selecting the turn-on and turn-off (or conduction) angles. While this is challenging for the VRM as well, the optimization is more straightforward because only the turn-on angle dictates the peak phase current. For the VRG, both turn-on and turn-off angles contribute to peak phase current above base speed.

For operation below base speed, VRG operation is very similar to that of the VRM. In particular, the phase currents would be regulated to follow a prescribed value as a function of rotor position. The value is nominally constant unless very low torque ripple is part of the objective. Conduction would be carried out over the region of decreasing inductance. The determination of the turn-on angle is based on providing adequate time to build up the phase current before entering the torque production region. Current regulation for the VRG is usually based on unison operation of both controllable switches in Fig. 13.2.

To illustrate the optimization problem, Fig. 13.21 shows the average dc link current as a function of excitation angles for the VRG of Fig. 13.1 for operation at 3000 rpm and a fixed 300 V dc bus. The data suggest that there are multiple combinations of excitation parameters that provide the same average dc link current. Typical performance metrics such as efficiency, torque ripple, RMS dc link current, etc. can be dramatically different for the various combinations of excitation parameters that yield the same power output. Figure 13.22 shows the RMS phase current for the VRG of Fig. 13.1 as a function of average dc link current for operation at 3000 rpm[14]. Each point corresponds to a different combination of turn-on and turn-off angle; the circled points correspond to the minimum RMS phase current for a

---

[14] See E. Mese, Y. Sozer, J. M. Kokernak, and D. A. Torrey, "Optimal excitation of a high speed switched reluctance generator," *Proc. of the IEEE Applied Power Electronics Conf*, pp. 362-368, 2000 for additional details.

13.4. Control of VR Generators

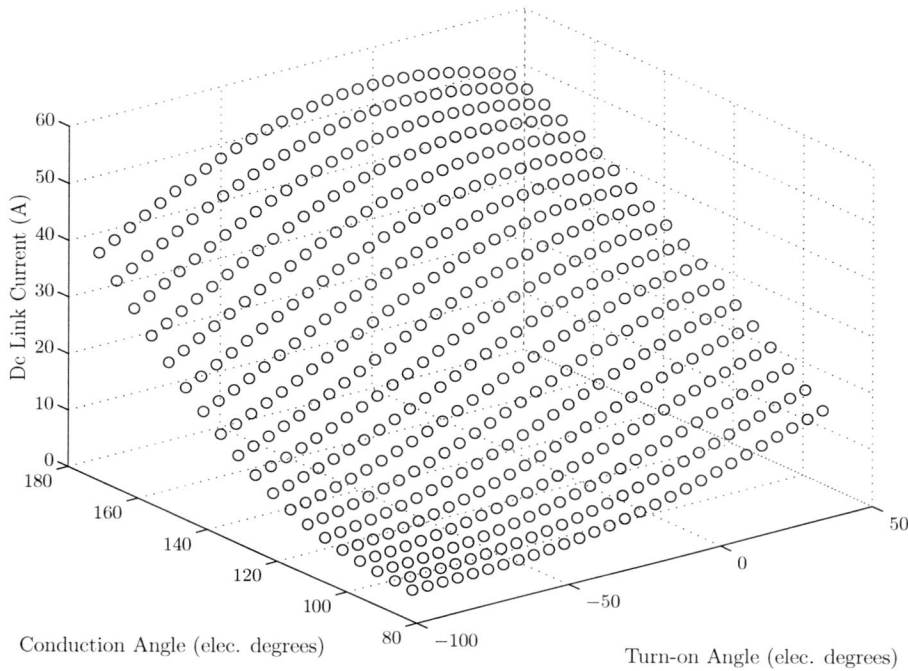

Figure 13.21: The average dc link current as a function of excitation parameters for the VRG of Fig. 13.1 for operation at 3000 rpm.

specific value of dc link current.

Where maximization of efficiency at each operating point is the optimization objective, minimization of RMS phase current is a reasonably pragmatic approach. Minimizing the RMS phase current implies minimizing the resistive losses in the VRG, minimizing the peak flux in the VRG thereby minimizing core losses, and minimizing conduction losses within the inverter. A more precise optimization would require very detailed models for every loss mechanism within the VRG system and would be difficult to enforce in practice[15]. Minimization of RMS phase currents is a reasonable objective of, for example, self-tuning control[16].

---

[15] See also the discussion in D. A. Torrey and J. H. Lang, "Optimal-efficiency excitation of variable-reluctance motor drives," *IEE Proc.*, Vol. 138, pt. B, pp. 1-14, 1991.

[16] See, for example, B. Fahimi, G. Suresh, J. P. Johnson, M. Ehsani, M. Arefeen, and I. Panahi, "Self-tuning control of switched reluctance motors for optimized torque per Ampere at all operating points," *Proc. of the IEEE Applied Power Electronics Conf.*, pp. 778-783, 1998, and K. Russa, I. Husain and M. Elbuluk, "A self-tuning controller for switched reluctance motors," *IEEE Trans. on Power Electronics*, Vol. 15, pp. 545-552, 2000.

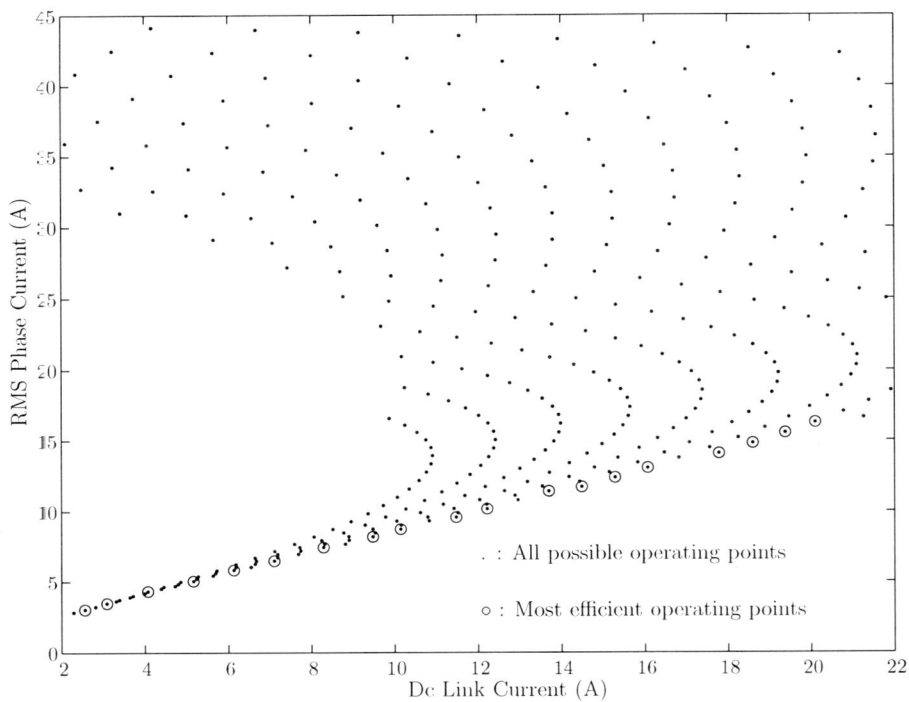

Figure 13.22: The RMS phase current as a function of average dc link current for the VRG of Fig. 13.1 for operation at 3000 rpm.

It is certainly possible to perform the optimization experimentally in order to build up a table of excitation parameters based on speed and required dc bus current. Figures 13.23 and 13.24 show the turn-on and conduction angles for the VRG of Fig. 13.1 that result in maximum system efficiency as a function of operating speed and average dc link current. It is possible to fit functions to the data of Figs. 13.23 and 13.24 to calculate the excitation parameters, rather than use a table[17]. Figures 13.23 and 13.24 show the tendency to advance turn-on and extend conduction for higher output current (power).

Other optimization criteria might include minimization of torque ripple or RMS dc link current. While these criteria could lead to a different choice of excitation parameters than minimization of RMS phase currents, minimization of RMS phase currents would tend to reduce both torque ripple

---

[17]This is the approach taken in S. R. MacMinn and J. W. Sember, "Control of a switched-reluctance aircraft starter-generator over a very wide speed range," *Proc. of the Intersociety Energy Conversion Engineering Conf.*, pp. 631-638, 1989 to handle operation in the motor mode.

13.4. Control of VR Generators

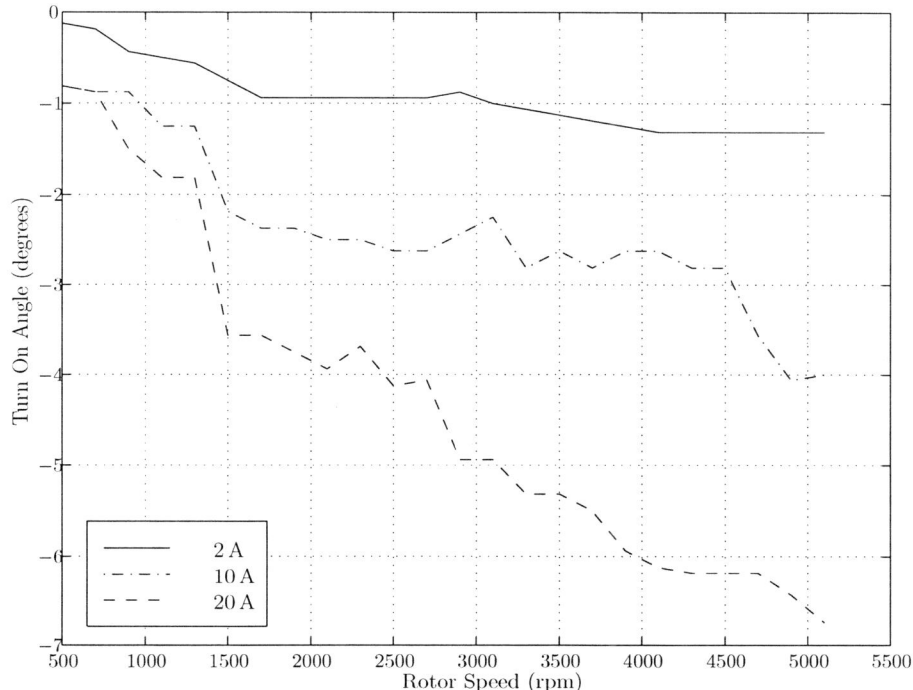

Figure 13.23: The efficiency-optimal turn-on angle as a function of speed and average dc link current for the VRG of Fig. 13.1.

and RMS dc link current because minimizing RMS phase currents would reduce the current peaks and positive torque production that lead to large RMS dc link current and large torque ripple.

In applications where the VRG system is used to regulate the voltage used to excite the VRG, linearization of the VRG control characteristic may be the preferred optimization goal. These applications include aerospace power systems[18]. In one approach, linearizing the control characteristic of the VRG is achieved by fixing the phase current at turn-off. Advances in turn-on angle then produced increased dc bus current. At low power levels the turn-on angle is fixed and the turn-off angle is used to control the average current in order to increase system efficiency. Another approach is to fix the turn-on angle and use a map of average dc link current as a function of

---

[18]Control linearization is discussed in S. R. MacMinn and J. W. Sember, "Control of a switched-reluctance aircraft starter-generator over a very wide speed range," *Proc. of the Intersociety Energy Conversion Engineering Conf.*, pp. 631-638, 1989, and D. E. Cameron and J. H. Lang, "The control of high-speed variable-reluctance generators in electric power systems," *IEEE Trans. on Industry Applications*, Vol. 29, pp. 1106-1109, 1993.

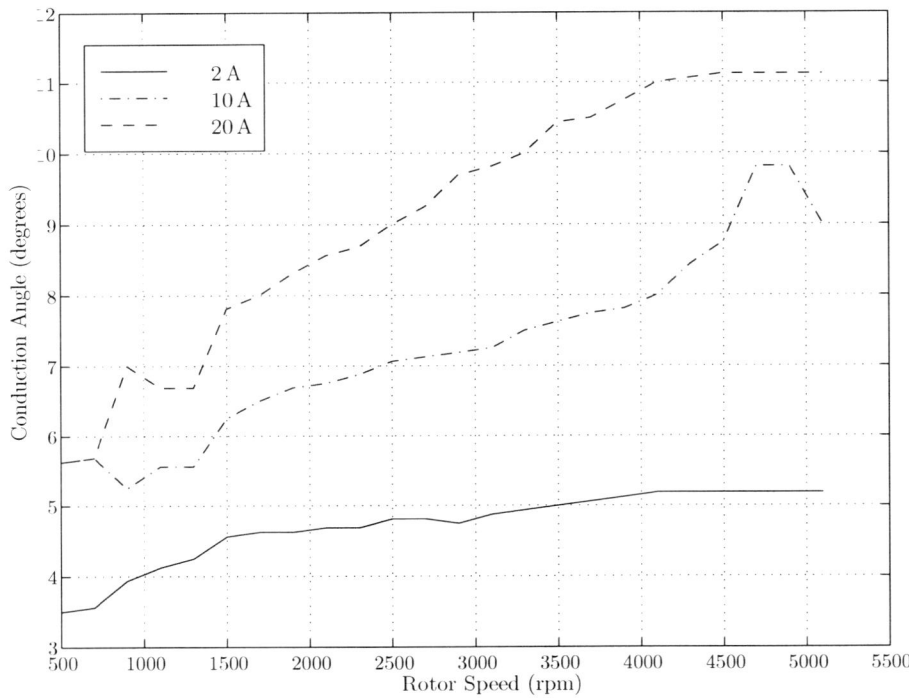

Figure 13.24: The efficiency-optimal conduction angle as a function of speed and average dc link current for the VRG of Fig. 13.1.

conduction angle to globally linearize the control characteristic.

13.4.4 Control Implementation

Control of the VRG is usually accomplished via a computer, whether it be a microcontroller, digital signal processor (DSP), or something more sophisticated. Consistent with Figs. 13.19 and 13.20, there are several tasks that must be handled by the controller. Implementation of the excitation parameters and current regulation are the most time-critical because small implementation errors in excitation angles can have a significant impact on the electromechanical performance. The commutator and current regulator typically operate on very short time scales. The outer control loop generally operates on a much longer time scale. A separation of time scales can often be employed to aid with the design of the outer control loop as we have seen elsewhere.

For a VRG that must operate at very high speed, it may be necessary to implement the commutator in hardware based on parameters determined

13.4. Control of VR Generators

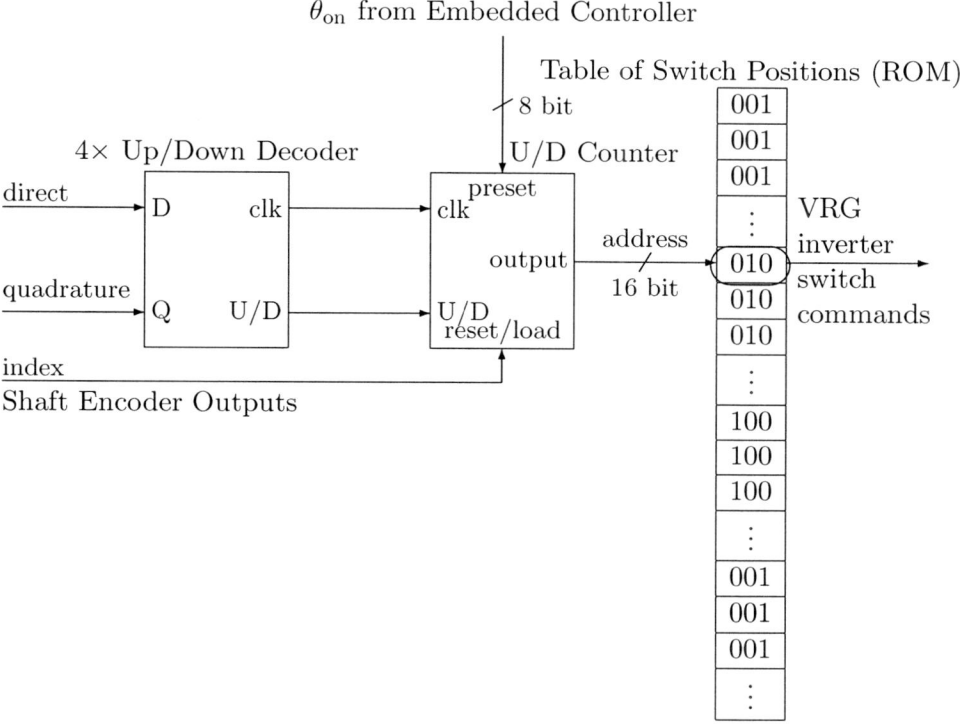

Figure 13.25: A method of implementing excitation angles using a ROM-based table of switch positions. Multiple tables are used to accommodate different conduction angles.

by software. This approach can take the rotor position as an index through a ROM-based table of switch states. Adjustment of the turn-on angle is accomplished by shifting the base pointer to the table. The essence of the concept is shown in Fig. 13.25 for a fixed conduction angle. Adjustment of conduction angle is accomplished by having multiple tables in the ROM, each based on a different conduction angle. Because of the spatial periodicity of the VRG the ROM table needs only to be large enough to service one electrical cycle with each table entry being accessed N_r times per revolution.

With the DSPs optimized for motor control it is possible to integrate commutation and current regulation algorithms directly into the processor thereby simplifying the system hardware. These DSPs generally have pulse-width modulation (PWM) generators, analog to digital (A/D) converters,

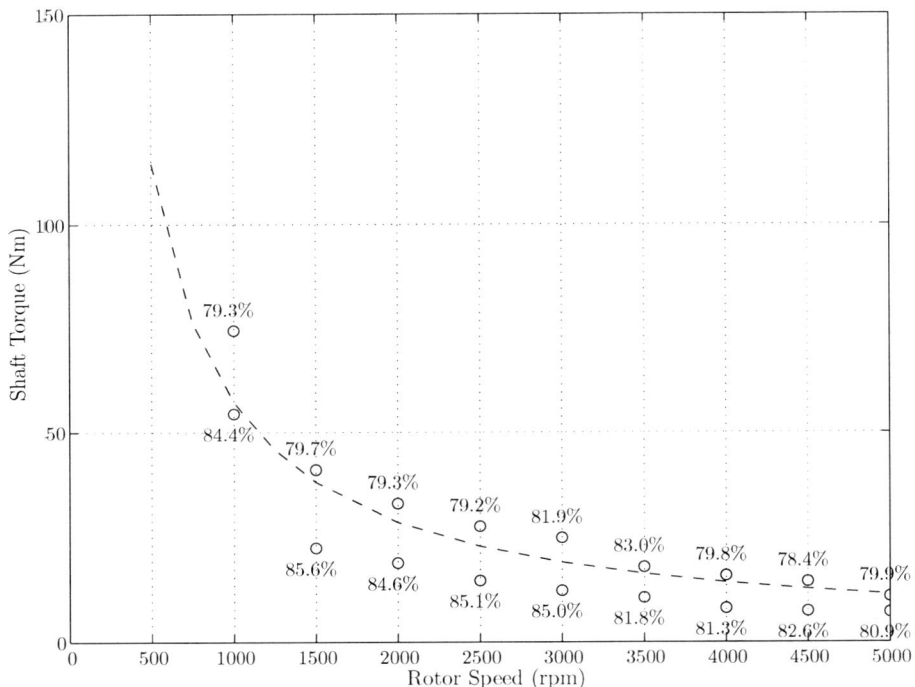

Figure 13.26: The system efficiency as a function of torque and speed, based upon the VRG of Fig. 13.1.

and incremental encoder interface units available on-chip. For this type of implementation, commutation decisions can be implemented with lookup tables and if-then-else comparisons. This is the approach taken for the VRG of Fig. 13.1. Acceptable efficiency is achieved with a rotor position resolution of 5° (electrical). System (combined VRG and inverter) efficiency for the VRG of Fig. 13.1 is given in Fig. 13.26.

13.5 Summary

This chapter has concerned itself with the VRM and its control. While motor and generator operation share many common features, the control implementation is sufficiently different that separate consideration is worthwhile.

Switched-reluctance motor efficiency is characterized in terms of operating speed, torque production and excitation angles. An efficient and easily implementable control algorithm for the excitation angles is developed. The

13.5. Summary

approach provides for automatic turn-on angle adjustment without the need for motor parameters or self-tuning techniques. The algorithm monitors the peak phase current and where the peak current occurs. It places the position of the first peak of phase current at θ_m in order to maximize $T_{\text{avg}}/I_{ph\text{rms}}$ produced by the VRM. The controller also ensures that the peak phase current is equal to the reference current. Turn-off angles are a function of peak phase current level and speed, so they can be represented through a curve fitting function. The turn-off angle is easily optimized by determining the efficiency-optimal excitation parameters at only four operating points.

Most applications of the VRG are oriented toward constant power output, thereby placing substantial interest in VRG operation above base speed. For operation above base speed the VRG system produces increasing average dc bus current for increasing excitation voltage. In certain systems this suggests open-loop instability, for which closed loop control must be used. Operation above base speed results in two excitation parameters impacting the peak phase current. This allows satisfying the required electromechanical performance along with another control objective. For VRG systems that deliver energy to a dc bus of fixed voltage, efficiency is commonly optimized. For VRG systems that deliver energy to a dc bus of variable voltage, control linearity is commonly optimized. This can be accomplished through use of a fixed turn-off current or global linearization through inverse mapping to achieve a known relationship between the average dc link current and a $(\theta_{\text{on}}, \theta_{\text{off}})$ combination.

The VRM is capable of supporting high system efficiencies over a wide speed range. Essentially constant system efficiency for constant power output is typically achievable over at least a three to one speed range, with acceptable system efficiency over a ten to one speed range. The VRM can offer high efficiency over a very wide speed range with a rotor structure that is compatible with high speeds, and an overall structure that is tolerant of extreme environments.

Appendix A

Symbols Used

The following symbols are used. Unfortunately, some symbols take on more than one meaning. Hopefully the correct meaning is obvious based on context. The page number indicates where the symbol is first introduced.

Alphabetic Symbols

Symbol	Units	Page	Quantity
a		291	Magnet magnetization reduction factor
a_n	m	287	Fourier coefficient of shape function
A	A/m	122	Electric loading
B	T	2	Magnetic flux density
B_r	T	14	Residual flux density
B	kg	25	Damper coefficient
c	m/s	5	Speed of light in free space
c	J/kgK	133	Heat capacity
d		166	Duty ratio
D	C/m^2	2	Electric displacement
D	m	123	Diameter
e	A	176	Current error
e_k	V	109	Back emf of the kth phase
E	V/m	2	Electric field intensity
E_b	W/m^2	136	Blackbody emissive flux
f	m	287	Shape function
f^e	N	37	Force of magnetic origin
f_n	At	71	Fourier coefficient of MMF
$\mathcal{F}$	At	9	Magnetomotive force

Symbol	Units	Page	Description
F_R	At	98	MMF of rotor back iron
g_m	T	291	Magnetization function
G	Ω^{-1}	10	Conductance
$G(s)$		180	Plant transfer function
h	W/m^{2}K	135	Heat transfer coefficient
h_{bi}	m	94	Height of back iron
h_g	m	94	Air gap length
h_g'	m	94	Effective air gap length
h_{sh}	m	94	Height of pole shoe
h_{st}	m	94	Height of pole shoe taper
h_t	m	94	Height of tooth
H	A/m	3	Magnetic field intensity
H_c'	A/m	19	Effective coercive force
H	Nms^2	25	Rotational moment of inertia
$H(s)$		180	Compensator transfer function
i	A	5	Current
I		143	The identity matrix
$\hat{I}$	A	215	Sinusoidal steady state (phasor) current
J	A/m^2	2	Current density
J		143	A matrix analogous to j; $J^2 = -I$
k	N/m	25	Spring constant
k		73	Coil span in slots, used for integral windings
k	W/mK	133	Thermal conductivity
k	Wb	269	Flux linkage due to magnet flux
k_α		88	Skew factor
k_C		94	Carter coefficient
k_d		73	Winding distribution factor
k_G		96	Slotting correction factor
k_p		71	Winding pitch factor
k_{sl}		86	Phase offset in slots
k_t	Nms^2	229	Load torque constant used in simulation
k_{w1}		124	Fundamental winding factor
K	Wb	271	Flux linkage amplitude due to magnet flux
K	T	292	Constant used to enforce flux conservation
K_i	Ω/s	177	Integral gain
K_o		80	Phase offset in slots
K_p	Ω	177	Proportional gain
$\vec{K}_s$	A/m	122	Surface current density
l	m	5	Characteristic length
ℓ	m	8	Magnetic path length
L	H	6	Inductance, could be a matrix
L_s	m	87	Active stack length

M	A/m	7	Magnetization density
M	kg	25	Mass
M	H	57	Mutual inductance, could be a matrix
$\hat{n}$		123	Normal unit vector
N		9	Number of turns
$N_{c\phi}$		83	Number of coils per phase
N_p		65	Number of magnetic poles
N_s		68	Number of winding slots
N_ϕ		69	Number of phases
N_{sp}		69	Number of winding slots per pole
$N_{sp\phi}$		69	Number of winding slots per pole per phase
p	W	116	Instantaneous power
$\mathcal{P}$	H	10	Permeance
q		72	Number of coils in winding phase belt
$\hat{q}$	W/m^2	133	Heat flux
$\dot{q}$	W/m^3	133	Internal heat generation
$\mathcal{R}$	1/H	9	Reluctance
R	Ω	10	Resistance, could be a matrix
R	m	85	Radius
R_ψ		144	Rotation transformation through angle ψ
s		215	Slip, the ratio of rotor to stator frequency
S		80	Coil span in slots, used for fractional windings
S		158	Inverter switch, usually carrying a number
S_k		98	Stator tooth mmf direction
t	s	3	Time
T	K	133	Temperature
T	s	183	Duration of inverter switching cycle
T		141	Transformation from $abc \to \alpha\beta0$
T_{23}		142	Transformation from $abc \to \alpha\beta$
T_{32}		142	Transformation from $\alpha\beta \to abc$
T_{mn}	N/m^2	39	Maxwell stress tensor
T^e	Nm	48	Torque of magnetic origin
u	J/m^3	133	Intensive internal energy
v^*	V	176	Voltage command
V	m^3	3	Volume
V	V	240	Voltage amplitude
$\vec{V}$	V	166	Voltage vector, produced by an inverter
V_{dc}	V	149	Inverter dc bus voltage
V_2	V	190	$\|\vec{V}^*\|$ when transition to six-step begins
V_6	V	190	$\|\vec{V}^*\|$ when transition to six-step ends
$\hat{V}$	V	215	Sinusoidal steady state (phasor) voltage
w'_m	J/m^3	49	Magnetic field coenergy density

Symbol	Units	Page	Quantity
w_m	J/m^3	49	Magnetic field energy density
w_s	m	94	Tooth pitch
w_{sh}	m	94	Tooth shoe width
w_{sl}	m	85	Slot width
w_{sb}	m	94	Tooth width at bottom (near air gap)
w_{st}	m	94	Tooth width at top (near back iron)
W_m	J	43	Magnetic field energy
W_m'	J	46	Magnetic field coenergy
x	As	177	State variable, the integral of current
$\overline{x}$		177	Local average of a variable
X	Ω	218	Inductive reactance

Greek Symbols

Symbol	Units	Page	Quantity
α		86	Skew angle normalized to slot pitch
α	m^2/s	134	Thermal diffusivity
β	rad	289	Angle of shoe tooth
χ_m		7	Magnetic susceptibility
χ	rad	85	Electrical angle of slot opening
δ	m	24	Skin depth
δ		39	Kronecker delta function
δ		70	Coil side angle
ϵ	F/m	3	Permittivity
ε_g	m	289	Magnitude of air gap eccentricity
γ'	rad	183	Angle of desired voltage, relative to $\vec{V}_a$
γ_g	rad	289	Phase of eccentricity
λ	Vs	10	Flux linkage
$\hat{\lambda}$	Vs	215	Sinusoidal steady state (phasor) flux linkage
μ	H/m	3	Permeability
ω	rad/s	24	Radian frequency, as in a current source
ω	rad/s	5	Angular velocity
ω_c	rad/s	181	Crossover frequency
ω_{syn}	rad/s	215	Synchronous speed
ϕ	Vs	9	Magnetic flux
ρ	C/m^3	2	Free charge density
ρ	kg/m^3	133	Mass density
σ	S	10	Conductivity
σ	W/m^{2}K^4	136	Stefan-Boltzmann constant

σ		253	Coupling factor between stator and rotor
σ_r		257	Rotor leakage factor
$\vec{\tau}$	N/m²	42	Traction
τ_l	Nm	212	Load torque
τ_r	s	228	Induction machine rotor time constant
θ	rad	35	Mechanical rotor position
θ_e	rad	71	Electrical angle
θ_s	rad	69	Slot pitch
θ_{se}	rad	69	Slot pitch in electrical measure

Subscripts

Subscripts are sometimes used in combination. For example, λ_{sa} is the flux linking the phase a stator winding.

Symbol   **Quantity**

a	Phase a
b	Phase b
c	Phase c
c	Core
C	Contour
d	Direct axis
g	Air gap
I	Fictitious winding
II	Fictitious winding
ℓ	Leakage
m	Magnet
m	Mechanical
m	Mutual, as in air gap
n	Normal
q	Quadrature axis
r	Rotor
R	Recoil
s	Stator
sl	Slip
sl	Stator slot
S	Surface
t	Tangential
x	Component in x direction

V	Volume
y	Component in y direction
z	Zero vector
α	Phase α
β	Phase β
0	Phase 0
$0-7$	Voltage vector number
$1-4$	Inverter switch number

Superscripts

Symbol	Quantity
s	Stator reference frame
r	Rotor reference frame
$*$	Commanded value, as in a closed-loop controller
$*$	Complex conjugate

Appendix B

Signal Analysis

B.1 Introduction

This Appendix provides summary information about signal analysis concepts that are useful for understanding the decomposition of waveforms, whether they be spatial or temporal. Since electric machines and their excitation do not contain pure distributions of mmf or single frequency currents, it is inevitable that consideration of interaction among harmonics must be considered.

We start with describing periodic waveforms, and move to other signal analysis concepts such as root mean square (rms) and total harmonic distortion.

B.2 Fourier Series

This section gives a thorough, albeit brief, explanation of the Fourier series. See a text on signal analysis for more detail and some of the mathematical subtleties[1]. This development is based on signals that depend on time, but it is a straightforward extension to consider signals that depend on space. Essentially, integration over time gets replaced by integration over space, with the period becoming an angular measure rather than time.

---

[1] See, for example, B. P. Lathi, *Signals, Systems and Communication*, John Wiley & Sons, 1965, Chapter 3 or S. S. Soliman and M. D. Srinath, *Continuous and Discrete Signals and Systems*, Prentice-Hall, 1990, Chapter 3.

B.2.1 The Basic Series

A signal $f(t)$ is said to be periodic with period T if it repeats itself every T seconds. Examples of periodic signals include sine waves and square waves. Any periodic signal can be broken down into fundamental components, called basis functions. These basis functions are chosen to be orthogonal, just as we use orthogonal unit vectors when representing an arbitrary vector.

By definition, two periodic functions $f(t)$ and $g(t)$ are considered to be orthogonal if

$$\int_0^T f(t)g(t)\,dt = 0 \quad . \tag{B.1}$$

A very useful set of orthogonal basis functions is formed by sinusoids and cosinusoids. If we note that

$$\int_0^T \sin m\omega t \sin n\omega t\,dt = \begin{cases} T/2 & \text{for } m = n \\ 0 & \text{for } m \neq n \end{cases}, \tag{B.2}$$

$$\int_0^T \cos m\omega t \cos n\omega t\,dt = \begin{cases} T/2 & \text{for } m = n \\ 0 & \text{for } m \neq n \end{cases}, \tag{B.3}$$

and

$$\int_0^T \sin m\omega t \cos n\omega t\,dt = 0 \tag{B.4}$$

where $\omega = 2\pi/T$, we have done all of the ground work for analytically describing any periodic signal.

The concept of orthogonality is significant in any discipline that involves power, where we are often describing the conversion of energy from one form to another. Our work generally involves dealing with signals that contain many frequency components. In terms of the average power represented by these signals, Eq. B.3 implies that if a particular frequency component is to contribute to average power flow, then that frequency must be present in both the voltage and the current. Exploitation of signal orthogonality can dramatically simplify circuit analysis in some cases. Thinking about the structure of the waveform before beginning the analysis can save significant time and effort. Similar arguments hold for functions of space.

The trigonometric version of the Fourier series representation of $f(t)$ is

$$f(t) = a_0 + \sum_{n=1}^{\infty} [a_n \cos n\omega t + b_n \sin n\omega t] \quad . \tag{B.5}$$

Equation B.5 states that any periodic signal can be decomposed into a constant (dc) component and the superposition of an infinite number of orthogonal basis functions.

B.2. Fourier Series

So far we have rationalized a functional form for $f(t)$, but we have not addressed how to find the coefficients a_0, a_n and b_n. We determine these coefficients by using the very properties we used in choosing the basis functions in the first place. To determine a_0, we multiply both sides of Eq. B.5 by dt and integrate over one period. This gives

$$\int_0^T f(t)\, dt = \int_0^T a_0\, dt + \int_0^T \left\{ \sum_{n=1}^\infty [a_n \cos n\omega t + b_n \sin n\omega t] \right\} dt \quad . \tag{B.6}$$

By recognizing that every sine and cosine term in the series will integrate to zero (because we are integrating over one or more full periods), we get

$$a_0 = \frac{1}{T} \int_0^T f(t)\, dt \quad . \tag{B.7}$$

In other words, a_0 is the dc component of $f(t)$.

We follow a similar process to determine a_n. Multiplying both sides of Eq. B.5 by $\cos m\omega t$ and integrating over a period gives

$$\int_0^T f(t) \cos m\omega t\, dt = \int_0^T a_0 \cos m\omega t\, dt +$$

$$\int_0^T \left\{ \sum_{n=1}^\infty [a_n \cos n\omega t + b_n \sin n\omega t] \right\} \cos m\omega t\, dt \quad . \tag{B.8}$$

By inspection, the first term on the right side integrates to zero. In addition, every cross product between a sine and cosine will integrate to zero; it does not matter whether the frequencies are the same or not. The orthogonality relationship for cosinusoids of different frequencies (Eq. B.3) indicates that the only term of the cosine series which contributes to the integral is the term $n = m$. Using this, the expression for a_n becomes

$$a_n = \frac{2}{T} \int_0^T f(t) \cos n\omega t\, dt \quad . \tag{B.9}$$

The determination of b_n exactly parallels the determination of a_n. The starting point is the multiplication of both sides of Eq. B.5 by $\sin m\omega t$ before integrating over the period. The result is

$$b_n = \frac{2}{T} \int_0^T f(t) \sin n\omega t\, dt \quad . \tag{B.10}$$

An alternative representation of function $f(t)$ is the exponential version of the Fourier series:

$$f(t) = \sum_{n=-\infty}^\infty f_n \exp(\jmath n\omega t) \quad , \tag{B.11}$$

where f_n is potentially complex and the complex exponential replaces the trigonometric functions. Thinking about the relationship between complex exponentials and circular trigonometric functions suggests we can determine f_n by multiplying both sides of Eq. B.11 by $\exp(-\jmath m\omega t)dt$ and integrating over a period. Noting that the complex exponential will only contribute a non-zero result if $m = n$, we conclude

$$f_n = \frac{1}{T} \int_0^T f(t) \exp(-\jmath n\omega t)\, dt \quad . \tag{B.12}$$

The power of the exponential version of the Fourier series is with how naturally phase shift can be incorporated. In fact, the form of the series in Eq. B.11 should be reminiscent of phasor representation.

For example, $f(t - \beta)$ is given by

$$f(t - \beta) = \sum_{n=-\infty}^{\infty} f_n \exp(\jmath n\omega (t - \beta)) \quad , \tag{B.13}$$

which can also be written as

$$f(t - \beta) = \sum_{n=-\infty}^{\infty} f_n \exp(-\jmath n\omega \beta) \exp(\jmath n\omega t) \quad . \tag{B.14}$$

This natural representation of shifting is why the exponential version of the Fourier series is used when describing windings in Chapter 3.

From Eq. B.12 we conclude that

$$f_{-n} = f_n^* \quad , \tag{B.15}$$

where * denotes complex conjugation. By comparing Eqs. B.5 and B.11, we see that

$$a_n = \Re\{f_n + f_{-n}\} \quad ; \tag{B.16}$$

$$b_n = \Im\{f_n - f_{-n}\} \quad ; \tag{B.17}$$

$$f_n = \frac{a_n - \jmath b_n}{2} \quad . \tag{B.18}$$

B.2.2 Symmetry Conditions

There are several symmetry conditions which sometimes occur in periodic signals. These conditions can be exploited to simplify our work in describing the signal. This subsection considers these conditions. Note that there are many practical examples of waveforms which contain more than one type of symmetry.

B.2. Fourier Series

In order to exploit the symmetry conditions, the symmetry must first be recognized. One method of determining whether or not symmetry exists involves mentally trying to fit some different basis functions to the signal in question. Another method mentally pictures the time origin in different locations. A simple time shift may mask the symmetry. You should practice these and any other techniques you discover. The objective is not to be able carry out nasty integrals, but to recognize and exploit symmetry conditions in order to enhance our insight and reduce the time we spend performing computations. That is, work smarter, not harder.

Even Symmetry

Even symmetry is defined as

$$f(t) = f(-t) \ . \tag{B.19}$$

A simple cosine is a perfect example of a waveform with even symmetry. For an even waveform,

$$a_n = \frac{4}{T} \int_0^{T/2} f(t) \cos n\omega t \, dt \ , \tag{B.20}$$

and

$$b_n = 0 \ . \tag{B.21}$$

Even symmetry can easily be masked by a time shift. For example, a sinusoid can be considered to be a cosine with a time shift of $T/4$ seconds. In the exponential form,

$$f_n = \frac{2}{T} \int_0^{T/2} f(t) \cos n\omega t \, dt \ . \tag{B.22}$$

Odd Symmetry

Odd symmetry is defined as

$$f(t) = -f(-t) \ . \tag{B.23}$$

A simple sinusoid is a perfect example of a waveform with odd symmetry. For an odd waveform,

$$b_n = \frac{4}{T} \int_0^{T/2} f(t) \sin n\omega t \, dt \ , \tag{B.24}$$

and

$$a_n = 0 \ . \tag{B.25}$$

In exponential form,

$$f_n = \frac{-j\,2}{T} \int_0^{T/2} f(t) \sin n\omega t\, dt \quad . \tag{B.26}$$

Half-Wave Symmetry

Half-wave symmetry is defined as

$$f(t) = -f\left(t \pm \frac{T}{2}\right) \quad . \tag{B.27}$$

Half-wave symmetry is sometimes referred to as rotation symmetry. A waveform which possesses half-wave symmetry has half cycles which are identical in shape but opposite in sign. The waveform shown in Fig. B.1 gives an example of half-wave symmetry.

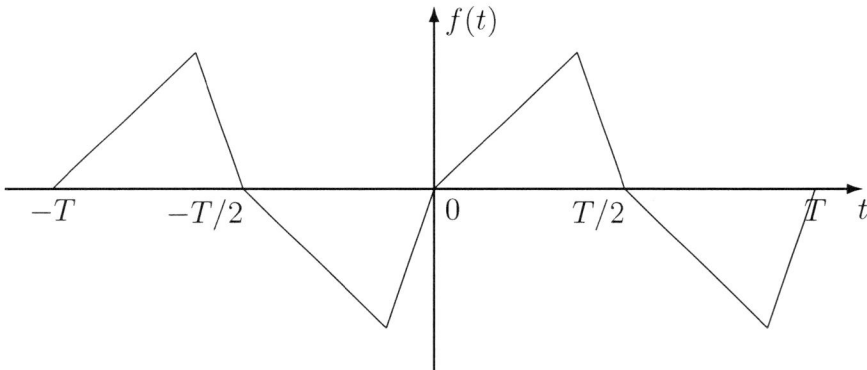

Figure B.1: A waveform with half-wave symmetry.

It can be shown that the Fourier series of a signal with half-wave symmetry contains only odd harmonics, that is, frequencies ω, 3ω, 5ω, ... etc. Trying to mentally superimpose candidate basis functions on a signal with half-wave symmetry will result in the conclusion that the even harmonics are not consistent with the waveform, so their coefficients should be zero.

Hidden Symmetry

Often one comes across a periodic function without any apparent form of symmetry, yet it possesses a Fourier series containing only cosine, sine, or odd harmonic terms. The reason for this is that a dc component can obscure the symmetry of the waveform. When looking for even, odd, or half-wave symmetry, one must be careful to remove any dc components. Another way of looking at it is that the ac component of the waveform (see the section below) dictates the symmetry, irrespective of whether the waveform contains dc or not. Even, odd and half-wave symmetry can be hidden by a dc component. Time shifts, as discussed above, can also hide symmetry.

B.3 The Decomposition of Waveforms

The last section discussed the representation of a periodic signal in terms of a Fourier series. This section examines the relationship between the Fourier series representation given by Eqs. B.5 and B.11 and other waveform attributes.

To begin, a periodic signal $f(t)$ can be considered to be the superposition of two components: an ac component and a dc component. Mathematically, we have

$$f(t) = F_{\text{dc}} + f_{\text{ac}}(t) \quad . \tag{B.28}$$

This representation is connected to the Fourier series such that

$$F_{\text{dc}} = a_0 = f_0 \quad , \tag{B.29}$$

and

$$f_{\text{ac}}(t) = \sum_{n=1}^{\infty} [a_n \cos n\omega t + b_n \sin n\omega t] = \sum_{\substack{n=-\infty \\ n \neq 0}}^{\infty} f_n \exp jn\omega t \quad . \tag{B.30}$$

We are frequently interested in the rms value of a waveform. It is important to understand how the rms value of each component contributes to the rms value of the total waveform. Applying the definition of the rms value to $f(t)$,

$$F_{\text{rms}} = \sqrt{\frac{1}{T} \int_0^T [f(t)]^2 \, dt} \quad . \tag{B.31}$$

Bringing in the Fourier series representation of $f(t)$, we have

$$F_{\text{rms}} = \sqrt{\frac{1}{T} \int_0^T \left\{ a_0 + \sum_{n=1}^{\infty} [a_n \cos n\omega t + b_n \sin n\omega t] \right\}^2 dt} \quad . \tag{B.32}$$

This certainly looks like an unpleasant equation to evaluate! If we think about what we have, however, we will find it is not quite as bad as it looks.

The series expression which forms the integrand in Eq. B.32 is comprised of the self and cross products of sinusoids, cosinusoids and dc terms. The integral is being performed over an integer number of periods, so we can exploit orthogonality once more. Specifically, the only terms that will contribute to the integral are a_0^2, $a_n^2 \cos^2 n\omega t$ and $b_n^2 \sin^2 n\omega t$. This allows us to simplify Eq. B.32 to

$$F_{\text{rms}} = \sqrt{\frac{1}{T}\left\{\int_0^T a_0^2\, dt + \sum_{n=1}^{\infty} \int_0^T [a_n^2 \cos^2 n\omega t + b_n^2 \sin^2 n\omega t]\, dt\right\}} \quad . \quad \text{(B.33)}$$

The elimination of the cross product terms has allowed us to interchange the summation and integral signs. This reduces still further to

$$F_{\text{rms}} = \sqrt{a_0^2 + \sum_{n=1}^{\infty} \frac{a_n^2 + b_n^2}{2}} \quad . \quad \text{(B.34)}$$

One final piece of manipulation should pull all of this together. If we recognize that

$$a_n \cos n\omega t + b_n \sin n\omega t = A_n \cos(n\omega t - \phi_n) \quad , \quad \text{(B.35)}$$

where

$$A_n = \sqrt{a_n^2 + b_n^2} \quad , \quad \text{(B.36)}$$

and

$$\phi_n = \tan^{-1} \frac{b_n}{a_n} \quad , \quad \text{(B.37)}$$

then we see that the summation term in Eq. B.34 is superimposing the squares of the rms values of each harmonic. Said another way,

$$F_{\text{rms}} = \sqrt{a_0^2 + \sum_{n=1}^{\infty} \left[\frac{A_n}{\sqrt{2}}\right]^2} = \sqrt{a_0^2 + \sum_{n=1}^{\infty} A_{n\,\text{rms}}^2} \quad . \quad \text{(B.38)}$$

In exponential form,

$$F_{\text{rms}} = \sqrt{\sum_{n=-\infty}^{\infty} f_n f_{-n}} \quad . \quad \text{(B.39)}$$

It follows that the total rms value of a waveform is related to the rms values of the ac and dc components through

$$F_{\text{rms}} = \sqrt{F_{\text{dc}}^2 + F_{\text{ac rms}}^2} \quad , \quad \text{(B.40)}$$

where

$$F_{\text{ac rms}} = \sqrt{\sum_{n=1}^{\infty} A_{n\,\text{rms}}^2} \quad . \tag{B.41}$$

Physically, this can be interpreted as something analogous to an n dimensional application of the Pythagorean Theorem and is known as Parseval's theorem. There is a lot of physical insight to be gained by thinking about the analysis which has been presented here, and relating it to the decomposition and magnitude determination of arbitrary vectors from basic mechanics.

B.4 Measures of Distortion

An important figure of merit for many waveforms is the ability of a waveform to approximate a desired shape. In particular, synthesis of sinusoidal waveforms is difficult to do efficiently. In this case, it is useful to have a measure of the distortion introduced by approximating the desired waveform. A common measure of this distortion is the Total Harmonic Distortion (THD). Formally, the THD is defined as

$$\text{THD} = \sqrt{\dfrac{\sum_{\substack{n=-\infty \\ n \neq x}}^{\infty} f_n f_n^*}{f_x f_x^*}} \quad , \tag{B.42}$$

where x is the frequency component of interest. Often x is the fundamental, but sometimes it is the dc or another component. The difficulty of evaluating the series in the numerator motivates a much more pragmatic calculation of the THD using Parseval's Theorem:

$$\text{THD} = \sqrt{\dfrac{F_{\text{rms}}^2 - f_x f_{-x}}{f_x f_{-x}}} \quad , \tag{B.43}$$

or

$$\text{THD} = \sqrt{\dfrac{F_{\text{rms}}^2}{f_x f_{-x}} - 1} \quad . \tag{B.44}$$

B.5 Summary

This Appendix has presented a brief discussion of orthogonal signals followed by the development of the Fourier series representation of a periodic signal. In addition, the general decomposition of periodic waveforms was discussed from the perspective of relating the ac and dc components of a waveform to the Fourier series of the waveform, and ultimately how the different components contribute to the rms value of the waveform.

The concepts presented here are very important to power electronics and detailed analysis of electric machines. A strong command of signal analysis techniques allows for greater insight into the energy conversion properties of electronic converters and electric machines.

Appendix C

Matlab Simulation Code

This Appendix provides the Matlab code used for various simulations.

C.1 Current Regulator

The code that follows was used to implement the simulation of a current regulator in Sec. 8.6. This code generated Figs. 8.22, 8.23, and 8.24. The simulation is based on three files. The first file performs the functions of the modulator. The second file carries out the simulation, using the modulator function and the file that contains the system state equations.

```
function [d_a,d_b,d_z,V_a,V_b,V_z,v_alpha_pr,v_beta_pr] = ...
    sv_modulator(v_alpha,v_beta,V_2,V_6,V_dc);
%
% David Torrey 4/30/2008
%
% Space Vector Modulator
%
% This is a two-level space vector modulator.
%
% This space vector modulation function performs the following
% operations:
%    1. It determines the sector in which the commanded voltage
%       sits.
%    2. It determines the angle the commanded voltage makes
%       relative to vector V_a.
%    3. It determines the duty ratios of the three voltage
%       vectors that are nearest to the desired vector.
%    4. It compares the magnitude of the commanded voltage to
%       two voltage levels.  The first level (V_2) is the level
```

```
%       where the transition to six-step operation takes place.
%       The second level (V_6) is the level where the transition
%       to six-step operation is completed.
%   5.  It modifies the commanded voltage if six-step operation
%       is encountered, preserving the commanded angle.  This is
%       effectively used for prevention of integrator windup.
%
% The function returns the duty ratios associated with the
% three switching vectors.
%
% A reference for this function is Chapter 8: Current Regulators.

% Convert the commanded voltage into polar form.

V_mag = sqrt(v_alpha^2 + v_beta^2);
gamma = atan2(v_beta,v_alpha);   % Gives gamma on [-pi,pi]
if gamma < 0                     % Determine gamma on [0,2pi]
    gamma = gamma + 2*pi;
end

% Determine the sector and the relative sector angle.

sector = 1 + fix(3*gamma/pi);
gamma_pr = mod(gamma,pi/3);

% Select vectors V_a, V_b, and V_z based on sector.  The numerical
% value of the vectors corresponds to the vector number.

switch sector
    case 1
        V_a = 1;
        V_b = 2;
        V_z = 7;
    case 2
        V_a = 2;
        V_b = 3;
        V_z = 0;
    case 3
        V_a = 3;
        V_b = 4;
        V_z = 7;
    case 4
        V_a = 4;
        V_b = 5;
        V_z = 0;
    case 5
        V_a = 5;
        V_b = 6;
```

C.1. Current Regulator

```
                V_z = 7;
            case 6
                V_a = 6;
                V_b = 1;
                V_z = 0;
            case 7
                V_a = 1;
                V_b = 2;
                V_z = 7;
        end

        % Determine the duty ratios for the three vectors.  These may get
        % modified below depending on the results.

        d_a = sqrt(2)*(V_mag/V_dc)*sin(pi/3-gamma_pr);
        d_b = sqrt(2)*(V_mag/V_dc)*sin(gamma_pr);
        d_z = 1 - d_a - d_b;

        if (d_z < 0) || (V_mag > V_2)
        % We have some degree of overmodulation.
            % If d_z < 0 we are in at least overmodulation mode 1;
            % if V_mag > V_2 we are in overmodulation mode 2,
            % transitioning toward six-step operation.
            if V_mag > V_6   % We have six-step operation
                if gamma_pr <= pi/6
                % We apply vector V_a during the entire cycle
                    d_a = 1;
                    d_b = 0;
                    d_z = 0;
                    v_alpha_pr = sqrt(2/3)*V_dc * cos((sector-1)*pi/3);
                    v_beta_pr = sqrt(2/3)*V_dc * sin((sector-1)*pi/3);
                else
                % We apply vector V_b during the entire cycle
                    d_a = 0;
                    d_b = 1;
                    d_z = 0;
                    v_alpha_pr = sqrt(2/3)*V_dc * cos(sector*pi/3);
                    v_beta_pr = sqrt(2/3)*V_dc * sin(sector*pi/3);
                end
            elseif V_mag > V_2
            % We have transition to six-step operation
                if gamma_pr <= (pi/6)*((V_mag-V_2)/(V_6-V_2))
                % Dwell at vector V_a
                    d_a = 1;
                    d_b = 0;
                    d_z = 0;
                    v_alpha_pr = sqrt(2/3)*V_dc * cos((sector-1)*pi/3);
                    v_beta_pr = sqrt(2/3)*V_dc * sin((sector-1)*pi/3);
```

```
                elseif gamma_pr >= (pi/6)*((2*V_6-V_mag-V_2)/(V_6-V_2))
                % Dwell at vector V_b
                    d_a = 0;
                    d_b = 1;
                    d_z = 0;
                    v_alpha_pr = sqrt(2/3)*V_dc * cos(sector*pi/3);
                    v_beta_pr = sqrt(2/3)*V_dc * sin(sector*pi/3);
                else
                % Implement duty ratios placing V* on hexagon perimeter
                    d_b = 2*sin(gamma_pr)/(sin(gamma_pr)+ ...
                        sqrt(3)*cos(gamma_pr));
                    d_a = 1 - d_b;
                    d_z = 0;
                    v_v = (V_dc / sqrt(2))*d_b;
                    v_h = v_v / tan(gamma_pr);
                    V_mag_pr = sqrt(v_v^2 + v_h^2);
                    v_alpha_pr = (V_mag_pr/V_mag) * v_alpha;
                    % Adjust magnitude
                    v_beta_pr = (V_mag_pr/V_mag) * v_beta;
                end
        else
        % Implement duty ratios placing V* on hexagon perimeter
            d_b = 2*sin(gamma_pr)/(sin(gamma_pr)+ ...
                sqrt(3)*cos(gamma_pr));
            d_a = 1 - d_b;
            d_z = 0;
            v_v = (V_dc / sqrt(2))*d_b;
            v_h = v_v / tan(gamma_pr);
            V_mag_pr = sqrt(v_v^2 + v_h^2);
            v_alpha_pr = (V_mag_pr/V_mag) * v_alpha;
            % Adjust magnitude
            v_beta_pr = (V_mag_pr/V_mag) * v_beta;
        end
else
    v_alpha_pr = v_alpha;
    v_beta_pr = v_beta;
end
```

The following Matlab script is used to control the overall simulation of the current regulator.

```
% David Torrey 5/6/2008
%
% This m-file has been written to simulate the operation of a
% three-phase inverter driving a balanced load comprised of
% inductors, resistors, and series voltage sources.  The
% choice of inverter output voltages is based on space vector
```

C.1. Current Regulator

```
% modulation of the output voltage vector. The state equations
% are in file "inverter_ode.m".
%

close all;       % Close all open figure windows.
clear all;       % Clear workspace.

% First start with the circuit parameters:
global L R V_dc;        % Inverter system parameters
global V_s omega;       % Source parameters
global mu K_i K_p;      % Controller parameters

L = 0.005;              % Inverter output inductor.
R = 0.1;                % Parasitic resistance.
V_dc = 700;             % Dc bus voltage.
V_s = sqrt(2/3)*460;    % Peak ac line to neutral voltage.
omega = 377;            % Radian frequency of the ac source.
f_s = 15000;            % Switching frequency.
T = 1/f_s;              % Sampling period.
mu = 25 / V_s;          % Desired magnitude of phase current.
K_i = 44410;            % Integral gain
K_p = 47.12;            % Proportional gain

V_2 = V_dc/sqrt(2)+0.2*(sqrt(2/3)-1/sqrt(2))*V_dc;
% Voltage where the transition to six-step starts
V_6 = V_dc/sqrt(2)+0.8*(sqrt(2/3)-1/sqrt(2))*V_dc;
% Voltage where the transition to six-step ends

% Define the transformations between alpha/beta and abc frames

T_23 = sqrt(2/3) * [1 -0.5 -0.5; 0 sqrt(3)/2 -sqrt(3)/2];
T_32 = T_23';

% Simulation parameters

t_start = 0;            % Start time of the simulation.
t_final = 0.05;         % End time of the simulation.
tspan = [t_start t_start+T];  % Time span of first simulation.

% Initial conditions are based on periodic steady state operation
omegat = omega*t_start; % Electrical position at start
v_alphan = sqrt(3/2)*V_s*sin(omegat);
v_betan = -sqrt(3/2)*V_s*cos(omegat);
i_alpha0 = mu*v_alphan;
i_beta0 = mu*v_betan;

y_0 = [0 0 0 0];   % Compile initial conditions
```

```
% Set up ODE solver parameters.  We do not need the
% 'events' flag here.
options = odeset('RelTol',1e-9,'AbsTol',1e-12,'MaxStep', ...
         T/10,'events','off');

% Define a matrix of realizable space vectors
vectors = V_dc * [0 0; ...
                  sqrt(2/3) 0; ...
                  sqrt(1/6) sqrt(1/2); ...
                  -sqrt(1/6) sqrt(1/2); ...
                  -sqrt(2/3) 0; ...
                  -sqrt(1/6) -sqrt(1/2); ...
                  sqrt(1/6) -sqrt(1/2); ...
                  0 0];

%
% Now initialize arrays which are used to store the variables of
% the simulation.  These vectors are used to compile the results
% of the simulations within each topological state.  They are
% initialized as empty vectors here so that they can be available
% for use below.

Y = y_0;
% State variable vector: i_alpha, i_beta, v_alpha*, and v_beta*
TT = t_start;    % Time vector.
v = [];          % Matrix for implemented v_alpha and v_beta
v_alphabeta_pr = [];
% Matrix for commanded inverter output voltages

% Now perform the simulation by moving through as many sampling
% intervals as necessary to complete the simulation.  For each
% sampling interval we:
%    1. determine the duty ratios d_a, d_b, and d_z
%    2. determine the appropriate vectors to implement during
%       the sampling period
%    3. carry out the appropriate number of simulation pieces,
%       based on the duty ratios and implementation order
%    4. accumulate the state variables and other useful data

direction_flag = 1; % Start implementation with V_a, V_b, V_z
end_loop_flag = 0;
% Initialize flag that looks for the end of the simulation.
while end_loop_flag == 0
    t_s = TT(length(TT));   % Get start time of simulation
    % Start by determining the appropriate duty ratios, voltage
    % vectors, and updated voltage commands.  The updated
    % voltage commands reflect any limiting (saturation) that
    % might be necessary to keep the commanded voltages within
```

C.1. Current Regulator

```
% the confines of the hexagonal vector space.  These updated
% voltage commands are passed to the ODEFILE for
% implementation of integrator anti-windup.
i_alpha_c = sqrt(3/2)*V_s*mu*sin(omega*t_s);
i_beta_c = -sqrt(3/2)*V_s*mu*cos(omega*t_s);
i_alpha = Y(length(Y(:,1)),1);
i_beta = Y(length(Y(:,1)),2);
x_alpha = Y(length(Y(:,1)),3);
x_beta = Y(length(Y(:,1)),4);
v_alpha_pr_desired = K_p*(i_alpha_c-i_alpha)+K_i*x_alpha;
v_beta_pr_desired = K_p*(i_beta_c-i_beta)+K_i*x_beta;
[d_a,d_b,d_z,V_a,V_b,V_z,v_alpha_pr,v_beta_pr] = ...
    sv_modulator(v_alpha_pr_desired,v_beta_pr_desired,V_2, ...
              V_6,V_dc);
if d_a == 1
% We only implement vector V_a during the interval
    t_f = t_s + T;           % Finish time of simulation
    tspan = [t_s t_f];
    y_0 = Y(length(Y(:,1)),:);   % Initial condition vector
    [t,y] = ode45('inverter_ode',tspan,y_0,options,V_a, ...
              v_alpha_pr,v_beta_pr);
% Now accumulate data and set up the next piece of
% the simulation
    nt = length(t);
% Determine number of time points to accumulate.
    Y = [Y; y(2:nt,:)];
% Update inductor current and time vectors.  Start with the
    TT = [TT; t(2:nt)];
% second element to avoid redundant entries.
    v = [v; vectors(V_a+1,1)*ones(nt-1,1) ...
         vectors(V_a+1,2)*ones(nt-1,1)];
    v_alphabeta_pr = [v_alphabeta_pr; v_alpha_pr*ones(nt-1,1)...
              v_beta_pr*ones(nt-1,1)];
elseif d_b == 1
% We only implement vector V_b during the interval
    t_f = t_s + T;           % Finish time of simulation
    tspan = [t_s t_f];
    y_0 = Y(length(Y(:,1)),:);   % Initial condition vector
    [t,y] = ode45('inverter_ode',tspan,y_0,options,V_b, ...
              v_alpha_pr,v_beta_pr);
% Now accumulate data and set up the next piece of
% the simulation
    nt = length(t);
    Y = [Y; y(2:nt,:)];
    TT = [TT; t(2:nt)];
    v = [v; vectors(V_b+1,1)*ones(nt-1,1) ...
         vectors(V_b+1,2)*ones(nt-1,1)];
    v_alphabeta_pr = [v_alphabeta_pr; v_alpha_pr*ones(nt-1,1)...
```

```
                    v_beta_pr*ones(nt-1,1)];
    elseif (d_a+d_b) == 1
    % We use only V_a and V_b during the interval and there are
    %two simulations during the interval
        if direction_flag == 1      % Implement V_a then V_b
            t_f = t_s + d_a*T;      % Finish time of simulation
            tspan = [t_s t_f];
            y_0 = Y(length(Y(:,1)),:);   % Initial condition vector
            [t,y] = ode45('inverter_ode',tspan,y_0,options,V_a, ...
                           v_alpha_pr,v_beta_pr);
            % Now accumulate data and set up the next piece of
            % the simulation
            nt = length(t);
            Y = [Y; y(2:nt,:)];
            TT = [TT; t(2:nt)];
            v = [v; vectors(V_a+1,1)*ones(nt-1,1) ...
                 vectors(V_a+1,2)*ones(nt-1,1)];
            v_alphabeta_pr = [v_alphabeta_pr; ...
                              v_alpha_pr*ones(nt-1,1) ...
                              v_beta_pr*ones(nt-1,1)];
            t_s = TT(length(TT));
            % Get start time of second simulation
            t_f = t_s + d_b*T;    % Finish time of simulation
            tspan = [t_s t_f];
            y_0 = Y(length(Y(:,1)),:);   % Initial condition vector
            [t,y] = ode45('inverter_ode',tspan,y_0,options,V_b, ...
                           v_alpha_pr,v_beta_pr);
            % Now accumulate data and set up the next piece of
            % the simulation
            nt = length(t);
            Y = [Y; y(2:nt,:)];
            TT = [TT; t(2:nt)];
            v = [v; vectors(V_b+1,1)*ones(nt-1,1) ...
                 vectors(V_b+1,2)*ones(nt-1,1)];
            v_alphabeta_pr = [v_alphabeta_pr; ...
                              v_alpha_pr*ones(nt-1,1) ...
                              v_beta_pr*ones(nt-1,1)];
        else                          % Implement V_b then V_a
            t_f = t_s + d_b*T;       % Finish time of simulation
            tspan = [t_s t_f];
            y_0 = Y(length(Y(:,1)),:);   % Initial condition vector
            [t,y] = ode45('inverter_ode',tspan,y_0,options,V_b, ...
                           v_alpha_pr,v_beta_pr);
            % Now accumulate data and set up the next piece of
            % the simulation
            nt = length(t);
            Y = [Y; y(2:nt,:)];
            TT = [TT; t(2:nt)];
```

C.1. Current Regulator

```
            v = [v; vectors(V_b+1,1)*ones(nt-1,1) ...
                    vectors(V_b+1,2)*ones(nt-1,1)];
            v_alphabeta_pr = [v_alphabeta_pr; ...
                              v_alpha_pr*ones(nt-1,1) ...
                              v_beta_pr*ones(nt-1,1)];
            t_s = TT(length(TT));
            % Get start time of second simulation
            t_f = t_s + d_a*T;     % Finish time of simulation
            tspan = [t_s t_f];
            y_0 = Y(length(Y(:,1)),:);   % Initial condition vector
            [t,y] = ode45('inverter_ode',tspan,y_0,options,V_a, ...
                          v_alpha_pr,v_beta_pr);
            % Now accumulate data and set up the next piece of
            % the simulation
            nt = length(t);
            Y = [Y; y(2:nt,:)];
            TT = [TT; t(2:nt)];
            v = [v; vectors(V_a+1,1)*ones(nt-1,1) ...
                    vectors(V_a+1,2)*ones(nt-1,1)];
            v_alphabeta_pr = [v_alphabeta_pr; ...
                              v_alpha_pr*ones(nt-1,1) ...
                              v_beta_pr*ones(nt-1,1)];
     end
   else                 % We have three simulations during the interval
       if direction_flag == 1
       % Simulation through V_a, V_b, and V_z in sequence
           if d_a > 0
           % Do not execute this section if d_a=0
               t_f = t_s + d_a*T;     % Finish time of simulation
               tspan = [t_s t_f];
               y_0 = Y(length(Y(:,1)),:);
               % Initial condition vector
               [t,y] = ode45('inverter_ode',tspan,y_0,options, ...
                             V_a,v_alpha_pr,v_beta_pr);
               % Now accumulate data and set up the next piece of
               % the simulation
               nt = length(t);
               Y = [Y; y(2:nt,:)];
               TT = [TT; t(2:nt)];
               v = [v; vectors(V_a+1,1)*ones(nt-1,1) ...
                       vectors(V_a+1,2)*ones(nt-1,1)];
               v_alphabeta_pr = [v_alphabeta_pr; ...
                                 v_alpha_pr*ones(nt-1,1) ...
                                 v_beta_pr*ones(nt-1,1)];
               t_s = TT(length(TT));
               % Get start time of second simulation
           end
           if d_b > 0
```

```
            % Do not execute this section if d_b = 0
                t_f = t_s + d_b*T;     % Finish time of simulation
                tspan = [t_s t_f];
                y_0 = Y(length(Y(:,1)),:);
                % Initial condition vector
                [t,y] = ode45('inverter_ode',tspan,y_0,options, ...
                            V_b,v_alpha_pr,v_beta_pr);
                % Now accumulate data and set up the next piece of
                % the simulation
                nt = length(t);
                Y = [Y; y(2:nt,:)];
                TT = [TT; t(2:nt)];
                v = [v; vectors(V_b+1,1)*ones(nt-1,1) ...
                    vectors(V_b+1,2)*ones(nt-1,1)];
                v_alphabeta_pr = [v_alphabeta_pr; ...
                            v_alpha_pr*ones(nt-1,1) ...
                            v_beta_pr*ones(nt-1,1)];
                t_s = TT(length(TT));
                % Get start time of third simulation
            end
            if d_z > 0
            % Do not execute this section if d_z = 0
                t_f = t_s + d_z*T;     % Finish time of simulation
                tspan = [t_s t_f];
                y_0 = Y(length(Y(:,1)),:);
                % Initial condition vector
                [t,y] = ode45('inverter_ode',tspan,y_0,options, ...
                            V_z,v_alpha_pr,v_beta_pr);
                % Now accumulate data and set up the next piece of
                % the simulation
                nt = length(t);
                Y = [Y; y(2:nt,:)];
                TT = [TT; t(2:nt)];
                v = [v; vectors(V_z+1,1)*ones(nt-1,1) ...
                    vectors(V_z+1,2)*ones(nt-1,1)];
                v_alphabeta_pr = [v_alphabeta_pr; ...
                            v_alpha_pr*ones(nt-1,1) ...
                            v_beta_pr*ones(nt-1,1)];
            end
        else
        % Simulation through V_z, V_b, and V_a in sequence
            if d_z > 0
            % Do not execute this section if d_z = 0
                t_f = t_s + d_z*T;     % Finish time of simulation
                tspan = [t_s t_f];
                y_0 = Y(length(Y(:,1)),:);
                % Initial condition vector
                [t,y] = ode45('inverter_ode',tspan,y_0,options, ...
```

C.1. Current Regulator

```
                       V_z,v_alpha_pr,v_beta_pr);
    % Now accumulate data and set up the next piece of
    % the simulation
    nt = length(t);
    Y = [Y; y(2:nt,:)];
    TT = [TT; t(2:nt)];
    v = [v; vectors(V_z+1,1)*ones(nt-1,1) ...
            vectors(V_z+1,2)*ones(nt-1,1)];
    v_alphabeta_pr = [v_alphabeta_pr; ...
                      v_alpha_pr*ones(nt-1,1) ...
                      v_beta_pr*ones(nt-1,1)];
    t_s = TT(length(TT));
    % Get start time of second simulation
end
if d_b > 0
% Do not execute this section if d_b = 0
    t_f = t_s + d_b*T;  % Finish time of simulation
    tspan = [t_s t_f];
    y_0 = Y(length(Y(:,1)),:);
    % Initial condition vector
    [t,y] = ode45('inverter_ode',tspan,y_0,options, ...
                  V_b,v_alpha_pr,v_beta_pr);
    % Now accumulate data and set up the next
    % piece of the simulation
    nt = length(t);
    Y = [Y; y(2:nt,:)];
    TT = [TT; t(2:nt)];
    v = [v; vectors(V_b+1,1)*ones(nt-1,1) ...
            vectors(V_b+1,2)*ones(nt-1,1)];
    v_alphabeta_pr = [v_alphabeta_pr; ...
                      v_alpha_pr*ones(nt-1,1) ...
                      v_beta_pr*ones(nt-1,1)];
    t_s = TT(length(TT));
    % Get start time of third simulation
end
if d_a > 0
% Do not execute this section if d_a = 0
    t_f = t_s + d_a*T;  % Finish time of simulation
    tspan = [t_s t_f];
    y_0 = Y(length(Y(:,1)),:);
    % Initial condition vector
    [t,y] = ode45('inverter_ode',tspan,y_0, ...
            options,V_a,v_alpha_pr,v_beta_pr);
    % Now accumulate data and set up the next
    % piece of the simulation
    nt = length(t);
    Y = [Y; y(2:nt,:)];
    TT = [TT; t(2:nt)];
```

```matlab
                        v = [v; vectors(V_a+1,1)*ones(nt-1,1) ...
                                vectors(V_a+1,2)*ones(nt-1,1)];
                        v_alphabeta_pr = [v_alphabeta_pr; ...
                                          v_alpha_pr*ones(nt-1,1) ...
                                          v_beta_pr*ones(nt-1,1)];
                end
            end
        end
        direction_flag = 1 - direction_flag;
        % Set implementation order for next interval
        if t(nt) >= t_final
        % Check to see if simulation should continue.
            end_loop_flag = 1;
        end
end

% Convert the alpha beta currents back into three-phase
% currents.

i_alphabeta = Y(:,1:2); % Extract currents from state vector
i_abc = T_32 * i_alphabeta';
i_abc = i_abc';

% Calculate source voltages

v_abc = V_s * [sin(omega*TT) sin(omega*TT-2*pi/3) ...
               sin(omega*TT+2*pi/3)];
v_alphabeta = (T_23 * v_abc')';

% Augment the alpha beta voltages so that the matrix is of the
% same size as the others

v = [v(1,:); v];

% Now the results can be plotted.

TT = 1000*TT;         % Put time into milliseconds
figure;
plot(TT,i_abc(:,1),'b','LineWidth',3);
hold;
grid;
plot(TT,i_abc(:,2),'r','LineWidth',3);
plot(TT,i_abc(:,3),'g','LineWidth',3);
xlabel('Time (ms)','FontWeight','Bold');
ylabel('Phase Currents (A)','FontWeight','Bold');
plot(TT,mu*v_abc(:,1),'r','LineWidth',2);
legend('i_a','i_b','i_c','\mu v_{an}',0);
```

C.1. Current Regulator

```
figure;
plot(i_alphabeta(:,1),i_alphabeta(:,2),'b','LineWidth',3);
grid;
axis square;
xlabel('i_\alpha (A)','FontWeight','Bold');
ylabel('i_\beta (A)','FontWeight','Bold');
hold;
plot(mu*v_alphabeta(:,1),mu*v_alphabeta(:,2),'r','LineWidth',3);

figure;
plot(v_alphabeta_pr(:,1),v_alphabeta_pr(:,2),'b','LineWidth',3);
grid;
axis square;
xlabel('v_\alpha (V)','FontWeight','Bold');
ylabel('v_\beta (V)','FontWeight','Bold');
```

The following function contains the state equations for the current regulator.

```
function [out1,out2,out3] = inverter_ode(t,x,flag,V_vector, ...
                                        v_alpha_pr_sat, ...
                                        v_beta_pr_sat);
% David Torrey 5/10/2008
%
% This file contains the state equations for a three-phase
% inverter driving currents into a balanced set of voltages.
% Four states are considered:
% x(1) is the alpha component of the inverter output current
% x(2) is the beta component of the inverter output current
% x(3) is the alpha component of the integral of the error
% (with suitable correction for anti-windup)
% x(4) is the beta component of the integral of the error
%
% This file assumes that parameters that describe the system
% are shared with the main function through global declarations.
%

global L R V_dc;    % Inverter system parameters
global V_s omega;   % Source parameters
global mu K_i K_p;  % Controller parameters

% Source voltages in alpha, beta coordinates
v_alpha = sqrt(3/2)*V_s*sin(omega*t);
v_beta = -sqrt(3/2)*V_s*cos(omega*t);

% Generate current commands
i_alpha_c = mu*v_alpha;
```

```
i_beta_c = mu*v_beta;

% Define a matrix of realizable space vectors
vectors = V_dc * [0 0; ...
                  sqrt(2/3) 0; ...
                  sqrt(1/6) sqrt(1/2); ...
                  -sqrt(1/6) sqrt(1/2); ...
                  -sqrt(2/3) 0; ...
                  -sqrt(1/6) -sqrt(1/2); ...
                  sqrt(1/6) -sqrt(1/2); ...
                  0 0];

v_alpha_pr = vectors(1+V_vector,1);
v_beta_pr = vectors(1+V_vector,2);

% Put state variables into useful names

i_alpha = x(1);
i_beta = x(2);
x_alpha = x(3);
x_beta = x(4);

% Calculate useful quantities to simplify the structure of
% the state equations

e_alpha = i_alpha_c - i_alpha;            % Current errors
e_beta = i_beta_c - i_beta;
% Voltages before saturation
v_alpha_pr_desired = K_p*e_alpha+K_i*x_alpha;
v_beta_pr_desired = K_p*e_beta+K_i*x_beta;
% Inputs to integral controllers.  If there is no saturation
% of the commanded voltages, only the error gets integrated.
% Otherwise, the anti-windup term gets put to use.
integrand_alpha = e_alpha-(K_p/K_i)*(v_alpha_pr_desired- ...
                                    v_alpha_pr_sat);
integrand_beta = e_beta-(K_p/K_i)*(v_beta_pr_desired- ...
                                   v_beta_pr_sat);

if nargin < 3 | isempty(flag)

   % Return dx/dt = f(t,x); this is the state equation.
   out1 = [(v_alpha_pr - v_alpha)/L - (R/L)*i_alpha; ...
           (v_beta_pr - v_beta)/L - (R/L)*i_beta; ...
           integrand_alpha;...
           integrand_beta];

else
   switch(flag)
```

```
    otherwise
        error(['Unknown flag ''' flag '''.']);
    end
end
```

C.2 Induction Machine

The code that follows was used to simulate an induction machine being line-started. This simulation code was used to create Figs. 9.12 through 9.19. Two files are used to simulate the induction motor. The first file provides the parameters and the structure of the simulation. It also provides post-processing of the simulation results and generation of the graphs.

The second file contains the system state equations that are solved by the Matlab ordinary differential equation solver ode45.

```
% David Torrey 4/30/2001
%
% This file manages the simulation of an induction motor starting
% on line.  The state equations for the simulation are in
% "im_ode.m."
%

% Start with motor parameters
global P R_s R_r M L_S L_R D; % Shared parameters with ode file
P = 3;              % Number of pole pairs
R_s = 0.294;        % Stator resistance (abc, dq frames)
R_r = 0.156;        % Rotor resistance (abc, dq frames)
L_sl = 0.00139;     % Stator leakage inductance (abc frame)
L_rl = 0.00074;     % Rotor leakage inductance (abc frame)
L_sr = 0.041;       % Mutual inductance (abc frame)

M = 1.5 * L_sr;     % Mutual inductance (dq frame)
L_S = M + L_sl;     % Stator inductance (dq frame)
L_R = M + L_rl;     % Rotor inductance (dq frame)
D = L_S*L_R - M^2;  % Leakage factor

% Now we have parameters associated with the source and load
global V_s omega_e H k_t; % Shared parameters with ode file
V_s = sqrt(2/3) * 220;   % The amplitude of the phase voltage
omega_e = 377;           % The source radian frequency
H = 0.5;                 % Rotor inertia
k_t = 0.0038;            % The fan load torque constant

t_start = 0;             % Start time of the simulation
t_final = 1.0;           % End time of the simulation
```

```matlab
tspan = [t_start t_final];    % Time span of simulation

% Initial conditions are based on starting from standstill
% with the motor deenergized
y_0 = [0; 0; 0; 0; 0; 0];     % State variables are flux
%                               linkages, rotor velocity and
%                               position

% Set up ODE solver parameters.  We do not need to 'events' flag
% here nor is there any need to limit the maximum step size
options = odeset('RelTol',1e-9,'AbsTol',1e-12);

% Now perform the simulation.  The ode file is "im_ode.m."

[t,y] = ode45('im_ode',tspan,y_0,options);    % Solve system

% Now it is time to post-process the results.  For the post-
% processing shown here, we have the following state variables
% with which to work:
% y(:,1) is the stator direct axis flux
% y(:,2) is the stator quadrature axis flux
% y(:,3) is the rotor direct axis flux
% y(:,4) is the rotor quadrature axis flux
% y(:,5) is the rotor position
% y(:,6) is the rotor velocity
%
% The variables of interest are:
% electromagnetic torque
% stator currents in the dq reference frame
% rotor currents in the dq reference frame
% stator currents in the abc reference frame

% Compute the electromagnetic torque and the load torque
em_torque = (P*M/D)*(y(:,3).*y(:,2)- y(:,4).*y(:,1));
ld_torque = k_t * y(:,6) .* y(:,6);

% Now the currents in the dq reference frame, using the
% inverted flux linkage relations
i_sd = (1/D) * (L_R * y(:,1) - M * y(:,3));
i_sq = (1/D) * (L_R * y(:,2) - M * y(:,4));
i_rd = (1/D) * (-M * y(:,1) + L_S * y(:,3));
i_rq = (1/D) * (-M * y(:,2) + L_S * y(:,4));

% Now we take the currents from the dq reference frame into
% the alpha beta reference frame

i_salpha = i_sd .* cos(omega_e*t) - i_sq .* sin(omega_e*t);
i_sbeta  = i_sd .* sin(omega_e*t) + i_sq .* cos(omega_e*t);
```

C.2. Induction Machine

```
i_ralpha = i_rd .* cos(omega_e*t-P*y(:,5)) - i_rq .* ...
           sin(omega_e*t-P*y(:,5));
i_rbeta = i_rd .* sin(omega_e*t-P*y(:,5)) + i_rq .* ...
          cos(omega_e*t-P*y(:,5));

% Now we can take the currents from the alpha beta frame
% back to the abc frame.  Only the currents from phase a
% are computed.

i_sa = sqrt(2/3) * i_salpha;
i_ra = sqrt(2/3) * i_ralpha;

% Convert speed from rad/s to rev/min
speed = (30/pi) * y(:,6);

% Now the results can be plotted.

figure;                           % Rotor speed
plot(t,speed,'b','LineWidth',2);
grid;
xlabel('Time (s)','FontWeight','Bold');
ylabel('Rotor Speed (rpm)','FontWeight','Bold');

figure;                           % Torques
subplot(2,1,1)
plot(t,em_torque,'b','LineWidth',2);
grid;
xlabel('Time (s)','FontWeight','Bold');
ylabel('\tau_m (Nm)','FontWeight','Bold');
subplot(2,1,2)
plot(t,ld_torque,'g','LineWidth',2);
grid;
xlabel('Time (s)','FontWeight','Bold');
ylabel('\tau_l (Nm)','FontWeight','Bold');

figure;                           % dq stator fluxes
subplot(2,1,1)
plot(t,y(:,1),'b','LineWidth',2);
grid;
xlabel('Time (s)','FontWeight','Bold');
ylabel('\lambda_{sd} (Wb)','FontWeight','Bold');
subplot(2,1,2)
plot(t,y(:,2),'g','LineWidth',2);
grid;
xlabel('Time (s)','FontWeight','Bold');
ylabel('\lambda_{sq} (Wb)','FontWeight','Bold');

figure;                           % dq rotor fluxes
```

```
subplot(2,1,1)
plot(t,y(:,3),'b','LineWidth',2);
grid;
xlabel('Time (s)','FontWeight','Bold');
ylabel('\lambda_{rd} (Wb)','FontWeight','Bold');
subplot(2,1,2)
plot(t,y(:,4),'g','LineWidth',2);
grid;
xlabel('Time (s)','FontWeight','Bold');
ylabel('\lambda_{rq} (Wb)','FontWeight','Bold');

figure;                         % dq stator currents
subplot(2,1,1)
plot(t,i_sd,'b','LineWidth',2);
grid;
xlabel('Time (s)','FontWeight','Bold');
ylabel('i_{sd} (A)','FontWeight','Bold');
subplot(2,1,2)
plot(t,i_sq,'g','LineWidth',2);
grid;
xlabel('Time (s)','FontWeight','Bold');
ylabel('i_{sq} (A)','FontWeight','Bold');

figure;                         % dq rotor currents
subplot(2,1,1)
plot(t,i_rd,'b','LineWidth',2);
grid;
xlabel('Time (s)','FontWeight','Bold');
ylabel('i_{rd} (A)','FontWeight','Bold');
subplot(2,1,2)
plot(t,i_rq,'g','LineWidth',2);
grid;
xlabel('Time (s)','FontWeight','Bold');
ylabel('i_{rq} (A)','FontWeight','Bold');

figure;                         % phase a stator current
plot(t,i_sa,'b','LineWidth',2);
grid;
xlabel('Time (s)','FontWeight','Bold');
ylabel('Phase a Stator Current (A)','FontWeight','Bold');

figure;                         % phase a rotor current
plot(t,i_ra,'b','LineWidth',2);
grid;
xlabel('Time (s)','FontWeight','Bold');
ylabel('Phase a Rotor Current (A)','FontWeight','Bold');
```

This next function is used by the differential equation solver during sys-

C.2. Induction Machine

tem simulation. This function evaluates the system state variables. It might also include conditions that would cause the simulation to stop prematurely, but those are not needed here.

```
function [out1,out2,out3] = im_ode(t,x,flag);
% David Torrey 4/30/2001
%
% This file contains the state equations for an induction
% motor in dq coordinates.  Six states are considered:
% x(1) is the stator direct axis flux
% x(2) is the stator quadrature axis flux
% x(3) is the rotor direct axis flux
% x(4) is the rotor quadrature axis flux
% x(5) is the rotor position
% x(6) is the rotor velocity
%
% This file assumes that parameters that describe the
% induction motor, the source and the load are shared
% with the main function through global declarations.
%

global P R_s R_r M L_S L_R D;      % motor parameters
global V_s omega_e H k_t;          % Source and load parameters

% Source voltages in abc coordinates
v_a = V_s * sin(omega_e*t);
v_b = V_s * sin(omega_e*t-2*pi/3);
v_c = V_s * sin(omega_e*t+2*pi/3);

% Convert source voltages to alpha, beta coordinates
v_alpha = sqrt(2/3) * (v_a - 0.5 * (v_b + v_c));
v_beta = sqrt(1/2) * (v_b - v_c);

% Convert source voltages to dq coordinates
v_sd = v_alpha * cos(omega_e * t) + v_beta * sin(omega_e * t);
v_sq = - v_alpha * sin(omega_e * t) + v_beta * cos(omega_e * t);

if nargin < 3 | isempty(flag)

    % Return dx/dt = f(t,x); this is the state equation for the
    % induction machine.
    out1 = [v_sd - (R_s/D)*(L_R*x(1) - M*x(3)) + omega_e*x(2); ...
        v_sq - (R_s/D) * (L_R*x(2) - M*x(4)) - omega_e*x(1); ...
        - (R_r/D)*(L_S*x(3) - M*x(1)) + (omega_e-P*x(6))*x(4); ...
        - (R_r/D)*(L_S*x(4) - M*x(2)) - (omega_e-P*x(6))*x(3); ...
        x(6); ...
        (1/H)*( (P*M/D)*(x(3)*x(2)-x(4)*x(1)) - k_t*x(6)*x(6) ) ];
```

```
else
   switch(flag)
   otherwise
      error(['Unknown flag ''' flag '''.']);
   end
end
```

C.3 Simulation of a Brushless Dc Machine System

The code that follows was used to implement the simulation of a brushless dc motor in Sec. 12.3. The Simulink environment was used to develop this simulation using a graphical approach rather than a programming approach. This code generated Figs. 12.9 through 12.14. The top level of the model is given in Fig. C.1. The model for each block follows. Two "S-functions" are used to complete the simulation. The first synthesizes the inverter output voltages within the inverter block. The second computes $\partial\lambda/\partial\theta$ for each phase to facilitate calculation of back emfs and phase torques.

```
function [sys,x0,str,ts] = voltage_lookup_table(t,x,u,flag);

switch flag,
  %%%%%%%%%%%%%%%%%%%%
  % Initialization %
  %%%%%%%%%%%%%%%%%%%%
  % Initialize the states, sample times, and state
  % ordering strings.
  case 0
    [sys,x0,str,ts]=mdlInitializeSizes;

  %%%%%%%%%%%%%
  % Outputs %
  %%%%%%%%%%%%%
  % Return the outputs of the S-function block.
  case 3
    sys=mdlOutputs(t,x,u);

  %%%%%%%%%%%%%%%%%%%%%%
  % Unhandled flags %
  %%%%%%%%%%%%%%%%%%%%%%
  % There are no termination tasks (flag=9) to be handled.
  % Also, there are no continuous or discrete states,
  % so flags 1,2, and 4 are not used, so return an emptyu
  % matrix
  case { 1, 2, 4, 9 }
    sys=[];
```

```
%%%%%%%%%%%%%%%%%%%%%%%%%%%%%%%%%%%%
% Unexpected flags (error handling)%
%%%%%%%%%%%%%%%%%%%%%%%%%%%%%%%%%%%%
% Return an error message for unhandled flag values.
otherwise
   error(['Unhandled flag = ',num2str(flag)]);

end
%=================================================================
% mdlInitializeSizes
% Return the sizes, initial conditions, and sample times for
% the S-function.
%=================================================================
%
function [sys,x0,str,ts] = mdlInitializeSizes()

sizes = simsizes;
sizes.NumContStates  = 0;
sizes.NumDiscStates  = 0;
sizes.NumOutputs     = 3;   % dynamically sized
sizes.NumInputs      = 13;  % dynamically sized
sizes.DirFeedthrough = 1;   % has direct feedthrough
sizes.NumSampleTimes = 1;

sys = simsizes(sizes);
str = [];
x0  = [];
ts  = [-1 0];   % inherited sample time

% end mdlInitializeSizes

%
%=================================================================
% mdlOutputs
% Return the output vector for the S-function
%=================================================================
%
function sys = mdlOutputs(t,x,u)

q_a  = u(1);      % Phase a switching function
ph_a = u(2);      % Phase a phase current
i_a  = u(3);      % Phase a current
e_a  = u(4);      % Phase a back emf
q_b  = u(5);      % Phase b switching function
ph_b = u(6);      % Phase b phase function
i_b  = u(7);      % Phase b current
e_b  = u(8);      % Phase b back emf
```

```
q_c = u(9);      % Phase c switching function
ph_c = u(10);    % Phase c phase function
i_c = u(11);     % Phase c current
e_c = u(12);     % Phase c back emf
V_dc = u(13);    % Dc bus voltage

I_zero = 0.1;    % Current below this value is taken to be zero
v_n = -(e_a+e_b+e_c)/3;    % Voltage at motor neutral point

if (ph_a > 0) && (ph_b < 0) % Phases a and b conducting
    if q_a == 1             % Phase a used for regulation
        if i_c > I_zero     % Phase c lower device conducts
            v_a = 2*V_dc/3;
            v_b = -V_dc/3;
            v_c = -V_dc/3;
        else                % Phase c devices off
            v_a = V_dc/2-e_c/2-v_n/2;
            v_b = -V_dc/2-e_c/2-v_n/2;
            v_c = e_c+v_n;
        end
    else
        if i_c > I_zero     % Phase c lower device conducts
            v_a = 0;
            v_b = 0;
            v_c = 0;
        else                % Phase c devices off
            v_a = -e_c/2-v_n/2;
            v_b = -e_c/2-v_n/2;
            v_c = e_c+v_n;
        end
    end
elseif (ph_a < 0) && (ph_b > 0) % Phases a and b conducting
    if q_b == 1             % Phase b used for regulation
    if i_c < -I_zero        % Phase c upper device conducts
            v_a = -2*V_dc/3;
            v_b = V_dc/3;
            v_c = V_dc/3;
        else                % Phase c devices off
            v_a = -V_dc/2-e_c/2-v_n/2;
            v_b = V_dc/2-e_c/2-v_n/2;
            v_c = e_c+v_n;
        end
    else
        if i_c < -I_zero    % Phase c upper device conducts
            v_a = -V_dc/3;
            v_b = -V_dc/3;
            v_c = 2*V_dc/3;
        else                % Phase c devices off
```

C.3. Simulation of a Brushless Dc Machine System 395

```
                    v_a = -e_c/2-v_n/2;
                    v_b = -e_c/2-v_n/2;
                    v_c = e_c+v_n;
                end
            end
    elseif (ph_a > 0) && (ph_c < 0)  % Phases a and c conducting
        if q_a == 1                  % Phase a used for regulation
            if i_b < -I_zero         % Phase b upper device conducts
                v_a = V_dc/3;
                v_b = V_dc/3;
                v_c = -2*V_dc/3;
            else                     % Phase b devices off
                v_a = V_dc/2-e_b/2-v_n/2;
                v_b = e_b+v_n;
                v_c = -V_dc/2-e_b/2-v_n/2;
            end
        else
            if i_b < -I_zero         % Phase b upper device conducts
                v_a = -V_dc/3;
                v_b = -V_dc/3;
                v_c = 2*V_dc/3;
            else                     % Phase b devices off
                v_a = -e_b/2-v_n/2;
                v_b = e_b+v_n;
                v_c = -e_b/2-v_n/2;
            end
        end
    elseif (ph_a < 0) && (ph_c > 0)  % Phases a and c conducting
        if q_c == 1                  % Phase c used for regulation
            if i_b > I_zero          % Phase b lower device conducts
                v_a = -V_dc/3;
                v_b = -V_dc/3;
                v_c = 2*V_dc/3;
            else                     % Phase b devices off
                v_a = -V_dc/2-e_b/2-v_n/2;
                v_b = e_b+v_n;
                v_c = V_dc/2-e_b/2-v_n/2;
            end
        else
            if i_b > I_zero          % Phase b lower device conducts
                v_a = 0;
                v_b = 0;
                v_c = 0;
            else                     % Phase b devices off
                v_a = -e_b/2-v_n/2;
                v_b = e_b+v_n;
                v_c = -e_b/2-v_n/2;
            end
```

```
            end
        elseif (ph_b > 0) && (ph_c < 0)  % Phases b and c conducting
            if q_b == 1                  % Phase b used for regulation
                if i_a > I_zero          % Phase a lower device conducts
                    v_a = -V_dc/3;
                    v_b = 2*V_dc/3;
                    v_c = -V_dc/3;
                else                     % Phase a devices off
                    v_a = e_a+v_n;
                    v_b = V_dc/2-e_a/2-v_n/2;
                    v_c = -V_dc/2-e_a/2-v_n/2;
                end
            else
                if i_a > I_zero          % Phase a lower device conducts
                    v_a = 0;
                    v_b = 0;
                    v_c = 0;
                else                     % Phase a devices off
                    v_a = e_a+v_n;
                    v_b = -e_a/2-v_n/2;
                    v_c = -e_a/2-v_n/2;
                end
            end
        elseif (ph_b < 0) && (ph_c > 0)  % Phases b and c conducting
            if q_c == 1                  % Phase c used for regulation
                if i_a < -I_zero         % Phase a upper device conducts
                    v_a = V_dc/3;
                    v_b = -2*V_dc/3;
                    v_c = V_dc/3;
                else                     % Phase a devices off
                    v_a = e_a+v_n;
                    v_b = -V_dc/2-e_a/2-v_n/2;
                    v_c = V_dc/2-e_a/2-v_n/2;
                end
            else
                if i_a < -I_zero         % Phase a upper device conducts
                    v_a = 2*V_dc/3;
                    v_b = -V_dc/3;
                    v_c = -V_dc/3;
                else                     % Phase a devices off
                    v_a = e_a+v_n;
                    v_b = -e_a/2-v_n/2;
                    v_c = -e_a/2-v_n/2;
                end
            end
        else       % Should not get into this state; set voltages to zero
            v_a = 0;
            v_b = 0;
```

C.3. Simulation of a Brushless Dc Machine System

```
      v_c = 0;
end

sys(1) = v_a;    % Inverter output phase voltages
sys(2) = v_b;
sys(3) = v_c;
```

The code that follows is the "S-function" that computes $\partial\lambda/\partial\theta$ for each phase. From this, back emf and phase torques are computed within the Simulink block diagrams.

```
function [sys,x0,str,ts] = ...
         partial_lambda_partial_theta(t,x,u,flag);

switch flag,
%%%%%%%%%%%%%%%%%%
% Initialization %
%%%%%%%%%%%%%%%%%%
% Initialize the states, sample times, and state
% ordering strings.
  case 0
    [sys,x0,str,ts]=mdlInitializeSizes;

%%%%%%%%%%%
% Outputs %
%%%%%%%%%%%
% Return the outputs of the S-function block.
  case 3
    sys=mdlOutputs(t,x,u);

%%%%%%%%%%%%%%%%%%
% Unhandled flags %
%%%%%%%%%%%%%%%%%%
% There are no termination tasks (flag=9) to be handled.
% Also, there are no continuous or discrete states,
% so flags 1,2, and 4 are not used, so return an emptyu
% matrix
  case { 1, 2, 4, 9 }
    sys=[];

%%%%%%%%%%%%%%%%%%%%%%%%%%%%%%%%%%%
% Unexpected flags (error handling)%
%%%%%%%%%%%%%%%%%%%%%%%%%%%%%%%%%%%
% Return an error message for unhandled flag values.
  otherwise
```

```
        error(['Unhandled flag = ',num2str(flag)]);

end
%===============================================================
% mdlInitializeSizes
% Return the sizes, initial conditions, and sample times for
% the S-function.
%===============================================================
%
function [sys,x0,str,ts] = mdlInitializeSizes()

sizes = simsizes;
sizes.NumContStates  = 0;
sizes.NumDiscStates  = 0;
sizes.NumOutputs     = 3;   % dynamically sized
sizes.NumInputs      = 1;   % dynamically sized
sizes.DirFeedthrough = 1;   % has direct feedthrough
sizes.NumSampleTimes = 1;

sys = simsizes(sizes); str = []; x0  = [];
ts  = [-1 0];    % inherited sample time

% end mdlInitializeSizes

%
%===============================================================
% mdlOutputs
% Return the output vector for the S-function
%===============================================================
%
function sys = mdlOutputs(t,x,u)

theta = u(1);      % Mechanical rotor position
N_p = 6;           % Number of magnetic poles
N_h = 25;          % Number of (odd) harmonics used in series
E_max = 2.2;       % Peak back emf at omega_max (Vs/rad)
omega_max = 1;     % Phase angular velocity
delta = pi/6;      % Angle over which back emf increases from zero
                   % to its maximum value

K = E_max/omega_max; A = zeros(2*N_h,1);
for n = 1:2:(2*(N_h+1)), % Use a loop to compute the coefficients
    A(n) = (4*K*sin(n*delta))/(n^2*pi*delta);
end

p_lambda_a = 0;
p_lambda_b = 0;
p_lambda_c = 0;
```

C.3. Simulation of a Brushless Dc Machine System

```
for n = 1:2:(2*(N_h+1)), % Use a loop to compute the partials
    p_lambda_a = p_lambda_a + A(n)*sin(0.5*n*N_p*theta);
    p_lambda_b = p_lambda_b + A(n)*sin(0.5*n*N_p*theta-2*n*pi/3);
    p_lambda_c = p_lambda_c + A(n)*sin(0.5*n*N_p*theta+2*n*pi/3);
end

sys(1) = p_lambda_a;    % Partial of phase flux linkages with respect
sys(2) = p_lambda_b;    % to theta
sys(3) = p_lambda_c;
```

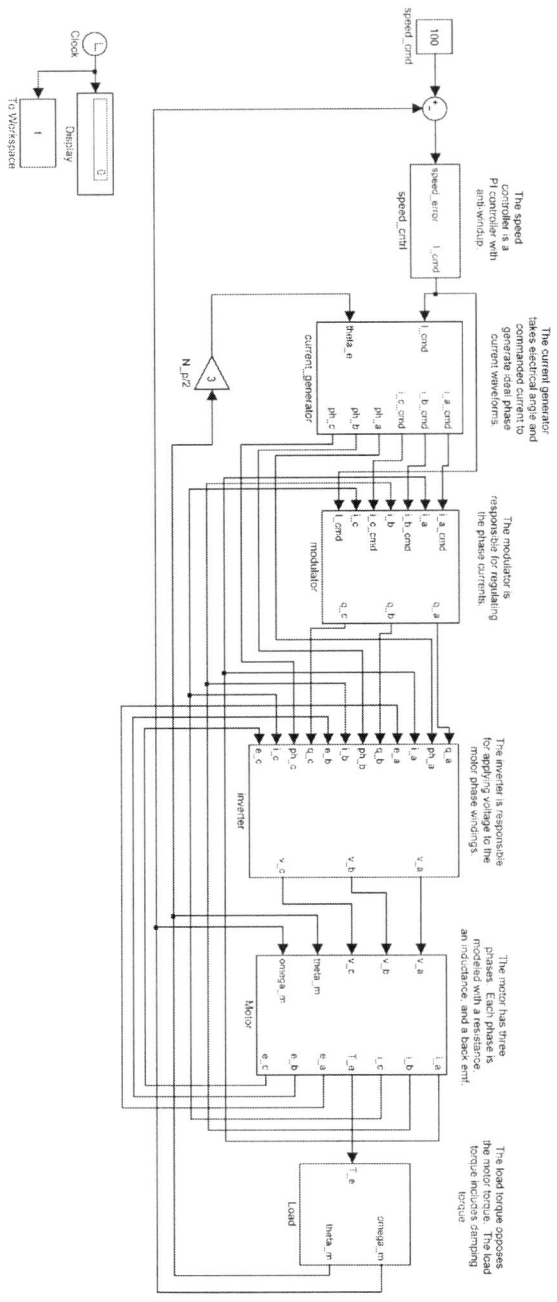

Figure C.1: The top level model used to simulate a brushless dc machine drive.

C.3. Simulation of a Brushless Dc Machine System

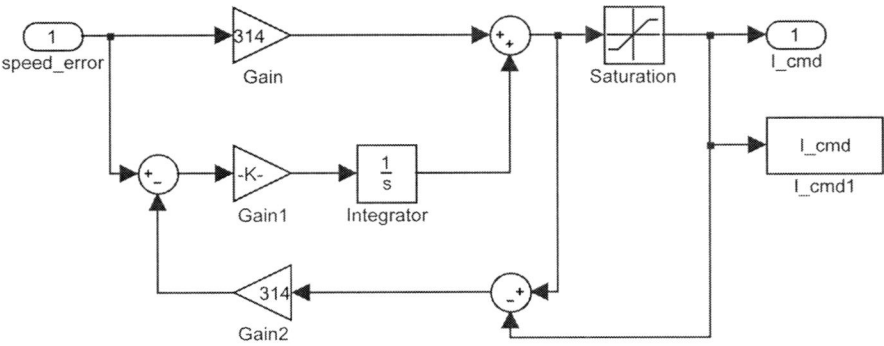

Figure C.2: The details of the speed control block used to simulate a brushless dc machine drive.

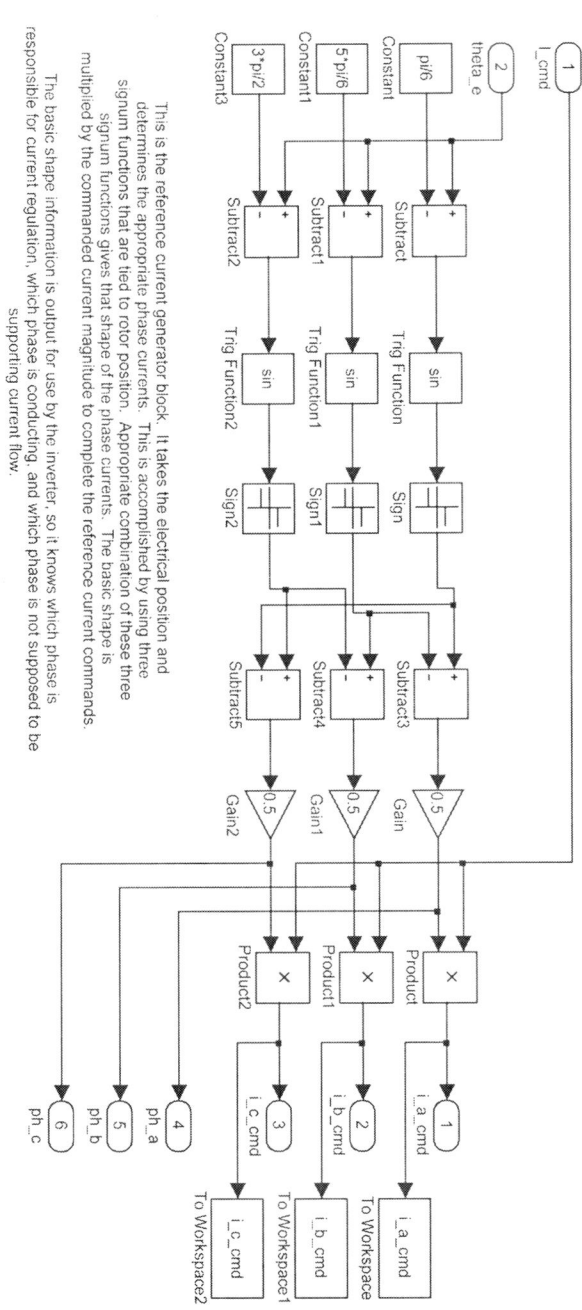

Figure C.3: The details of the reference current generator block used to simulate a brushless dc machine drive.

C.3. Simulation of a Brushless Dc Machine System

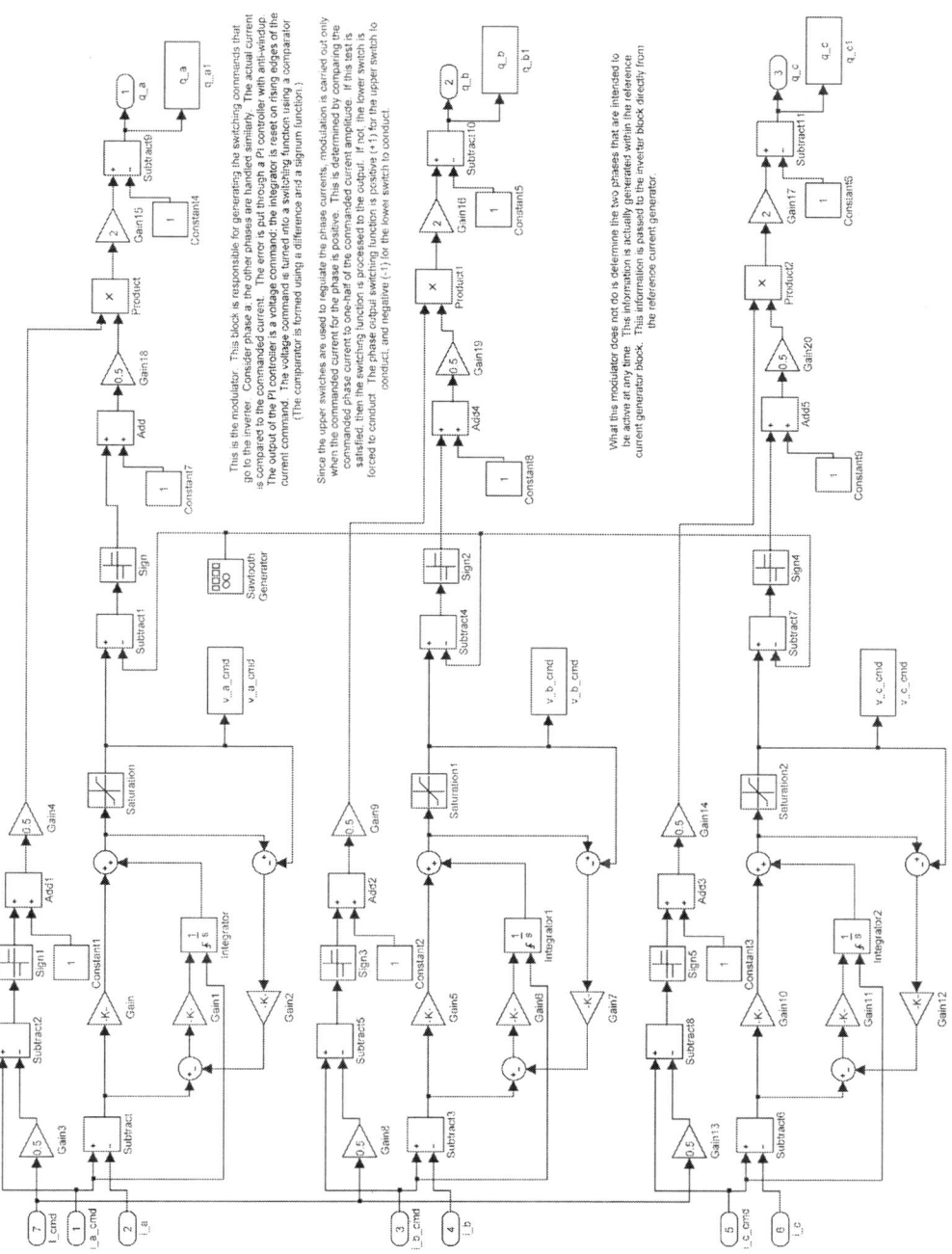

Figure C.4: The details of the modulator block used to simulate a brushless dc machine drive.

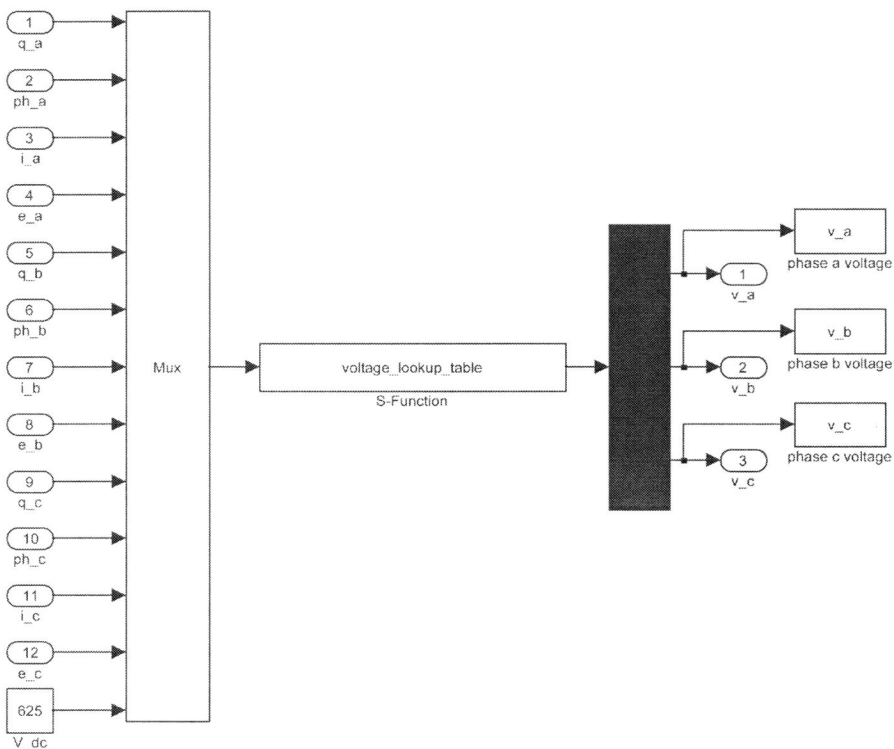

Figure C.5: The details of the inverter block used to simulate a brushless dc machine drive. This block relies heavily on the S-function "voltage_lookup_table" to determine the inverter output voltages.

C.3. Simulation of a Brushless Dc Machine System

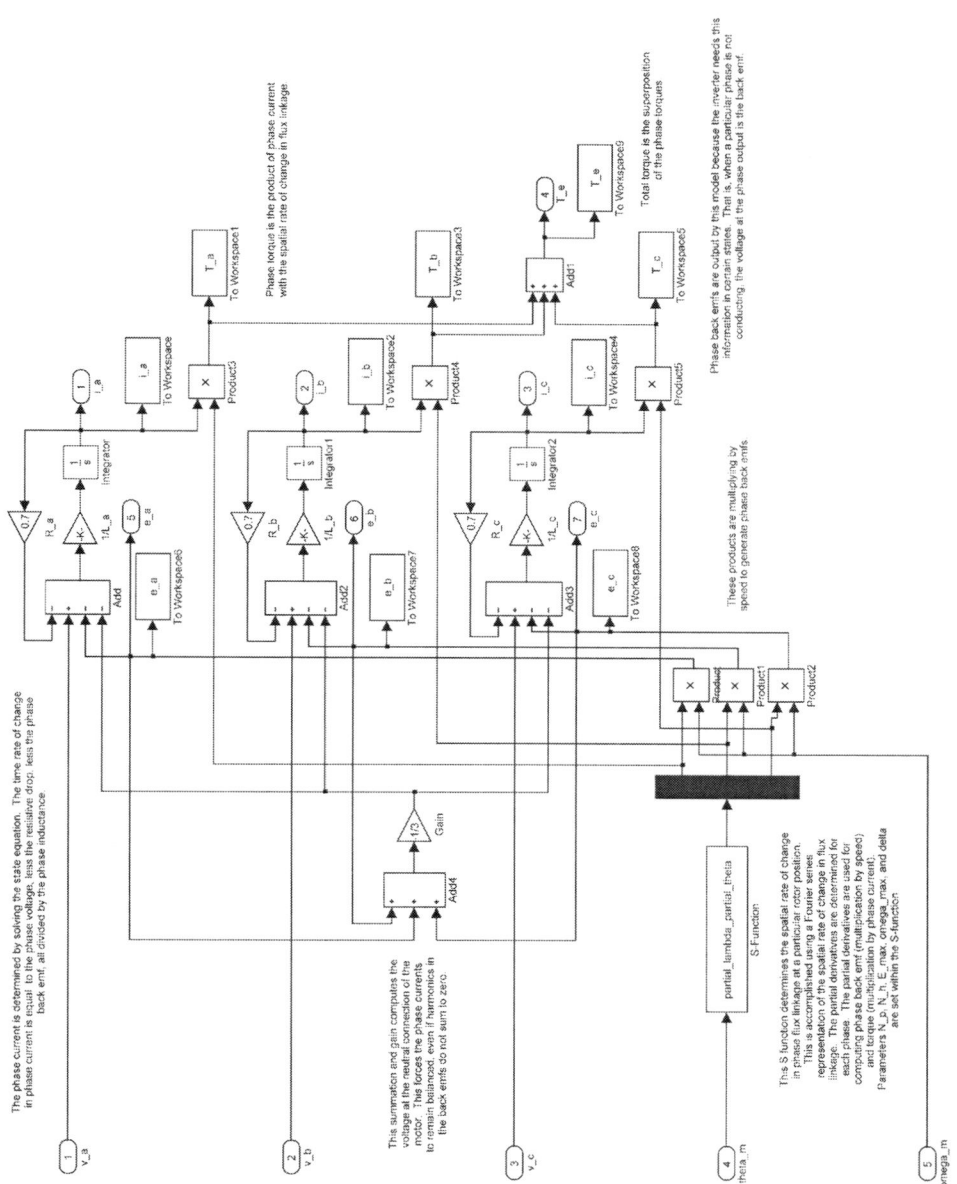

Figure C.6: The details of the motor block used to simulate a brushless dc machine drive. This block relies heavily on the S-function "partial_lambda_partial_theta" to determine the value of the functions underlying back emf and phase torque.

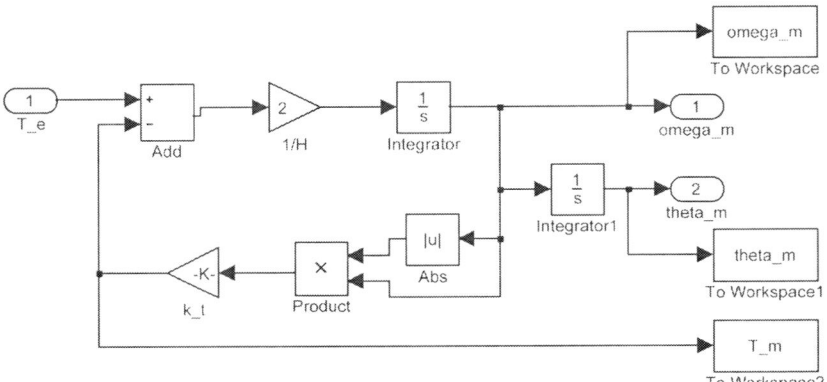

The mechanical load is modeled with a torque that is proportional to the square of the speed. The difference of the electromagnetic and load torque is multiplied by the reciprocal of the inertia and integrated to get the speed. The speed is integrated to get the mechanical position.

The load torque is computed using the absolute value so that load torque sign will follow the sign of the speed and not always be positive.

Figure C.7: The details of the load block used to simulate a brushless dc machine drive.

Index

$\alpha\beta 0 \to abc$ transformation, 141, 174, 183, 196, 221, 248, 275
$\alpha\beta \to dq$ transformation, 143, 203, 204, 222, 223, 225, 250, 278
$\alpha\beta 0$ reference frame, 140
abc reference frame, 140
$abc \to \alpha\beta 0$ transformation, 141, 166, 174, 183, 196, 221, 248, 275, 306
Ackermann, B., 283
acoustic noise, 115, 129–131, 323
    and number of poles, 130
adaptive control, 263
air gap, 9
air gap field orientation, 255
    slip condition, 257
air gap inductance, 98
air gap mmf
    and electric loading, 124
air gap orientation, 244
Al-Zamel, A. M., 170
Ampere's law, 3, 7, 9
    along flux tube, 286, 291
    boundary condition for, 123
Arefeen, M., 330, 349
Asher, G. M., 324, 346
asymmetrical bridge inverter, 326
average power conversion, 116
averaged circuit model, 177

back emf, 108, 265, 340
    and motional voltage, 32
    and space harmonics, 115, 266
    and torque, 109
    and transformations, 142

    of simple coil, 109
    simulation of, 392, 397, 405
    sinusoidal, 270
back iron, 36
back iron thickness
    and number of poles, 127
ball bearings, 131
base speed, 327
basis functions, 364
bearing currents, 131
bearing seats, 132
bearings, 131
Besbes, M., 324
Bianchi, N., 289
blackbody radiation, 136
Blaschke, F., 239
Bode magnitude plot, 181, 198
Bode plot, 198, 199
Bolognani, S., 289
Bolte, E., 283
Borisenko, A. I., 38
Bose, B. K., 148, 187
boundary layer, 134
    momentum, 134
    thermal, 134
breadloaf magnets, 284
Brockett, R., 143
Brod, D. M., 169, 170
brushless dc machine, 303, 310
brushless dc machines
    idealized waveforms, 311
    reference current generator, 402
    sensorless operation, 316
    simulation of, 400
    speed control, 312

torque ripple, 314, 317

Cameron, D. E., 324, 345, 351
Cardenas, R., 324, 346
Carter coefficient, 94
characteristic dimension, 5
characteristic polynomial, 180
charge conservation, 3
    and Kirchhoff's current law, 29
coenergy, 46
    and current, 46
    and energy, 46
    and field quantities, 49
    and flux linkage, 46
    and force, 46
    and permanent magnets, 50
    and torque, 212
    calculation, 48
    integration order, 48
coenergy density, 49, 104
coenergy equivalence, 49, 104, 107
coercive force, 15
cogging torque, 125, 270, 283
    and fractional slot winding, 296
    and magnet placement, 289
    reduction using skew, 294
coil pitch, 65
coil span, 70, 80
common mode voltage, 159
compensator, 176, 181
concentrated windings, 66, 325
concentric winding, 67
concentricity, 129, 132
conduction current density, 2
conduction currents, 3
conduction heat transfer, 132
conductors
    proximity effect, 23
    skin effect, 23, 24
conductors in hand, 23
conformal mapping, 94
conservation of energy, 43, 51
    and lossless system, 43

vector notation, 211
conservation of thermal energy, 133
conservative system
    and reciprocity, 50
constant torque loci, 305
constituitive relations, 7, 285
control error, 176
controller saturation, 198
convection heat transfer, 133, 134
convolution, 116
core loss
    and lamination design, 127
crest factor, 344
crossover frequency, 180, 181, 199
current density
    winding slot, 137
current regulator, 173, 175, 272
    α axis equivalent circuit, 178
    $\alpha\beta$ block diagram, 181
    $\alpha\beta$ state equations, 177, 385
    averaged circuit model of, 177
    d axis equivalent circuit, 204, 205
    dq block diagram, 204, 206
    dq state equations, 204
    modulator simulation, 373
    q axis equivalent circuit, 204, 205
    simulation control, 376
    simulation of, 373

dc machines
    armature current, 246
    field current, 246
    field flux, 246
demagnetization, 16
digital signal processor, 352
    A/D converters, 353
    encoder interface, 354
    PWM generator, 353
diode-clamped multilevel inverter, 158
direct axis, 142, 204, 225, 227, 240, 245, 246, 249, 252, 255, 263, 278, 282, 306, 307
direct field oriented control, 263

direct torque control, 240, 259, 261, 263, 306
    signal calculator, 262
displacement currents, 3
distributed winding, 66, 72
distribution factor, 73, 75, 114
Divan, D. M., 149, 150
divergence theorem, 3, 40
double-layer winding, 66, 127
$dq \rightarrow \alpha\beta$ transformation, 143, 203, 204, 222, 223, 225, 250, 278, 306
dq reference frame, 142, 144, 204, 225, 227, 244, 281, 306
DSP, 352
Dwight, H. B., 75
dynamic balance, 129

eddy current loss, 22
    laminations, 23, 27
eddy currents, 27
effective coercive force, 19
Ehsani, M., 330, 349
Elbuluk, M., 330, 349
electric and magnetic circuit analogs, 9
    limitations, 11
    table, 10
electric displacement, 2
electric field intensity, 2
electric loading, 121
    and air gap mmf, 124
    and forced convection, 137
    and liquid cooling, 137
    and winding mmf, 122
    ways to increase, 123
electrical and mechanical frequencies
    relationship between, 55, 57, 65, 116
electrical and mechanical measure, 64, 209
electrical and thermal analogy, 134
    summary table, 134

electromotive force, 9
electroquasistatic limit, 6
embedded permanent magnets, 304
empirical heat transfer correlations, 135
encoder, 251, 306
end bells, 132
end turn inductance, 98, 107
energy
    and coenergy, 46
    and current, 45
    and field quantities, 49, 285
    and flux linkage, 46
    and flux tubes, 292
    and force, 45
    and torque, 212
energy conversion
    and magnetic saturation, 54
energy conversion cycle, 51
    and generator operation, 53, 340
    and motor operation, 53
    and recirculated energy, 53
Enjeti, P. N., 161
EQS limit, 6
even symmetry, 367

Fahimi, B., 330, 349
fan load, 229
Faraday's law, 3, 31
    and Kirchhoff's voltage law, 29
feed-forward control, 251, 333
ferromagnetic materials, 13
    coercive force, 15
    core loss data, 26
    eddy current losses, 22
    hard, 13
    hard, table of, 16
    hysteresis, 15
    hysteresis loss, 15
    minor hysteresis loop, 17
    permanent magnets, 16
    residual flux density, 14
    saturation, 14

soft, 13, 208
soft, table of, 16
field orientation, 239, 242
    air gap, 255
    air gap decoupling network, 258
    analogy between ac and dc machines, 246
    direct method, 248, 249
    indirect method, 251
    rotor flux calculator, 250
    stator decoupling network, 252, 254, 255
    torque calculator, 250
field oriented control, 263, 307
field weakening, 267, 303–305
    in permanent magnet machines, 309
finite element analysis, 128, 131
finite slot width, 84
    slot factor, 86
Fitzgerald, A. E., 208
flux, 9
    and flux linkage, 11
flux focusing, 19
flux linkage, 10, 31
flux sensors, 248
flux tube analysis, 12, 94, 283
force density, 37
    field description, 38
forced convection
    and electric loading, 137
forward loop gain, 180
Fourier series, 71, 290, 295, 363
    even symmetry, 367
    exponential version, 365
    half-wave symmetry, 368
    hidden symmetry, 369
    odd symmetry, 367
    trigonometric version, 364
Fourier's law of heat conduction, 133
fractional pitch, 65
fractional slot winding, 69, 79
    and cogging torque, 296
    harmonic cancellation in, 83, 92
    mmf offset, 100
free charge density, 2
free convection, 135
full-bridge inverter, 156
full-pitch coil, 65

Gabsi, M., 324
gate drive circuit, 153
    high voltage integrated circuit in, 153
    optocouplers in, 154
    transformer-coupled, 154
Gauss's law, electric version, 3
Gauss's law, magnetic version, 3, 9
    along flux tube, 286, 291
generalized inductance, 30
generalized inductor, 30
generator operation, 53
Grioni, B., 324

H-bridge inverter, 156
Habetler, T. G., 168, 169
half-bridge inverter, 154, 156
half-wave symmetry, 368
Hall sensors, 248
Hanselman, D. C., 119
harmonic cancellation, 92, 160, 161
harmonic elimination, 160
    and programmed PWM, 161
He, J., 150
heat capacity, 133
heat transfer coefficient, 135
Hendershot, J. R., 119
hidden symmetry, 369
high voltage integrated circuit, 153
Hoang, E., 324
Hoft, R. G., 160
Holmes, D. G., 167, 187
Holtz, J., 161, 166
Howe, D., 283
Husain, I., 330, 349
HVIC, 153

hysteresis, 15
hysteresis current control, 170
hysteresis loss, 15

IGBT, 152
incompressible substance, 133
index notation, 38
    Kronecker delta function, 39
    summation convention, 39
indirect field oriented control, 263
induced voltage, 31
    motional, 108
    transformer, 108
inductance, 10
    air gap, 98
    and energy equivalence, 49, 104
    bottom coil side, 104
    end turn, 98, 107
    leakage, 104
    mutual, 98, 101, 266
    self, 98, 266
    slot, 98, 104
    tooth, 100
    top coil side, 106
    zig-zag, 98, 108
inductance matrix, 210
    electromechanical coupling, 211
induction machine, 208
    line starting, simulation of, 229
    rotor circuit, 217
    rotor slot design, 127
induction machines
    α axis equivalent circuit, 224
    α axis equivalent circuit, 223
    $\alpha\beta$ and dq reference frames, 226
    $\alpha\beta$ model, 220, 225
    $\alpha\beta$ reference frame, 222
    $\alpha\beta$ torque, 225
    $\alpha\beta$ voltage equations, 223
    air gap field orientation, 255
    air gap flux, 227, 243
    air gap orientation decoupling network, 258
    β axis equivalent circuit, 224
    block diagram, 227, 229
    breakdown torque, 219
    design Class A, 220
    design classes, 219
    direct axis equivalent circuit, 227, 228, 253, 256
    direct axis flux and quadrature axis voltage, 226
    direct field oriented control, 248, 249, 263
    direct torque control, 240, 259, 261, 263
        signal calculator, 262
    dq model, 220, 225
    electrical dynamics, 211
    field oriented control, 239, 242, 263
    field-oriented state, 230
    indirect field orientation, 251
    indirect field oriented control, 263
    inductance matrix, 210
    leakage inductance, 216
    mechanical dynamics, 212
    mutual flux, 227, 243
    phase equivalent model, 213
    quadrature axis equivalent circuit, 227, 228, 253, 256
    quadrature axis flux and direct axis voltage, 226
    relationship between voltage and flux, 240
    rotor flux calculator, 250
    rotor leakage factor, 257, 259
    rotor leakage inductance, 250
    rotor power, 217
    rotor time constant, 228, 260
    scalar control, 239, 240, 262
    scalar control speed range, 241
    simulation control, 387
    simulation of, 387
    single phase equivalent model, 216
    slip, 215

slip and torque, 217
slip condition, air gap orientation, 257
slip condition, rotor orientation, 246, 251
slip condition, stator orientation, 254
slip frequency, 215, 226, 252
slip speed, 251
squirrel cage winding, 208, 209
state equations, 390
stator field orientation, 252
stator field orientation restrictions, 254
stator orientation decoupling network, 252, 254, 255
stator rotor leakage factor, 253, 255
Thevenin equivalent circuit, 218
torque calculator, 250
torque for air gap orientation, 245
torque for rotor orientation, 244
torque for stator orientation, 245
torque in dq reference frame, 227
wound rotor, 208
insulated gate bipolar transistor, 152
insulation classes
    table of, 132
integral control, 179
    windup, 198
integral slot winding, 69
integrator windup, 198, 312
intensive internal energy, 133
International Rectifier, 154
inverter
    averaged circuit model of, 177
    diode-clamped multilevel, 158
    full-bridge, 156
    H-bridge, 156
    half-bridge, 154, 156
    multilevel, 157, 173
    simulation of, 404
    single-phase, 154, 156
    three-phase, 157, 173
    voltage source, 248
inverter phase leg, 148

Johnson, J. P., 330, 349
Joos, G., 170

Kailath, T., 143
Kaplan, M., 324
Kassakian, J. G., 148, 177, 181
Kazerani, M., 170
Kingsley, C., Jr., 208
Kirchhoff's current law
    and charge conservation, 29
Kirchhoff's voltage law
    and Faraday's law, 29
Kokernak, J. M., 324, 348
Krein, P. T., 148
Kronecker delta function, 39

Lai, J.-S., 158
lamination design
    and core loss, 127
laminations, 27, 325
    and eddy currents, 23, 27
Lammert, P., 166
Lang, J. H., 324, 328, 345, 349, 351
lap winding, 67, 68
Lateb, R., 289
Lathi, B. P., 363
Lawrenson, P. J., 325
leakage factor, 253
    and torque limit, 255
leakage flux, 104, 210
leakage inductance, 104, 211
Lecrivain, M., 324
Legowski, S., 166, 185
Li, W., 170
Lindsay, J. L., 161
Lipo, T. A., 98, 119, 149, 187, 242
liquid cooling
    and electric loading, 137
list of symbols, 357
local average, 177

Lorentz force, 37
   on conductors, 131
lossless electromechanical system, 43
lossless system, 43
   and reciprocity, 50
Lotzkat, W., 166

M-19 core loss data, 26
MacMinn, S. R., 324, 350, 351
magnet torque, 270, 271, 283
magnetic and electric circuit analogs, 9
   limitations, 11
   table, 10
magnetic circuit
   assumptions, 8
magnetic core, 9
magnetic diffusion, 23
magnetic domains, 14
magnetic field intensity, 3, 7, 9, 17, 18, 104, 121, 285
   and air gap mmf, 126
   boundary conditions, 122
magnetic flux density, 2, 7, 9, 18, 285
magnetic loading, 121, 125
   and tooth saturation, 125
magnetic poles, 64, 209, 268
   and acoustic noise, 130
   and back iron thickness, 127
   and field velocity, 65, 215
   and regular spacing, 289
   and torque, 127
magnetic saliency, 51, 267, 270, 325
magnetic saturation
   and energy conversion, 54
magnetization density, 7
magnetomotive force, 9, 140
magnetoquasistatic limit, 6
magnetostriction, 40
Majmudar, H., 208
Matlab, 373
   differential equation solver, 387
   function `ode45`, 387

   S-function, 392, 397, 404, 405
   Simulink, 392
maximum torque per Ampere, 304
Maxwell stress tensor, 39
Maxwell's equations
   differential form, 2
   electroquasistatic limit, 6
   EQS limit, 6
   integral form, 3
   magnetoquasistatic limit, 6
   MQS limit, 6
   quasistatic, 3
   wave equation, 5
McGilp, M. I., 329
McGrath, B. P., 167
McMurray snubber, 151
McMurray, W., 150
mechanical and electrical frequencies
   relationship between, 55, 57, 65, 116
mechanical and electrical measure, 64, 209
mechanical elements
   table of, 25
Meibody-Tabar, F., 289
Melcher, J. R., 24, 41
MEMS, 1
Mese, E., 324, 325, 348
microcontroller, 352
microelectromechanical systems, 1
Miller, T. J. E., 119, 325, 326, 329
mmf, 9, 21, 72, 98, 122, 140
   air gap, 66
   and current distribution, 124
   and dc offset, 79, 100
   and finite slot width, 86
   and short-pitching, 71
   and skew, 87, 294
   and spatial harmonics, 69
   and tangential field, 122, 124, 126
mmf distribution, 71
mmf offset, 100
mmf space vectors, 73, 74, 81

modulator, 173
    simulation of, 403
Mohan, N., 148, 150
momentum boundary layer, 134
motional voltage, 32, 108
    and back emf, 32
motor operation, 53
moving average, 177, 178
MQS limit, 6
multilevel inverter, 157, 173
    space-vector PWM, 167
Murai, Y., 149
mutual inductance, 37, 98, 101, 106, 107, 112, 122, 210, 217, 222, 266, 271, 272

National Electrical Manufacturers Association, 219
natural convection, 135
natural frequencies, 130
natural modes, 130
NEMA, 219
Newton's law of cooling, 135
Novotny, D. W., 169, 170, 242
number of poles
    and acoustic noise, 130
    and field velocity, 65
number of slots per pole, 69
number of slots per pole per phase, 69

odd symmetry, 367
optocouplers, 154
output equation, 126
overall heat transfer coefficient, 135
overmodulation, 185, 186

Panahi, I., 330, 349
parallel coil connection, 90
Parseval's theorem, 371
Patel, H. S., 160
Peng, F. Z., 158, 167
permanent magnet machines
    α axis equivalent circuit, 278

$\alpha\beta$ model, 274
β axis equivalent circuit, 279
brushless dc machine, 303, 310
constant torque loci, 305
current limitation, 309, 310
d axis equivalent circuit, 282, 308
direct torque control, 306
dq model, 278
electrical dynamics, 271
embedded magnets, 304
field oriented control, 303, 307
field weakening, 303–305, 309
maximum torque per Ampere, 304
mechanical dynamics, 271
q axis equivalent circuit, 282
spoke configuration, 268
thermal limitation, 309
trapezoidal back emf, 273, 310
trapezoidal back emf waveforms, 311
voltage limitation, 309, 310
with embedded magnets, 267–269
with surface magnets, 268
permanent magnet materials, 16
    bonded, 16
    demagnetization, 16
    effective coercive force, 19
    energy product, 19
    equivalence with coil, 19
    Norton model, 22
    recoil permeability, 17, 286
    sintered, 16
    table of, 16
    Thevenin model, 22
permanent magnets, 31
    and coenergy, 50
permeability, 7
    free space, 3
permeance, 10, 95, 97
    and slot inductance, 105
    and winding slots, 95
    incremental, 95, 105
    irregular shapes, 11

permanent magnet, 21
permittivity
    free space, 3
phase, 64
phase 0, 141, 142, 144, 222, 275
phase belt, 72
phase margin, 181
phase offset, 80
phase windings, 61, 64
phase-locked loop, 205
phasor analysis, 213
phasor representation, 73, 213, 215, 217, 366
pitch, 65
pitch angle, 73
pitch factor, 71
Plasse, C., 324
position sensing
    encoder, 205, 251, 306
        observer, 306
        resolver, 205, 306
power invariance, 139, 141, 143
Prantl number, 135
predictive current regulator, 171, 202, 203
proportional-integral control, 176
programmed PWM
    and harmonic elimination, 161
propagation time, 5
proportional-integral control, 178, 180, 199
proximity effect, 23
pulse-width modulation, 160, 161, 248
PWM, 161, 248
    sinusoidal, 162
        sinusoidal PWM and space vector, 165
    space-vector modulation, 166

quadrature axis, 142, 204, 225, 227, 245, 246, 249, 252, 255, 282, 283, 307

radiation heat transfer, 136
Radun, A., 324, 326, 345
Ray, W. F., 324, 346
reciprocity, 50, 210
recirculated energy, 53
recoil permeability, 17, 286
reference frames
    $\alpha\beta$, 204, 220, 222, 271, 274, 278, 283, 306
    $\alpha\beta$, 141, 174, 183, 196, 203
    $\alpha\beta 0$, 144
    abc, 140, 141, 174, 183, 196, 204, 211
    direct axis, 142, 204
    dq, 142, 144, 204, 220, 225–227, 240, 244, 271, 274, 278, 281, 283, 306
    quadrature axis, 142, 204
    rotating, 229
    rotor, 144, 212, 213, 215, 222, 225
    stationary, 229
    stator, 144, 212, 213, 220, 222, 223, 225
reluctance, 9, 97, 98, 107
    irregular shapes, 11
    permanent magnet, 21
reluctance torque, 52, 267, 270, 271, 283, 323
residual flux density, 14
resistance matrix, 211
resolver, 306
resonant dc link, 149, 150
resonant snubber, 151
Reynolds number, 135
Robbins, W. P., 148
roller bearings, 131
rotor, 29, 208
rotor eccentricity
    and cogging torque, 289
rotor field orientation, 244
    slip condition, 246, 251
rotor leakage factor, 257

and torque limit, 259
rotor losses, 137
rotor time constant, 260
Russa, K., 330, 349

saliency, 51, 267, 270, 325
Salon, S. J., 128
scalar control, 240, 262
    speed range, 241
Schlecht, M. F., 148, 177, 181
self inductance, 266
Sember, J. W., 324, 350, 351
series coil connection, 89
shaft power, 43, 109
shock load
    and bearing selection, 131
short-pitched, 65
short-pitched windings, 70
signal decomposition
    ac component, 369
    dc component, 369
    Fourier series, 71, 363, 364
    Parseval's theorem, 371
signal orthogonality, 364
silicon steel, 22, 208
Silverman, R. A., 38
Simonelli, J. M., 150
Simulink, 392
    S-function, 392, 397, 404, 405
single-layer winding, 66, 127
single-phase inverter, 154, 156
single-turn solenoid, 4
sinusoidal PWM, 162, 182
    and space-vector modulation, 165
six-step operation, 185, 190, 193
sizing equation, 121
skew, 86, 90, 293
    and reducing cogging torque, 294
skew factor, 88, 114, 294
Skibinski, G., 150
skin depth, 24
skin effect, 23, 24
Skudelny, H. C., 166

Slemon, G., 208
sliding-mode control, 170
slip, 215
slip frequency, 215, 226, 252
slip speed, 251
slot factor, 86, 114
slot inductance, 98, 104
    bottom coil side, 104
    top coil side, 106
slot pitch
    electrical measure, 69
    mechanical measure, 69
Slotine, J. J., 170
snubber circuits, 150
    McMurray, 151
    resonant, 151
    Undeland, 152
Soliman, S. S., 363
Sozer, Y., 318, 324, 325, 331, 348
space harmonics, 37, 115
    and torque ripple, 115, 117
space vectors, 140
    table of, 183
space-vector modulation, 166, 182
    and sinusoidal PWM, 165
    multilevel inverter, 167
spatial harmonics, 115
    and back emf, 115
    and flux linkage, 115
    and rotor flux, 115
speed control
    simulation of, 401
speed of light, 5
Srinath, M. D., 363
Stanke, G., 166
state equations, 197, 204
stator, 29, 208
stator deformation, 130
stator field orientation, 244
    restrictions, 254
    slip condition, 254
stator resistance compensation, 241
Stefan-Boltzmann law, 136

Steinmetz, C. P., 23
Stoke's law, 3
Stratton, J. A., 41
Straughen, A., 208
summation convention, 39
Suresh, G., 330, 349
switching functions
    table of, 183
symbols, list of, 357
synchronous speed, 215
system state, 43

Takorabet, N., 289
Tarapov, I. E., 38
THD, 77
thermal and electrical analogy, 134
    summary table, 134
thermal boundary layer, 134
thermal conductivity, 133
thermal diffusivity, 134
thermodynamics, 133
Thevenin equivalent, 218
three-phase inverter, 157, 173
thrust loads, 131
Tolbert, L. M., 167
tooth deflection, 130
tooth inductance, 100
torque
    and back emf, 109, 274
    and coenergy, 212
    and electric loading, 126
    and electromechanical coupling, 211
    and energy, 212
    and machine size, 126
    and magnetic loading, 126
    and rotor volume, 126
    cogging, 270
    magnet, 270, 271, 283
    reluctance, 267, 270, 271, 283, 323, 327
    simulation of, 392, 397, 405

torque ripple, 37, 58, 115, 314, 317, 323, 351
    elimination of, 60
    harmonic orders of, 117
    harmonic orders, table, 118
Torrey, D. A., 150, 170, 318, 324, 328, 331, 346, 348, 349
total harmonic distortion, 77
traction, 42
    axial force, 42
    radial force, 42
    torque, 42
transfer function pole, 181
transfer function zero, 181
transformations
    $\alpha\beta 0 \to abc$, 141, 174, 183, 196, 221, 248, 275, 306
    $\alpha\beta \to dq$, 143, 204, 222, 223, 225, 250, 278
    $\alpha\beta$ reference frame, 140
    $abc \to \alpha\beta 0$, 141, 166, 174, 183, 196, 221, 248, 275
    abc reference frame, 140
    and induced voltage, 142
    direct axis, 142
    $dq \to \alpha\beta$, 278
    $dq \to \alpha\beta$, 143, 203, 204, 222, 223, 225, 250, 306
    dq reference frame, 144
    power invariance, 139, 141, 143
    quadrature axis, 142
    rotary, 142
    stator to rotor, 213
    unitary, 143
transformer voltage, 31, 108
transformer-coupled gate drive, 154
trapezoidal windings, 265, 271
traveling waves, 60
triplen harmonics, 117, 274, 315
Trzynadlowski, A. M., 166, 185, 242

Umans, S. D., 208
Undeland snubber, 152

Undeland, T. M., 148, 150
underdamped response, 180
unitary transformation, 143
United States Steel, 22

Van der Broeck, H. W., 166
variable reluctance machines, 129
    acoustic noise, 323
    back emf coefficient, 340
    base speed, 327
    conduction angle, 326
    control of conduction angle, 346
    control of excitation parameters, 330
    control of turn-off angle, 336, 348
    control of turn-on angle, 333, 346, 348
    current regulation, 352
    current regulation during generation, 348
    dwell angle, 326
    energy conversion cycle, 53, 340
    excitation parameter control, 338, 352
    excitation parameters, 329
    inverter during phase demagnetization, 343
    inverter during phase excitation, 343
    magnetic nonlinearities, 327
    maxiumum inductance region, 331
    minimum inductance region, 331
    optimum turn-off angle, 335
    power regulation during generation, 347
    reference current, 329
    spatial nonlinearities, 327
    speed regulation during generation, 347
    torque production, 52, 327
    torque ripple, 323, 351
    turn-off angle, 326, 329, 332
    turn-on angle, 326, 329, 332
    turn-on angle control, 333
Verghese, G. C., 148, 177, 181
vibration, 42, 115, 129–131
virtual work, 43, 211
viscosity, 135
voltage clamp, 150, 152, 153
voltage source inverter, 168, 248, 308
VRG, see variable reluctance machines
VRM, see variable reluctance machines

wave winding, 68
winding design, 127
winding mmf
    and electric loading, 122
winding slot
    current density, 137
winding slots, 64
winding span, 70
windings
    coil span, 80
    common assumptions, 70
    concentrated, 66, 325
    concentric, 67, 266
    distributed, 66, 72, 208, 266
    distribution factor, 73, 75, 114
    double-layer, 66
    double-layer winding, 266
    finite slot width, 84
    fractional slot, 69, 79, 266
    harmonic cancellation in, 83, 92
    integral slot, 69, 266
    lap winding, 67, 68
    mmf distribution, 71, 140
    mmf space vectors, 73, 74, 81
    number of slots per pole, 69
    number of slots per pole per phase, 69
    parallel coil connection, 90
    phase belt, 72
    phase offset, 80
    pitch angle, 73
    pitch factor, 71
    series coil connection, 89

    short-pitched, 70
    single-layer, 66
    single-layer winding, 266
    skew, 86, 90
    skew factor, 88, 114
    slot factor, 86, 114
    slot pitch, 69
    span, 70
    THD of mmf distribution, 77
    trapezoidal, 265, 271
    wave, 68
Woodson, H. H., 24, 41

Zhu, Z. Q, 283
zig-zag inductance, 98, 108
Ziogas, P. D., 161, 170